Artificial Intelligence and Machine Learning Applications for Sustainable Development

The book highlights how technologies including artificial intelligence and machine learning are transforming renewable energy technologies and enabling the development of new solutions. It further discusses how smart technologies are employed to optimize energy production and storage, enhance energy efficiency, and improve the overall sustainability of energy systems.

This book:

- Discusses artificial intelligence–based techniques, namely, neural networks, fuzzy expert systems, optimization techniques, and operational research.
- Showcases the importance of artificial intelligence and machine learning in the energy market, demand analysis, and forecasting of renewable energy applications.
- Illustrates strategies for sustainable development using artificial intelligence and machine learning applications.
- Presents applications of artificial intelligence in the domain of electronics transformation and development, smart cities, and renewable energy utilization.
- Highlights the role of artificial intelligence in solving problems such as image and signal processing, smart weather monitoring, smart farming, and distributed energy sources.

It is primarily written for senior undergraduates, graduate students, and academic researchers in diverse fields, including electrical, electronics and communications, energy, and environmental engineering.

Artificial Intelligence and Machine Learning Applications for Sustainable Development

Edited by
A. J. Singh, Nikita Gupta, Sanjay Kumar,
Sumit Sharma, Subho Upadhyay, and
Sandeep Kumar

CRC Press
Taylor & Francis Group
Boca Raton London New York

CRC Press is an imprint of the
Taylor & Francis Group, an **informa** business

First edition published 2025
by CRC Press
2385 NW Executive Center Drive, Suite 320, Boca Raton FL 33431

and by CRC Press
4 Park Square, Milton Park, Abingdon, Oxon, OX14 4RN

CRC Press is an imprint of Taylor & Francis Group, LLC

ISBN: 978-1-032-74214-4 (hbk)
ISBN: 978-1-032-94688-7 (pbk)
ISBN: 978-1-003-58124-6 (ebk)

DOI: 10.1201/9781003581246

Typeset in Sabon
by Deanta Global Publishing Services, Chennai, India

Contents

Foreword

In the ever-evolving landscape of technology and sustainability, the convergence of artificial intelligence (AI) and machine learning (ML) has emerged as a beacon of hope, offering innovative solutions to some of the most pressing challenges facing humanity. As we stand on the threshold of a new era defined by rapid technological advancement and unprecedented environmental concerns, the book *Artificial Intelligence and Machine Learning Applications for Sustainable Development* comes as a timely and significant contribution to the discourse on harnessing the power of AI and ML for the greater good.

In this comprehensive volume, readers are presented with a wealth of knowledge and insights curated by experts in the field. From the evolution of AI and ML to their practical applications in renewable energy, smart cities, and beyond, the book offers a nuanced exploration of the intersection between technology and sustainability. Through a series of meticulously crafted chapters, readers are guided through the theoretical frameworks, practical methodologies, and real-world case studies that illuminate the transformative potential of AI and ML in advancing sustainable development goals.

What sets this book apart is not only its breadth of coverage but also its depth of analysis. Each chapter offers a deep dive into a specific aspect of AI and ML applications, providing readers with a comprehensive understanding of the challenges, opportunities, and best practices associated with each domain. Whether it is optimizing renewable energy sources, managing big data, or enhancing urban infrastructure, the authors provide valuable insights and actionable recommendations that will undoubtedly inform and inspire future research and innovation.

As we navigate the complexities of the 21st century, it is clear that collaboration and interdisciplinary dialogue will be essential in addressing the multifaceted challenges of sustainable development. This book serves as a testament to the power of collaboration, bringing together diverse perspectives and expertise to explore the potential of AI and ML in shaping a more sustainable future. From academia to industry, from policymakers to practitioners, the insights contained within these pages have the potential to catalyze positive change and drive meaningful progress toward a more equitable and resilient world.

Preface

The fundamental goal of this book is to provide a comprehensive guide to help readers understand the technology under the complex field of sustainable development with their real-world challenges due to its growing significance and demand. These developments have opened up new research, innovation, and exploration avenues. In recent years, the fields of artificial intelligence (AI) and machine learning (ML) have surged to the forefront of innovation, presenting unparalleled opportunities to address the pressing challenges of sustainable development. This book, *Artificial Intelligence and Machine Learning Applications for Sustainable Development*, curated by experts in the field, offers a comprehensive exploration of how these technologies can be harnessed to catalyze positive change.

The journey begins with an insightful overview of the evolution of AI and ML, tracing their origins and highlighting the key milestones that have shaped their development. From theoretical foundations to practical applications, readers will gain a deep understanding of the underlying principles driving this transformative field. As we delve deeper, the book navigates through a series of chapters that delve into specific applications of AI and ML in the realm of sustainable development. From leveraging renewable energy sources to optimizing smart city infrastructures, each chapter offers practical insights and real-world case studies that illustrate the potential impact of these technologies on building a more sustainable future. One of the core themes explored throughout the book is the intersection of AI and ML with big data management, cloud computing, image processing, pattern recognition, and signal processing. These chapters shed light on how advanced data analytics techniques can be used to derive valuable insights and inform decision-making processes in various domains related to sustainable development. In addition to addressing technological advancements, this book also examines the broader strategies and frameworks for integrating AI and ML into sustainable development initiatives. From policy considerations to ethical implications, readers will gain a holistic understanding of the multifaceted challenges and opportunities associated with harnessing these technologies for the greater good.

We thank all contributing authors who helped us tremendously with their contributions, time, critical thoughts, and suggestions to assemble this peer-reviewed edited volume. Their dedication and hard work have made this book possible. We also extend our thanks to the reviewers who have provided valuable feedback to the authors and helped improve the quality of the articles. The editors are also thankful to CRC Press and their team members for the opportunity to publish this volume. Lastly, we thank our family members for their love, support, encouragement, and patience during this work.

We hope this book will be valuable for researchers, professionals, and students interested in artificial intelligence and machine learning applications for sustainable development. This book will inspire further research and innovation and contribute to developing new applications and technologies. We look forward to future advancements in artificial intelligence and machine learning applications for sustainable development and hope this book will play a small role in shaping the future of this exciting field.

Prof. A. J. Singh
Dr. Nikita Gupta
Dr. Sanjay Kumar
Dr. Sumit Sharma
Dr. Subho Upadhyay
Dr. Sandeep Kumar

About the editors

A. J. Singh is Professor in the Department of Computer Science at HPU. He has been in this department since 1992. He obtained Bachelor of Engineering in Computer Technology from NIT Bhopal, Master of Science in Distribute Information Systems from the University of East London, UK, under British Government ODASS Scholarship, and Ph.D. from Himachal Pradesh University, Shimla. He worked on deputation to Royal Government of Bhutan under Colombo Plan for three years. He has published more than 50 research papers, supervised 6 Ph.D. dissertations, and 9 students are doing Ph.D. under his supervision.

Nikita Gupta received B. Tech. degree in Electrical and Electronics Engineering from National Institute of Technology, Hamirpur, India, in 2011 and M. Tech. degree in Power System from Delhi Technological University, Delhi, India, in 2014. She has completed her Ph.D. degree in the Department of Electrical Engineering from Delhi Technological University, Delhi, India, in December 2018. Presently, she is working as Assistant Professor at the University Institute of Technology, Himachal Pradesh University, Shimla (H.P.), India. Her research interests include power system engineering, renewable energy systems, power quality, soft computing techniques, and microgrids. Dr. Nikita has published numerous research papers in reputed international journals, including IEEE and IET, 2 book chapters, and 22 research papers in international and national conferences. Additionally, she has also published three edited books and two authored books on B. Tech. machines courses. She has received funding from HIMCOSTE for a project on wind turbines. She is also an active reviewer of journals, including *IEEE Transactions on Power Electronics, IET Power Electronics, Journal of Engineering, IEEE Access, Wiley Journal of Energy*, and many more. Dr. Nikita received the "Commendable Research Award" for excellence in research in 2018 from Delhi Technological University, Delhi, India. She is the recipient of the outstanding paper award from IEEE-GPECOM in 2020. She is the recipient of the best paper award from IEEE-INDICON in 2015. She is a member of various technical communities, including member IEEE, member IAENG, member ISRD and Lifetime Member TERA.

Sanjay Kumar received B. Tech. degree in Electrical Engineering from Himachal Pradesh University, Shimla, India, in 2007 and M. Tech. degree in Power System from the National Institute of Technology, Hamirpur, India, in 2010. He has completed his Ph.D. from the Department of Electrical Engineering at Punjab Engineering College Deemed to be University, Chandigarh, India, in December 2019. Presently, he is working as Assistant Professor in the University Institute of Technology, Himachal Pradesh University, Shimla (H.P.), India. He worked on a project "Development and validation of technology for production of high energy density from rice straw and agri-biomasses" (funding agency PSA, GOI, and Sweden). He published many research papers in reputed international and national journals. His area of research includes Power System, Renewable Energy, Network Planning, Micro-grid, Optimization, GIS, and Machine Learning. He has published several research papers in various journals and conferences. Recently, his paper on renewable energy has been published in a reputed journal. He has to his credit one patent and another one is in process. He has six years of teaching experience at NITs and government institutes. He is reviewer of various conference papers and international journals.

Sumit Sharma received B. Tech. in Electrical Engineering from Maharishi Markandeshwar University Mullana, Haryana, India, in 2013 and M. Tech in Electrical Engineering (Power System) from PEC University of Technology, Chandigarh, India, in 2017. He has completed his Ph. D. in Electrical Engineering from National Institute of Technology Hamirpur, India, in 2022. Presently, he is working as Assistant Professor at KIET Group of Institutions, Ghaziabad. His research interest includes microgrid in rural environment, renewable energy sources, and power system deregulation. He is an active reviewer of various international journals and conference papers, including international publications.

Subho Upadhyay received his ME degree in Electrical and Electronics Engineering from the Birla Institute of Technology, Mesra, India, in 2012 and Ph. D. from the Indian Institute of Technology, Roorkee, India, in August 2017. He is currently working as Assistant Professor in Electrical Engineering Department at Dayalbagh Educational Institute, Agra. Dr. Upadhyay teaches Power Electronics, Electrical Machine, Electrical Measurements, and Fuzzy Systems. His research interests include integrated renewable energy system, power electronics applications in renewable energy, soft computing techniques, and optimization. He has edited the book *Renewable Energy System Modeling, Optimization and Applications*. He is an active reviewer of various international journals and conference papers. He is also working as NSS Program Officer in his institute.

Dr. Sandeep Kumar Professor, School of Computer Science and Artificial Intelligence, SR University, Warangal, India. He completed his postdoctorate from Pentagram Pvt. Ltd. in August 2021. He has good academics and research experience in various areas of electronics and communication. His area of research includes Embedded System, Image Processing, Biometrics, and Machine Learning. He has to his credit 25 patents (17 national and 8 international patents) and has filed 7 patents successfully (6 national and 1 international patents). He has been invited 20 times to be a Guest in Scopus Indexed IEEE/Springer Conferences. He has been invited 12 times to be an expert in various colleges/universities in India. He has published 150 research papers in various international/national journals (including IEEE, Springer, etc.) and proceedings of the reputed international/national conferences (including Springer and IEEE). He has been awarded "Best Paper Presentation" in Nepal and India, respectively, in 2017 and 2018. He was awarded "Best Performer Award" in Hyderabad, India, in 2018. He has also been awarded the "Young Researcher Award" in Thailand in 2018 and the "Best Excellence Award" in Delhi in 2019. He has also been awarded the "Excellence in Academics Award" in Chennai in 2020. He is an active member of 22 various professional international societies. He has been nominated on the board of editors/reviewers of 25 peer-reviewed and refereed journals. He has conducted three international conferences and six workshops. He has also attended 45 seminars, workshops, and short-term courses in IITs. He has supervised 23 M. Tech. and 2 Ph. D. scholars and is currently supervising four Ph. D. scholars and two M. Tech. scholars. His 14 books have been published at the international level and 2 books are in the press for publication.

List of contributors

Altaf Alam
Drone Laboratory
Department of Electrical
Engineering
Rajkiya Engineering College Banda
India

Vinay Anand
School of Electronics and Electrical
Engineering
Lovely Professional University
Phagwara, India

Paul Arévalo
Department of Electrical
Engineering
University of Jaen
Jaen, Spain

Ankit Bhatt
Department of Electrical
Engineering
Graphic Era (Deemed to be
University) Dehradun
India

Antonio Cano
Department of Electrical
Engineering
University of Jaen
Jaen, Spain

Anurag Chauhan
Department of Electrical
Engineering
Rajkiya Engineering College
Banda, India

Sushil Chauhan
Department of Electrical
Engineering
National Institute of Technology
Hamirpur, India

Dev Dhaked
Department of Electrical
Engineering
Faculty of Engineering
Dayalbagh Educational Institute
(Deemed to be University)
Agra, India

Sourav Diwania
KIET Group of Institutions
Ghaziabad, India

Mohammad Riahi Se Gonbad
Urban Planning & Management
University of Tehran
Iran

Varun Gupta
National Institute of Technology
Sikkim
Ravangla, India

Himesh Handa
Department of Electrical
Engineering
NIT Hamirpur
Hamirpur, India

A. K. M. Mahmudul Haque
Department of Political Science
Varendra University
Rajshahi, Bangladesh

Imran Hossain
Department of Political Science
Varendra University
Rajshahi, Bangladesh

Dharmendra K. Jhariya
Electrical Engineering Department
National Institute of Technology
New Delhi, India

Francisco Jurado
Department of Electrical
Engineering
University of Jaen
Jaen, Spain

Abdul Kadir
Department of Computer Science
and Engineering
Dhaka University of Engineering &
Technology
Gazipur, Bangladesh

Seema Kalonia
Banasthali Vidyapith
Tonk
India

Mohd Tauseef Khan
Drone Laboratory
Department of Electrical
Engineering
Rajkiya Engineering College
Banda
India

Amit Kumar
Electrical Engineering Department
National Institute of Technology
New Delhi, India

Aniket Kumar
Department of Electrical
Engineering
National Institute of Technology
Hamirpur, India

Anuj Kumar
Department of Thermal and Energy
Engineering
School of Mechanical Engineering
Vellore Institute of Technology
Vellore, India

Rajesh Kumar
Department of Electrical
Engineering
National Institute of Technology
Hamirpur, India

Rishabh Kumar
School of Mechanical Engineering
VIT University
Vellore, India

Ankur Maheshwari
Department of Electrical
Engineering
National Institute of Technology
Hamirpur, India

Sanjay Kumar Maurya
Department of Electrical
Engineering
GLA University Mathura, India

Meenakshi
Banasthali Vidyapith
Tonk, India

Sara Naghibizadeh
Urban Planning & Management
University of Tehran, Iran

Sahar Dehghan Niri
Urban Planning & Management
University of Tehran
Iran

Danny Ochoa-Correa
Department of Electrical
Electronics and
Telecommunications Engineering
Universidad de Cuenca
Cuenca, Ecuador

Pallav
Department of Electrical
Engineering
NIT Hamirpur
Hamirpur, India

Meenakshi Pareek
Banasthali Vidyapith
Tonk, India

Md. Sohel Rana
Department of Political Science
Barisal University
Barisal, Bangladesh

Salim
KIET Group of Institutions
Ghaziabad, India

Himanshu Sharma
School of Electronics and Electrical
Engineering
Lovely Professional University
Phagwara, India

Sumit Sharma
KIET Group of Institutions
Ghaziabad, India

Kamal Singh
Electrical Engineering Department
National Institute of Technology
New Delhi, India

Mukesh Singh
Department of Electrical
Engineering
National Institute of Technology
Hamirpur, India

Nitin Singh
Department of Electrical
Engineering
Faculty of Engineering
Dayalbagh Educational Institute
(Deemed to be University)
Agra, India

Nitin Singh Singha
Electrical Engineering Department
National Institute of Technology
New Delhi, India

Sonu
Department of Electrical
Engineering
National Institute of Technology
Hamirpur, India

Mahdi Suleimany
Urban Planning & Management
University of Tehran
Iran

Amrita Upadhyay
Banasthali Vidyapith
Tonk, India

Edisson Villa-Ávila
Department of Electrical
Engineering
University of Jaen
Jaen, Spain

Prem Prakash Vuppuluri
Department of Electrical
Engineering
Faculty of Engineering
Dayalbagh Educational Institute
(Deemed to be University)
Agra, India

Acknowledgments

This book has been made possible through the collaboration and support of numerous individuals and institutions. We extend our heartfelt gratitude to all those who have contributed to this endeavor, directly or indirectly. First and foremost, we express our sincere appreciation to the contributing authors whose expertise and dedication have enriched the content of this volume. Their insightful contributions, meticulous research, and innovative ideas have laid the foundation for a comprehensive exploration of artificial intelligence (AI) and machine learning (ML) applications in sustainable development. We are deeply grateful for their commitment to advancing knowledge in this rapidly evolving field.

We would also like to acknowledge the invaluable feedback provided by the reviewers who rigorously evaluated the chapters and offered constructive criticism to enhance the quality and clarity of the content. Their expertise and attention to details have played a crucial role in ensuring the scholarly rigor and integrity of this work. We are indebted to them for their time, expertise, and commitment to academic excellence.

Furthermore, we extend our gratitude to CRC Press and their team members for their unwavering support throughout the publication process. Their professionalism, guidance, and dedication to promoting scholarly work have been instrumental in bringing this book to fruition. We are grateful for the opportunity to collaborate with CRC Press and for their commitment to disseminating knowledge and fostering academic discourse.

In addition, we would like to thank our families for their unwavering support, understanding, and patience during the oft-demanding process of writing and editing this book. Their encouragement, love, and belief in our abilities have been a constant source of strength and motivation throughout this journey. We are profoundly grateful for their unwavering support and encouragement.

Lastly, we express our gratitude to the readers of this book, whether researchers, professionals, or students, for their interest in the intersection of artificial intelligence, machine learning, and sustainable development. It is our hope that this book will serve as a valuable resource and inspire further research, innovation, and collaboration in this important field. Together, we can harness the power of AI and ML to create a more sustainable and equitable world for future generations.

Chapter 1

Comprehensive framework and classification of advanced artificial intelligence and machine learning modelling techniques

Kamal Singh, Amit Kumar, Nitin Singh Singha, and Dharmendra K. Jhariya

1.1 INTRODUCTION

In the modern era, we find ourselves immersed in a technological land-scape where rapid transformations occur across various areas: technology, industries, societal norms, and operational procedures [1]. These shifts are driven by increased interconnectivity and the implementation of intelligent automation systems. The ongoing revolution profoundly affects industries worldwide, driving rapid and non-linear transformations. These changes occur at an unprecedented pace, with far-reaching implications for various fields, economies, and societies. Within this context, three essential criteria have emerged: automation, intelligence, and smart computing. Automation entails minimizing human intervention in operations, while intelligence refers to the capability to derive insights from data. Smart computing refers to systems or devices that can observe, analyse, and report on their own performance. These criteria are integral to the development of modern applications and systems, influencing various facets of our daily lives. This is more pronounced than ever before.

Nowadays, the adoption of advanced smart technologies empowers organizations to make intelligent and quick decisions in their business processes. This enhances overall productivity and profitability, with artificial intelligence (AI) playing a pioneering role in this domain. In our rapidly evolving technological landscape, the upheaval of AI parallels prior industrial upheavals that sparked significant economic operations across trade, transportation, manufacturing, and various other domains. AI research is a focal point within the scientific community, delving into both theoretical aspects and practical applications across society. Machine learning (ML) enables predictions of new data properties based on known training data. Within ML, deep learning (DL) has gathered notable interest in the past. These AI methods find practical applications in diverse fields, with computer science, engineering, and mathematics being dominant areas of study

DOI: 10.1201/9781003581246-1

1

[2–4]. In the realm of computer science, researchers explore a wide range of topics. These studies focus on areas such as image processing, analysis of signals [5, 6], processing of natural language [7, 8], security, and the creation of hardware and smart software (including fascinating innovations like brain–computer interfaces) [9, 10]. These investigations rely on various techniques, including classification methods [11, 12], data clustering [13], and visualization [14]. These approaches extend beyond computer science and find applications in diverse domains. AI plays a pivotal role in shaping modern industries. It significantly impacts fields like trade, logistics, automated manufacturing, and banking. Researchers also contemplate AI's potential in various sectors of the economy [15, 16]. However, while many studies discuss the feasibility of AI applications as shown in Figure 1.1, practical implementation faces challenges. We systematically analyse the difficulties and challenges encountered during this process. Sustainable

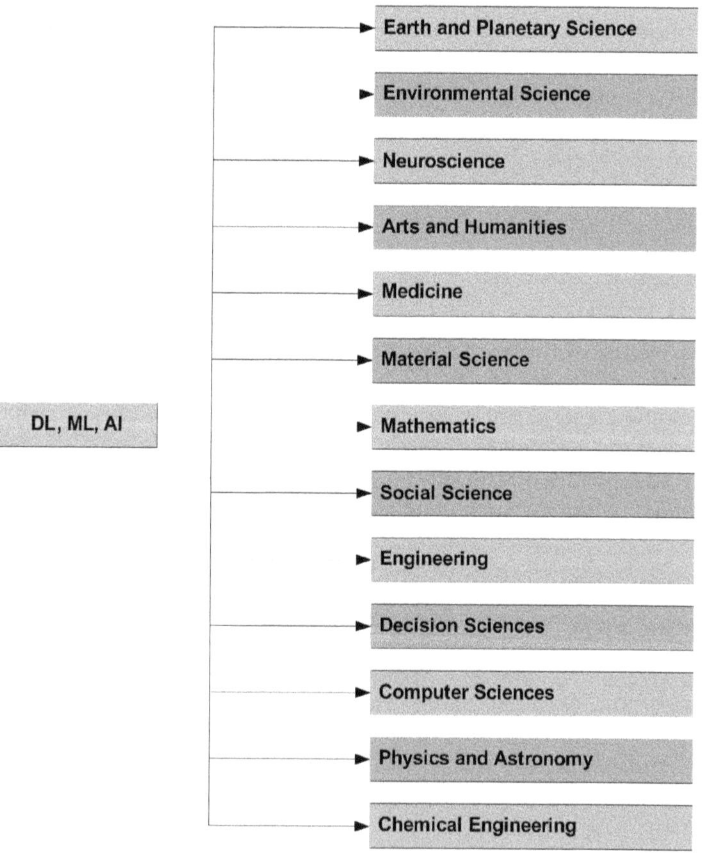

Figure 1.1 Area of applications for AI, ML, and DL.

development necessitates a shift towards novel management methods and technologies.

It also involves additional resources within the region, such as agricultural goods, livestock management, recently discovered mineral resources, and human resources. The lack of widespread adoption of modern technologies can reduce efficiency and promote the expansion of industries that offer little value addition.

1.2 AI AND ML TECHNOLOGIES CLASSIFICATION

As depicted in Figure 1.2, the environment of AI comprises five key components:

1. *Machines:* These are the hardware and software systems that process information and perform tasks.
2. *Human intelligence:* Our own cognitive abilities and decision-making processes play a crucial role in shaping AI systems.
3. *ML algorithms:* These are mathematical frameworks enabling machines to learn from data and enhance their performance progressively.
4. *Internet of Things (IoT):* A system comprising interconnected devices and sensors for collecting and exchanging data.
5. *Internet of Everything (IoE):* An extension of IoT, IoE includes not only devices but also people, processes, and data.

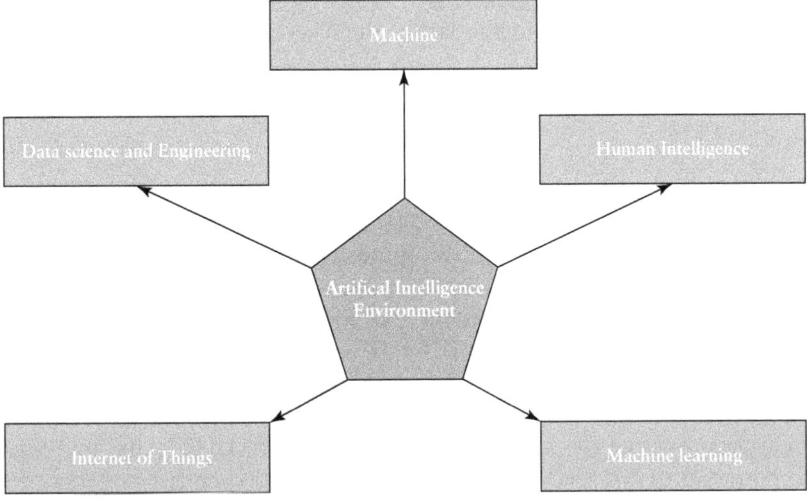

Figure 1.2 AI environment.

In both non-AI- and AI-based environments, machines serve as fundamental and implicit components. Human intelligence, a fascinating trait that distinguishes us from animals and even from other humans, is important in the field of AI. Identifying the most effective human intelligence for problem-solving is crucial. This intelligence is then embedded into machines, transforming them into smart entities capable of carrying out tasks using a set of instructions (commonly known as programs or code).

ML serves as an efficient platform for AI developers and programmers. ML algorithms enable self-learning within AI systems by leveraging their environment and accumulated experiences. These algorithms are pivotal in predicting future events and trends based on the data available.

The IoE and IoT have a close relationship with the world of AI. Here's how they connect:

1. *Real-time data and decision-making:*
   - IoE and IoT rely on sensor technology to generate real-time data. This data becomes crucial for decision-making.
   - An intelligent program, coded by humans, utilizes this sensor data. The program operates within the AI environment, making informed choices based on the incoming information.
2. *Smarter machines:*
   - By integrating sensor data, the machine becomes smarter. It can adjust, react, and optimize its behaviours according to the surroundings.
   - Imagine a factory where machines adjust production based on real-time sensor inputs – this is the power of combining IoE, IoT, and AI.
3. *Data science and engineering:*
   - Data analysis is a vital part of the AI ecosystem. It ensures efficient decision-making.
   - When the machine processes data effectively, it can optimize its operations through its programming.

Many AI applications heavily rely on ML techniques, which embody the foundational principles of AI [17]. ML is utilized to enhance outcomes in various domains such as speech recognition [18], emotional speech analysis [19], economic planning [20], and manufacturing control [21]. According to Ref. [22], ML serves as a potent tool for data analysis and finds applications in diverse expert systems [23]. Currently, ML stands out as a primary focus of research in the field of robotics [24], where it is frequently employed to tackle both scientific and applied challenges. For instance, researchers explore the applicability conditions of ML [25] and investigate the potential of DL methodologies [26] to address chemistry-related problems. ML finds extensive use across multiple disciplines, including medicine [27, 28], particularly in the realm of medical imaging as well as in astronomy [29], computational biology [30, 31], agriculture [32], urban planning [33], industrial processes [34], construction [35], environmental modelling [36], geo-ecological assessments [37], petrographic analysis [38], exploration [39], and mining forecasts [40], among others. ML serves as a vital component in contemporary research within the natural language processing field [41, 42], driving advancements and innovation.

1.3 MACHINE LEARNING

ML is a component of AI that uses optimized algorithms to extract valuable information from data, enabling machines to learn and make accurate predictions in real-world applications (Figure 1.3) [43, 44]. ML stands as a potent instrument for comprehensive data analysis, offering versatile applicability within expert systems. Presently, it constitutes a principal focus of research within the domain of robotics, reflecting its pivotal role in advancing the field's methodologies and capabilities [45, 46].

Figure 1.4 illustrates the classifications of AI algorithms, showcasing the diverse categorization methods within the field. ML operates through four distinct structures [47].

1.3.1 Supervised learning

This is a ML approach used when defined objectives need to be achieved with a provided set of inputs. This method employs labelled training data and examples to derive a function, employing a "task-driven strategy." By utilizing labelled data, algorithms are trained to classify information or predict outcomes. An illustrative example of a supervised learning application is the detection of spam-like emails. In supervised learning, the primary

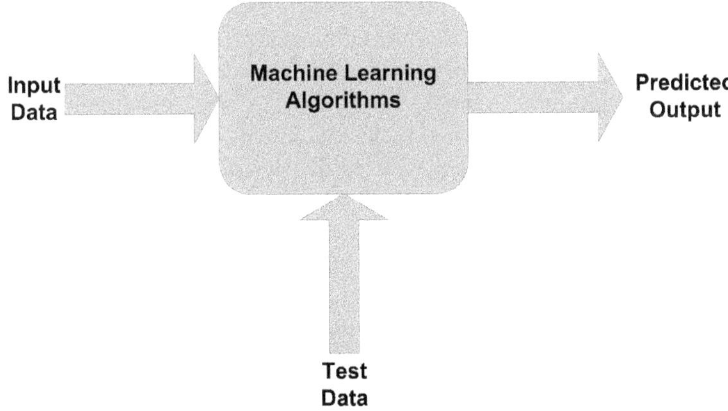

Figure 1.3 Basic ML model.

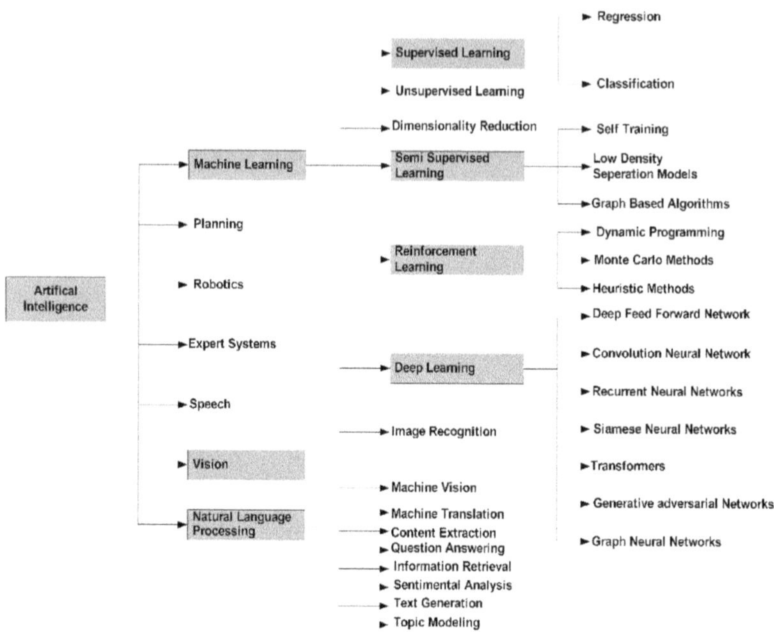

Figure 1.4 Classifications of AI algorithms.

tasks typically involve classification, where the objective is to predict a label assigned to each input, and regression, which entails predicting a numerical quantity [48, 49].

1.3.2 Unsupervised learning

This approach autonomously analyses datasets without labels, representing a data-driven process without human intervention [50]. It is often utilized for identifying significant trends, extracting features, grouping outcomes, and performing analysis. Typical tasks include feature learning, clustering, association rule discovery, estimation of density, reduction of dimensionality, and detection of anomaly, among others [51].

1.3.3 Semi-supervised learning

This approach is considered a hybrid approach, incorporating elements from both the aforementioned techniques by leveraging both labelled and unlabelled data for model training. This approach proves effective in enhancing model performance, especially in scenarios where data labelling needs to be automated without direct human interaction. An illustrative example is the classification of online content or textual material, where a semi-supervised learning model can demonstrate utility [52, 53]. The main aim of semi-supervised learning is to produce improved predictive outcomes compared to using only labelled data [54, 55].

1.3.4 Reinforcement learning

This is an ML algorithm that allows agents and machines to independently assess optimal behaviour within specific environments to enhance efficiency [56]. It functions as an environment-based approach, where the environment is commonly represented as a Markov decision process, and decisions are determined by a reward function [57]. This learning paradigm relies on the idea that involves rewards or penalties, intending to leverage insights acquired from interactions with the environment to optimize actions, maximizing rewards or minimizing risks.

1.4 ML CLASSIFICATION ALGORITHMS

Classification involves categorizing data into distinct groups based on labelled attributes. It is categorized into binary and multiclass classification. In binary classification, the algorithm assigns new data to one of the two classes present in the dataset, typically represented as 0 or 1, corresponding to real-world labels such as "Yes" or "No." Unlike regression, the output in binary classification is a class label rather than a numerical value. On the other hand, multiclass classification involves dividing a dataset into multiple classes, extending beyond the binary scenario, with each class identified by a specific label [58, 59]. Figure 1.5 illustrates various ML classification algorithms utilized in the field. The following are discussions on various ML classification algorithms:

1.4.1 Support vector machine (SVM)

SVM operates on the principle of generating a hyperplane that efficiently divides data points into separate categories. Serving as a linear model, SVM is versatile, capable of addressing both regression and classification problems and accommodating both linear and non-linear scenarios. SVM, a supervised ML technique [60], is utilized for solving regression and classification tasks, facilitating the identification of optimal boundaries between different outputs. The objective is to construct an n-dimensional hyperplane that effectively divides assigning instances or data points to possible classes, positioned as closely as possible to the data points [61]. Despite the absence of definitive principles for kernel selection, SVM exhibits robust performance across different applications.

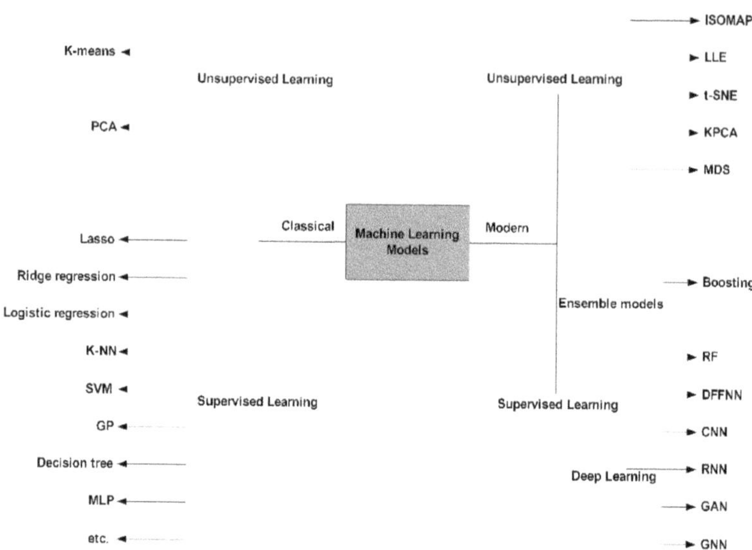

Figure 1.5 ML classification algorithms.

1.4.2 Logistic regression (LR)

LR is primarily employed for solving classification problems, utilizing a sigmoid function to predict values ranging between 0 and 1. It models the output as a categorical dependent variable, typically represented as binary values, such as 0 or 1, yes or no, true or false, and so on. LR serves as a supervised learning classification approach aimed at estimating the likelihood of a target variable. However, it is advisable to avoid LR when there is an insufficient amount of data relative to the number of features to prevent overfitting [62]. LR is utilized to classify data points into binary categories, whereas categorical classification is employed to classify outputs into two distinct classes (0 or 1). Within LR, two critical components are the sigmoid curve and the hypothesis. The sigmoid curve, derived from the hypothesis, enables the calculation of event probabilities. The hypothesis data can be fitted into a logarithmic function to generate a sigmoid curve, characterized by an S-shaped curve. This logarithmic function also facilitates the prediction of class categories [63].

1.4.3 Decision trees

Decision trees are flexible classification tools that construct a tree-shaped structure to make decisions using feature values. They are easy to interpret and commonly used for various classification tasks. They can map numerous outcomes using a set of inputs in ML [64–66].

1.4.4 Random forest

It is an ensemble learning method that improves classification accuracy and reliability by combining multiple decision trees. It constructs numerous trees during training and combines their predictions by taking the mode for classification tasks or the average for regression tasks. This classifier is widely used in ML and data science across various applications [67].

1.4.5 k-Nearest neighbours (KNN)

KNN is a lazy learning approach. Instead of building an internal model, it retains all training data instances within an n-dimensional space. When classifying new data points, KNN calculates their classification using similarity measures like the Euclidean distance function. It determines classification through a majority vote of the k-nearest neighbours for each point. While KNN demonstrates resilience to noisy training data, its accuracy is contingent on the quality of the data. However, choosing the optimal number of neighbours remains challenging. Notably, KNN can handle both regression and classification tasks [68, 69].

1.4.6 Naïve Bayes (NB)

The NB algorithm relies on Bayes' theorem, making the assumption of feature independence [70], and is widely used for binary and multiclass categorization tasks in various real-world scenarios like document classification and spam filtering. Utilizing the NB classifier [71] proves beneficial for accurately classifying noisy data instances and building resilient prediction models.

1.5 GRADIENT DESCENT METHOD

It is a fundamental optimization technique utilized in training ML models. It works on the basis of a convex function, where the model's weights are adjusted iteratively in the opposite direction of the gradients of the loss or cost function. These adjustments aim to gradually approach the optimal point of the objective function. The learning rate, represented by η, controls the size of each step in this process, affecting the speed of convergence and the number of iterations required to reach a minimum. The main goal of gradient descent is to minimize the loss or cost function, which measures the difference between predicted and actual outputs. The loss function, often denoted as $\ell\left(h\left(x;\beta\right),y\right)$, quantifies the deviation between the model's predictions and the true values. Gradient descent aims to find the minimum point on the convex function curve, which signifies either a local or a global minimum and is referred to as the point of convergence. The method plays a pivotal role in optimizing ML models [72, 73]. Gradient descent can be categorized into two main types for further analysis:

1.5.1 Batch gradient method

This is a standard gradient descent approach. In this approach, the algorithm computes all samples concurrently, instead of processing a single sample from the entire dataset per iteration. The rules for updating weights are as follows:

$$\beta_{n+1} = \beta_n + \delta\beta \tag{1.1}$$

where $=-\eta\nabla C(\beta)$: the loss/objective function.

$$F(\beta) = \frac{1}{k}\sum_{i=1}^{k}\nabla f_i(\beta) \tag{1.2}$$

The update rule for weights can be expressed as follows:

$$\beta_{n+1} \leftarrow \beta_n - \frac{\eta_n}{k}\sum_{i=1}^{k}\nabla f_i(\beta_t) \tag{1.3}$$

$$F(\beta) = \ell\big(h(x;\beta),y\big) \tag{1.4}$$

Here $h(x;\beta)$ represents the predicted output for input x, and y denotes the desired output.

1.5.2 Stochastic gradient method

Classical gradient methods become inefficient when handling large datasets in solving optimization problems in ML. However, the stochastic method can address this inefficiency. This approach selects one sample randomly per iteration instead of computing the complete gradient across the entire dataset each time [74].

$$\beta_{n+1} \leftarrow \beta_n - \eta_n\nabla f_{in}(\beta_n) \tag{1.5}$$

$n \in N := \{1,2,3,4...\}$, where i_n is randomly selected from the set {1, 2, 3, ..., n}. η_n represents a non-negative step size. $\nabla f_{in}(\beta_n)$ relates to computing a single sample.

The classification algorithms discussed here are merely a subset of the extensive array of techniques within the realm of ML tailored for classification tasks. Each algorithm boasts unique strengths and weaknesses, pivotal factors in determining the most suitable approach derived from the distinct attributes of the data, the problem domain, and computational constraints.

1.6 DEEP LEARNING

DL, a subset of ML, utilizes algorithms that gain knowledge from various hierarchical levels to represent complex data relationships. It is characterized by deep architecture where high-level

features are built upon lower-level ones. Mainly based on unsupervised learning, DL integrates neural networks, graphical modelling, and optimization, enabling significant advancements in information processing. Its popularity stems from enhancing computer processing capabilities, accommodating large datasets, and driving recent breakthroughs in ML [75]. DL, a subset of AI, utilizes computer optimization algorithms modelled after artificial neural networks (ANNs) to replicate human cognitive processes [76]. ANNs consist of interconnected nodes or artificial neurons that imitate the arrangement of human neural networks. Through learning mechanisms, such as reinforcement, synapses in ANNs strengthen, enabling the development of problem-solving capabilities. Within the scope of DL, the idea of transfer learning (TL) often arises. Figure 1.6 depicts various DL models utilized in the field. The following are discussions on the various DL architectures:

1. **Feedforward neural networks (FNNs)**

    FNNs are the basic structure of deep neural networks, comprising input, hidden, and output layers. Data flows unidirectionally from the input through the hidden layers to the output. FNNs are frequently applied in tasks like image classification and regression [77].

2. **Convolutional neural networks (CNNs)**

    CNNs are particularly effective for processing grid-like data, such as images. Figure 1.7 illustrates the basic CNN model, showcasing its fundamental architecture and capabilities in handling image data. CNNs have revolutionized tasks like image recognition, object detection, and image segmentation [78]. These architectures enable the identification of complex patterns within the input data, which remain consistent regardless of their position in the input signal vector (Table 1.1). Examples include horizontal or vertical lines, as well as other distinctive features present in images. The formation of convolutional filters, specifically the adjustment of neuron weights modelling such filters, occurs during the network training process. The utilization of

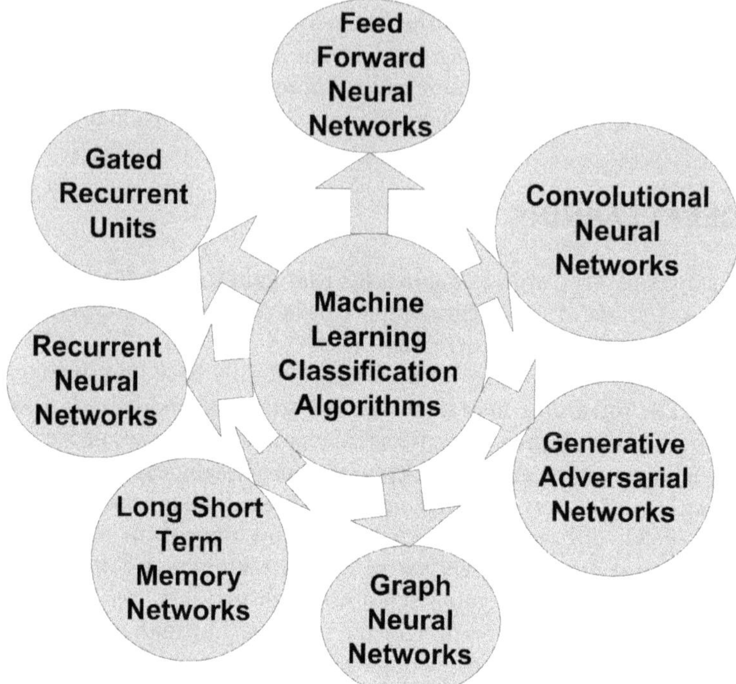

Figure 1.6 DL models.

convolution operations helps maintain a reasonable level of computational complexity during training.

3. **Recurrent neural networks (RNNs)**

RNNs are tailored for processing sequential data, where the order of input elements matters. They possess loops within their architecture, allowing them to retain information about past inputs. RNNs are widely used in time series analysis [79]. The one-to-many architecture is utilized when a short input sequence leads to the generation of longer data or signals. This is especially applicable in contexts like music generation [80, 82] or text generation [81].

RNNs can process sequential data, even of variable lengths. $x(t)$ is the signal that arrives at moment t and shapes the network's internal state; subsequent signals $x(t + 1)$ are added to it. As a result, the final output y is influenced by the entire sequence of signals. Recurrent networks are categorized into four main classes based on input and output sequences. It is worth noting that a FFNN can be viewed as an RNN with a one-to-one architecture (see Figure 1.8). The one-to-many architecture addresses scenarios where a short input sequence generates long output sequences, as seen in music [83] or text generation

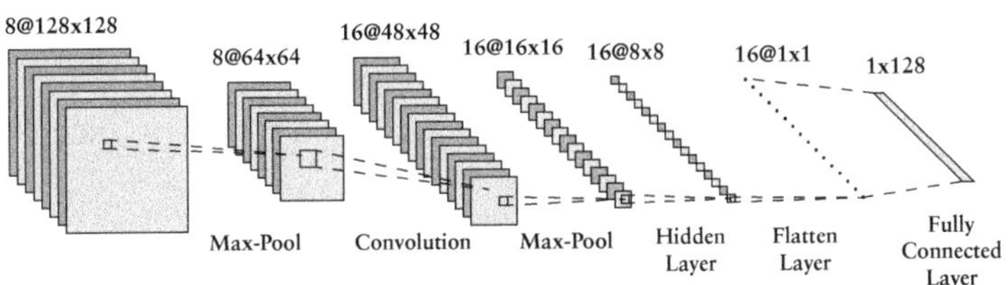

Figure 1.7 Basic CNN model.

Table 1.1 CNN parameters

Layer type	Kernel size	Number of kernels	Stride	Padding	Activation	Pooling	Pooling size	Normal-ization
Convolutional	128 × 128	8	1	Same	ReLU	MaxPooling	64 × 64	Batch-Norm
Convolutional	48 × 48	16	1	Same	ReLU	MaxPooling	16 × 16	Batch-Norm
Flatten	–	–	–	–	–	–	–	–
Fully connected	–	128	–	–	ReLU, Softmax	–	–	–

[84] tasks focused on setting style or theme. The many-to-one architecture is employed for classification purposes. For tasks like machine translation [85] and speech recognition [86], the many-to-many architecture is utilized. The evolution of RNNs led to the development of LSTM and transformer models like BERT [87], ELMO [88], GPT, and generative adversarial networks [89–92]. These models have gained popularity for their effectiveness in addressing natural language processing challenges.

1.7 APPLICATIONS

In today's competitive landscape, organizations across various industries strive to turn their data into actionable insights. AI and ML play an important role in achieving this transformation. By integrating AI and ML into their systems and strategic plans, leaders gain the ability to swiftly comprehend and act upon data-driven insights. The applications, as shown in Figure 1.9, of AI and ML are vast, offering opportunities for sustainable development. These technologies involve collaboration among stakeholders from diverse countries and sectors: some key areas where AI/ML make a significant impact:

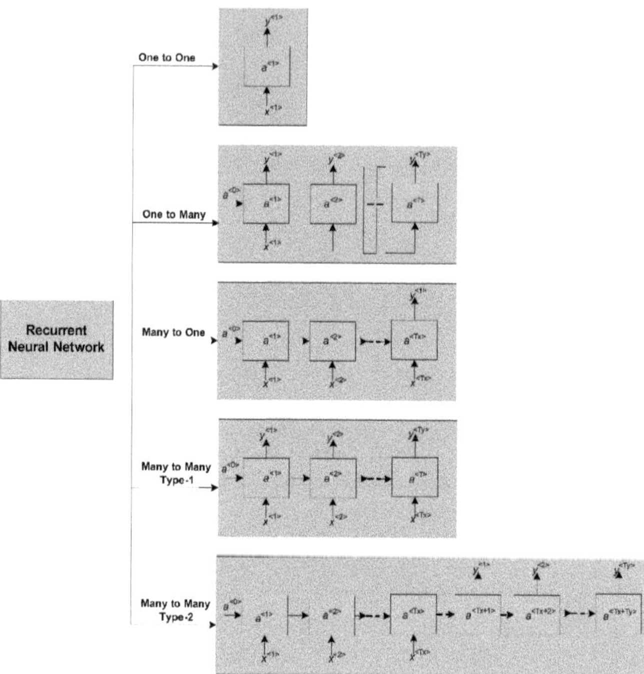

Figure 1.8 Basic RNN architecture.

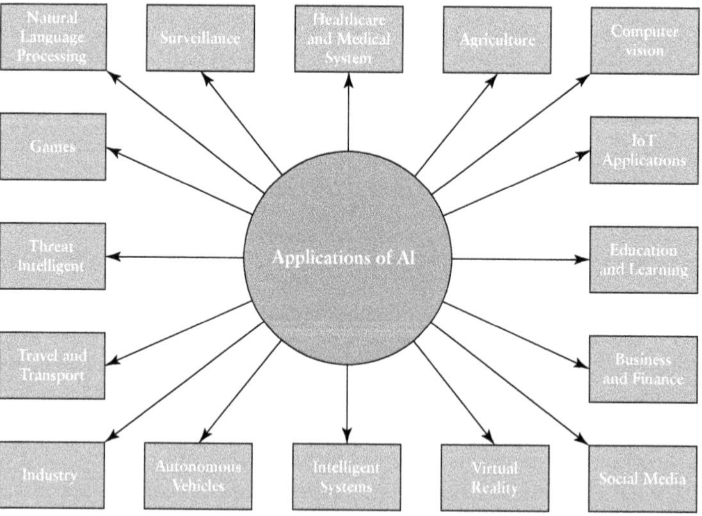

Figure 1.9 Various applications of AI.

- *Inventory and supply chain management:* AI algorithms optimize inventory levels, streamline logistics, and enhance supply chain efficiency.
- *Predictive maintenance:* By analyzing historical data, ML models predict equipment failures, allowing proactive maintenance and minimizing downtime.
- *Production optimization:* AI techniques fine-tune manufacturing processes, leading to improved product quality and resource utilization.

Researchers have extensively studied classification algorithms applied to renewable energy problems. These algorithms serve as powerful tools for predictive analysis, data preprocessing, and result interpretation, ultimately enhancing energy and resource management. Table 1.2 showcases a variety of AI techniques with application areas, demonstrating the broad scope and versatility of AI across different domains.

1.8 ADVANTAGES AND LIMITATIONS OF AI AND ML

AI and ML techniques offer numerous advantages, such as automation of repetitive tasks, improved decision-making based on data analysis, zero risks, unbiased decision, and the ability to handle large volumes of data efficiently. However, they also have limitations. For instance, they can sometimes produce biased results based on the data they were trained on, require substantial computing power and data to train effectively, and may lack the ability to understand context or make judgments like humans.

1.9 FUTURE CHALLENGES FOR DEVELOPING NEW ML

One future challenge for developing new ML techniques is ensuring robustness and reliability, especially in dynamic and complex environments. Another challenge is addressing ethical concerns related to privacy, fairness, and bias in ML algorithms. Additionally, advancing interpretability and explainability of ML models to foster trust and understanding is crucial.

1.10 CONCLUSION

AI and ML are rapidly transforming our world through automation and intelligent computing. AI's power comes from the collaboration between powerful machines, human expertise, and ML algorithms. These algorithms, like supervised learning for labelled data and unsupervised learning for unlabelled data, empower organizations to make data-driven decisions that promote sustainability. DL architectures take this a step further by

Table 1.2 Various AI techniques and application areas

AI techniques	Application areas	Tasks	Reference
Hybrid approach, searching, and optimization	Cybersecurity	Optimum feature selection	[93]
	Mobile application	Decision-making	[94]
	Recommendation systems	Hotel recommendation	[95]
	Sentiment analysis	Tweet sentiment accuracy analysis	[96]
Visual analytics, computer vision, and pattern recognition	Visual analytics	Navigation mark classification	[97]
	Healthcare	Cervical cancer diagnostics	[98]
	Computer vision	Human fall detection	[99]
Machine learning	Cybersecurity	Anomaly and attack detection	[100]
	Healthcare	COVID-19 aid	[101]
	Smartcity	Smart parking pricing system	[102]
Neural network and deep learning	Cybersecurity	Malware detection	[103]
	Healthcare	Diagnosis of COVID-19	[104]
	Smart agriculture	Plant disease detection	[105]
	Visual recognition	Facial expression analysis	[106]
Rule-based modelling and decision-making	Intelligent systems	Mining contextual rules	[107]
	Healthcare	Identifying risk factors	[108]
	Recommendation system	Web page recommendation	[109]
	Smart systems	Risk prediction	[110]
Visual analytics, computer vision, and pattern recognition	Healthcare	Cervical cancer diagnostics	[111]
	Computer vision	Human fall detection	[112]
	Visual analytics	Navigation mark classification	[113]

uncovering complex patterns in data, impacting everything from supply chains to healthcare.

Despite challenges like bias and job displacement, collaboration across sectors and continuous research offer a bright future. By harnessing AI's potential for efficiency and optimization, we can create a smarter and more sustainable world.

REFERENCES

1. Maynard, A.D. Navigating the fourth industrial revolution. *Nat Nanotechnol.* 2015, 10(12), 1005–1006.
2. Ghasemi, Y.; Jeong, H.; Choi, S.H.; Park, K.-B.; Lee, J.Y. Deep learning-based object detection in augmented reality: A systematic review. *Comput. Ind.* 2022, 139, 103661.

3. Widdows, D.; Kitto, K.; Cohen, T. Quantum mathematics in artificial intelligence. *J. Artif. Intell. Res.* 2021, 72, 1307–1341.

4. Panetto, H.; Iung, B.; Ivanov, D.; Weichhart, G.; Wang, X. Challenges for the cyber-physical manufacturing enterprises of the future. *Annu. Rev. Control.* 2019, 47, 200–213.

5. Izonin, I.; Tkachenko, R.; Peleshko, D.; Rak, T.; Batyuk, D. Learning-based image super-resolution using weight coefficients of synaptic connections. In *Proceedings of the 2015 Xth International Scientific and Technical Conference "Computer Sciences and Information Technologies" (CSIT)*, Lviv, Ukraine, 14–17 September 2015; pp. 25–29.

6. Shen, D.; Wu, G.; Suk, H.-I. Deep learning in medical image analysis. *Annu. Rev. Biomed. Eng.* 2017, 19, 221–248.

7. Barakhnin, V.; Duisenbayeva, A.; Kozhemyakina, O.Y.; Yergaliyev, Y.; Muhamedyev, R. The automatic processing of the texts in natural language. Some bibliometric indicators of the current state of this research area. In *Journal of Physics: Conference Series*, IOP Publishing: Bristol, UK, 2018; p. 012001.

8. Hirschberg, J.; Manning, C.D. Advances in natural language processing. *Science* 2015, 349, 261–266.

9. Kim, D.; Kim, S.-H.; Kim, T.; Kang, B.B.; Lee, M.; Park, W.; Ku, S.; Kim, D.; Kwon, J.; Lee, H. Review of machine learning methods in soft robotics. *PLoS ONE* 2021, 16, e0246102.

10. Torres, E.P.; Torres, E.A.; Hernández-Álvarez, M.; Yoo, S.G. EEG-based BCI emotion recognition: A survey. *Sensors* 2020, 20, 5083.

11. Kuchin, Y.; Mukhamediev, R.; Yakunin, K.; Grundspenkis, J.; Symagulov, A. Assessing the impact of expert labelling of training data on the quality of automatic classification of lithological groups using artificial neural networks. *Appl. Comput. Syst.* 2020, 25, 145–152.

12. Kotsiantis, S.B.; Zaharakis, I.; Pintelas, P. Supervised machine learning: A review of classification techniques. *Emerg. Artif. Intell. Appl. Comput. Eng.* 2007, 160, 3–24.

13. Hastie, T.; Tibshirani, R.; Friedman, J. Unsupervised learning. In *The Elements of Statistical Learning*, Springer: Berlin/Heidelberg, Germany, 2009; pp. 485–585.

14. Van der Maaten, L.; Hinton, G. Visualizing data using t-SNE. *J. Mach. Learn. Res.* 2008, 9, 2579–2605.

15. Zhao, H. Assessing the economic impact of artificial intelligence. ITU Trends. In *Emerging Trends in ICTs*, Morgan Kaufmann Publishers: Burlington, MA, USA, 2018.

16. Financial climate in the Republic of Kazakhstan. Available online: https://www2.deloitte.com/kz/ru/pages/about-deloitte/ articles/financial_climate _in_kazakhstan.html

17. Szczepanski, M. *Economic Impacts of Artificial Intelligence (AI)*. EPRS: European Parliamentary Research Service, 2019.

18. Watanabe, S.; Hori, T.; Karita, S.; Hayashi, T.; Nishitoba, J.; Unno, Y.; Soplin, N.E.Y.; Heymann, J.; Wiesner, M.; Chen, N. Espnet: End-to-end speech processing toolkit. *arXiv* 2018, arXiv:1804.00015.

19. Kerkeni, L.; Serrestou, Y.; Mbarki, M.; Raoof, K.; Mahjoub, M.A.; Cleder, C. Automatic speech emotion recognition using machine learning. In *Social Media and Machine Learning*, IntechOpen: London, UK, 2019.

20. An, J.; Mikhaylov, A.; Sokolinskaya, N. Machine learning in economic planning: Ensembles of algorithms. In *Journal of Physics: Conference Series*, IOP Publishing: Bristol, UK, 2019; p. 012126.

21. Usuga Cadavid, J.P.; Lamouri, S.; Grabot, B.; Pellerin, R.; Fortin, A. Machine learning applied in production planning and control: A state-of-the-art in the era of industry 4.0. *J. Intell. Manuf.* 2020, 31, 1531–1558.

22. Ogidan, E.T.; Dimililer, K.; Ever, Y.K. Machine learning for expert systems in data analysis. In *Proceedings of the 2018 2nd International Symposium on Multidisciplinary Studies and Innovative Technologies (ISMSIT)*, Ankara, Turkey, 19–21 October 2018; pp. 1–5.

23. Prasad, B.; Prasad, P.; Sagar, Y. An approach to develop expert systems in medical diagnosis using machine learning algorithms (asthma) and a performance study. *Int. J. Soft Comput.* 2011, 2, 26–33.

24. Mosavi, A.; Varkonyi, A. Learning in robotics. *Int. J. Comput. Appl.* 2017, 157, 8–11.

25. Artrith, N.; Butler, K.T.; Coudert, F.-X.; Han, S.; Isayev, O.; Jain, A.; Walsh, A. Best practices in machine learning for chemistry. *Nat. Chem.* 2021, 13, 505–508.

26. Mater, A.C.; Coote, M.L. Deep learning in chemistry. *J. Chem. Inf. Modeling* 2019, 59, 2545–2559.

27. Cruz, J.A.; Wishart, D.S. Applications of machine learning in cancer prediction and prognosis. *Cancer Inform.* 2006, 2, 59–77.

28. Miotto, R.; Wang, F.; Wang, S.; Jiang, X.; Dudley, J.T. Deep learning for healthcare: Review, opportunities and challenges. *Brief. Bioinform.* 2018, 19, 1236–1246.

29. Ball, N.M.; Brunner, R.J. Data mining and machine learning in astronomy. *Int. J. Mod. Phys.* D 2010, 19, 1049–1106.

30. Chicco, D. Ten quick tips for machine learning in computational biology. *BioData Min.* 2017, 10, 1–17.

31. Zitnik, M.; Nguyen, F.; Wang, B.; Leskovec, J.; Goldenberg, A.; Hoffman, M.M. Machine learning for integrating data in biology and medicine: Principles, practice, and opportunities. *Inf. Fusion* 2019, 50, 71–91.

32. Liakos, K.G.; Busato, P.; Moshou, D.; Pearson, S.; Bochtis, D. Machine learning in agriculture: A review. *Sensors* 2018, 18, 2674.

33. Mahdavinejad, M.S.; Rezvan, M.; Barekatain, M.; Adibi, P.; Barnaghi, P.; Sheth, A.P. Machine learning for Internet of Things data analysis: A survey. *Digit. Commun. Netw.* 2018, 4, 161–175.

34. Farrar, C.R.; Worden, K. *Structural Health Monitoring: A Machine Learning Perspective.* John Wiley & Sons: New York, NY, USA, 2012.

35. Lai, J.; Qiu, J.; Feng, Z.; Chen, J.; Fan, H. Prediction of soil deformation in tunnelling using artificial neural networks. *Comput. Intell. Neurosci.* 2016, 1–16

36. Recknagel, F. Applications of machine learning to ecological modelling. *Ecol. Model.* 2001, 146, 303–310.

37. Tatarinov, V.; Manevich, A.; Losev, I. A system approach to geodynamic zoning based on artificial neural networks. *Gorn. Nauk. I Tekhnologii Min. Sci. Technol.* 2018, 3, 14–25.

38. Kuchin, Y.I.; Mukhamediev, R.I.; Yakunin, K.O. One method of generating synthetic data to assess the upper limit of machine learning algorithms performance. *Cogent Eng.* 2020, 7, 1718821.
39. Chen, Y.; Wu, W. Application of one-class support vector machine to quickly identify multivariate anomalies from geochemical exploration data. *Geochem. Explor. Environ. Anal.* 2017, 17, 231–238.
40. Mukhamediev, R.I.; Kuchin, Y.; Amirgaliyev, Y.; Yunicheva, N.; Muhamedijeva, E. Estimation of filtration properties of host rocks in sandstone-type uranium deposits using machine learning methods. *IEEE Access* 2022, 10, 18855–18872.
41. Goldberg, Y. A primer on neural network models for natural language processing. *J. Artif. Intell. Res.* 2016, 57, 345–420.
42. Sadovskaya, L.L.; Guskov, A.E.; Kosyakov, D.V.; Mukhamediev, R.I. Natural language text processing: A review of publications. *Artif. Intell. Decis. Mak.* 2021, 95–115.
43. Deisenroth, M.P.; Faisal, A.A.; Ong, C.S. *Mathematics for Machine Learning.* Cambridge University Press, UK 2020.
44. Singh, K.; Singh, N.S. Performance analysis of large-scale machine learning optimization algorithms. In *2023 IEEE 12th International Conference on Communication Systems and Network Technologies (CSNT)*, Bhopal, India, 2023, pp. 226-230.
45. Prasadl, B., et al. An approach to develop expert systems in medical diagnosis using machine learning algorithms (a STHMA) and a performance study. *Int. J. Soft Comput.* 2011, 2, 26–33.
46. Mosavi, A.; Varkonyi, A. Learning in robotics. *Int. J. Comput. Appl.*, 2017, 157(1), 0975–8887.
47. Mohammed, M.; Khan, M.B.; Bashier Mohammed, B.E. *Machine Learning: Algorithms and Applications.* CRC Press; US, 2016.
48. Sarker, I.H.; Kayes, A.S.M.; Badsha, S.; Alqahtani, H.; Watters, P. Ng A. Cybersecurity data science: An overview from machine learning perspective. *J Big Data.* 2020, 7(1), 1–29.
49. Sarker, I.H. AI-based modeling: techniques, applications and research issues towards automation, intelligent and smart systems. *SN Comput. Sci.* 2022, 3(2), 158.
50. Han, J.; Pei, J.; Kamber, M. *Data Mining: Concepts and Techniques.* Elsevier: Amsterdam, 2011.
51. Sarker, I.H. Machine learning: algorithms, real-world applications and research directions. *SN Comput. Sci.* 2021, 2, 160.
52. Zhu, X.,; Goldberg, A. Introduction to semi-supervised learning. *Synthes. Lect. Artif. Intel. Mach. Learn.* 2009, 3(1), 1–130.
53. Chapelle, O.; Scholkopf, B.; Zien, A. *Semi-Supervised Learning,* Vol. 2. MIT Press: Cambridge, 2009.
54. Kaelbling, L.P.; Littman, M.L.; Moore, A.W. Reinforcement learning: A survey. *J. Artif. Intell. Res.* 1996, 4, 237–285.
55. Bellman, R. A Markovian decision process. *J. Math. Mech.* 1957, 2, 679–684.
56. Joy, J.; Selvan, M.P. A comprehensive study on the performance of different multi-class classification algorithms and hyperparameter tuning

techniques using optuna. In *2022 International Conference on Computing, Communication, Security and Intelligent Systems (IC3SIS)*, Kochi, India; 2022, pp. 1–5.

57. Hastie, T.; Tibshirani, R.; Friedman, J. *The Elements of Statistical Learning: Data Mining, Inference, and Prediction*. Springer Science & Business Media, 2009.

58. Guermeur, Y. Combining discriminant models with new multi-class SVMs. *Pattern Anal. Appl.* 2002, 5(2), 168–179.

59. *Learning with Kernels: Support Vector Machines, Regularization, Optimization, and Beyond*. MIT Press eBooks, IEEE Xplore. https://ieeexplore.ieee.org/book/6267332 (accessed Jun. 12, 2021).

60. Saw, M.; Saxena, T.; Kaithwas, S.; Yadav, R.; Lal, N. Estimation of prediction for getting heart disease using logistic regression model of machine learning. In *2020 International Conference on Computer Communication and Informatics (ICCCI), Coimbatore, India* 2020.

61. Ahsan, M.; Luna, S.; Siddique, Z. Machine-learning-based disease diagnosis: A comprehensive review. *Healthcare*, 2022, 10(3), 541.

62. Breiman, L.; Friedman, J.H.; Olshen, R.A.; Stone, C. J. *Classification and Regression Trees*. CRC Press US.

63. Quinlan, J.R. C4.5: Programs for machine learning. *Mach. Learn.* Volume 16, pages 235–240 1993.

64. Pal, M.; Mather, P. An assessment of the effectiveness of decision tree methods for land cover classification. *Remote Sens. Environ.* 2003, 86(4), 554–565.

65. Breiman, L. Random forests. *Mach. Learn.* 2001, 45(1), 5–32.

66. Aha, D.W.; Kibler, D.; Albert, M. Instance-based learning algorithms. *Mach. Learn.* 1991, 6(1), 37–66.

67. Pedregosa, F.; Varoquaux, G.; Gramfort, A.; Michel, V.; Thirion, B.; Grisel, O.; Blondel, M.; Prettenhofer, P.; Weiss, R.; Dubourg, V.; et al. Scikit-learn: Machine learning in python. *J. Mach. Learn. Res.* 2011, 12, 2825–2830.

68. John, G.H.; Langley, P. Estimating continuous distributions in Bayesian classifiers. In *Proceedings of the Eleventh Conference on Uncertainty in Artificial Intelligence*, Morgan Kaufmann Publishers Inc., 1995, pp. 338–345.

69. Sarker, I.H. A machine learning based robust prediction model for real-life mobile phone data. *Internet Things*. 2019, 5, 180–193.

70. Cho, Y.; Saul, L.K. Kernel methods for deep learning. *Proc. Adv. Neural Inf. Process. Syst. (NIPS)*, 2009, 22, 342–350.

71. Chauhan, R.; Ghanshala, K. K.; Joshi, R. Convolutional neural network (CNN) for image detection and recognition. In *2018 First International Conference on Secure Cyber Computing and Communication (ICSCCC)*, IEEE, 2018, pp. 278–282.

72. Goodfellow, I.; Bengio, Y.; Courville, A. *Deep Learning*. MIT Press, US, 2016.

73. LeCun, Y.; Bengio, Y.; Hinton, G. Deep learning. *Nature*. 521(7553), 436–444, 2015.

74. Schmidhuber, J. Deep learning in neural networks: An overview. *Neural Networks*, 61, 85–117, 2015.

75. Mukhamediev, R.I.; Kuchin, Y.; Amirgaliyev, Y.; Yunicheva, N.; Muhamedijeva, E. Estimation of filtration properties of host rocks in

sandstone-type uranium deposits using machine learning methods. *IEEE Access* 2022, 10, 18855–18872.

76. Nayebi, A.; Vitelli, M. Gruv: Algorithmic music generation using recurrent neural networks Course CS224D. *Deep. Learn. Nat. Lang. Processing (Stanf.)* 2015, 1–6. https://anayebi.github.io/files/projects/CS_224D_Final _Project_Writeup.pdf (accessed on 27 May 2022).

77. Lu, S.; Zhu, Y.; Zhang, W.; Wang, J.; Yu, Y. Neural text generation: Past, present and beyond. arXiv 2018.

78. Mukhamediev, R.I.; Popova, Y.; Kuchin, Y.; Zaitseva, E.; Kalimoldayev, A.; Symagulov, A.; Levashenko, V.; Abdoldina, F.; Gopejenko, V.; Yakunin, K.; et al. Review of artificial intelligence and machine learning technologies: classification, restrictions, opportunities and challenges. *Mathematics* 2022, 10, 2552.

79. Sun, S.; Cao, Z.; Zhu, H.; Zhao, J. A survey of optimization methods from a machine learning perspective.' *IEEE Trans. Cybernet.* 2020, 50(8), 3668–3681.

80. Han, S.; Ng, W.K.; Wan, L.; Lee, V.C.S. Privacy-preserving gradient-descent methods. *IEEE Trans. Know. Data Eng.* 2010, 22(6), 884–899.

81. Léon Bottou, F.E.C.; Nocedal, J. Optimization methods for large-scale machine learning.' *SIAM Rev.* 2018, 60(2), 221–487.

82. Robbins, H.; Monro, S. A stochastic approximation method. *Ann. Math. Stat.* 1951, 22(3), 400–407.

83. Nayebi, A.; Vitelli, M. Gruv: Algorithmic music generation using recurrent neural networks Course CS224D. *Deep. Learn. Nat. Lang. Processing (Stanf.)* 2015, 1–6. https://anayebi.github.io/files/projects/CS_224D_Final _Project_Writeup.pdf (accessed on 27 May 2022).

84. Lu, S.; Zhu, Y.; Zhang, W.; Wang, J.; Yu, Y. Neural text generation: Past, present and beyond. *arXiv* 2018, arXiv:1803.07133.

85. Wu, Y.; Schuster, M.; Chen, Z.; Le, Q.V.; Norouzi, M.; Macherey, W.; Krikun, M.; Cao, Y.; Gao, Q.; Macherey, K. Google's neural machine translation system: Bridging the gap between human and machine translation. *arXiv* 2016, arXiv:1609.08144.

86. Hannun, A.; Case, C.; Casper, J.; Catanzaro, B.; Diamos, G.; Elsen, E.; Prenger, R.; Satheesh, S.; Sengupta, S.; Coates, A. Deep speech: Scaling up end-to-end speech recognition. *arXiv* 2014, arXiv:1412.5567.

87. Devlin, J.; Chang, M.-W.; Lee, K.; Toutanova, K. Bert: Pre-training of deep bidirectional transformers for language understanding. *arXiv* 2018, arXiv:1810.04805.

88. Peters, M.; Neumann, M.; Iyyer, M.; Gardner, M.; Clark, C.; Lee, K.; Zettlemoyer, L. Deep contextualized word representations. *arXiv* 2018, arXiv:1802.05365.

89. Goodfellow, I.; Pouget-Abadie, J.; Mirza, M.; Xu, B.; Warde-Farley, D.; Ozair, S.; Courville, A.; Bengio, Y. Generative adversarial nets. *Adv. Neural Inf. Processing Syst.* 2014, 27, 1–9.

90. Creswell, A.; White, T.; Dumoulin, V.; Arulkumaran, K.; Sengupta, B.; Bharath, A.A. Generative adversarial networks: An overview. *IEEE Signal Processing Mag.* 2018, 35, 53–65.

91. Agnese, J.; Herrera, J.; Tao, H.; Zhu, X. A survey and taxonomy of adversarial neural networks for text-to-image synthesis. *Wiley Interdiscip. Rev. Data Min. Knowl. Discov.* 2020, 10, e1345.

92. LeCun, Y.; Bottou, L.; Bengio, Y.; Haffner, P. Gradient-based learning applied to document recognition. *Proc. IEEE* 1998, 86, 2278–2324.

93. Onah, J.O.; Abdullahi, M.; Hassan, I.H.; Al-Ghusham, A.; et al. Genetic algorithm-based feature selection and naïve Bayes for anomaly detection in fog computing environment. *Mach. Learn. Appl.* 2021, 6, 100156.

94. Sarker, I.H.; Khan, A.I.; Abushark, Y.B.; Alsolami, F. Mobile expert system: exploring context-aware machine learning rules for personalized decision-making in mobile applications. *Symmetry* 2021, 13(10), 1975.

95. Ramzan, B.; Bajwa, I.S.; Jamil, N.; Amin, R.U.; Ramzan, S.; Mirza, F.; Sarwar, N. An intelligent data analysis for recommendation systems using machine learning. *Sci Program.* 2019, 20, 19.

96. Phan, H.T.; Tran, V.C.; Nguyen, N.T.; Hwang, D. Improving the performance of sentiment analysis of tweets containing fuzzy sentiment using the feature ensemble model. *IEEE Access.* 2020, 8 14630–14641.

97. Pan, M.; Liu, Y.; Jiayi Cao, Y.; Li, C.L.; Chen, C.-H. Visual recognition based on deep learning for navigation mark classification. *IEEE Access.* 2020, 8, 32767–32775.

98. Elakkiya, R.; Subramaniyaswamy, V.; Vijayakumar, V.; Aniket, M. Cervical cancer diagnostics healthcare system using hybrid object detection adversarial networks. *IEEE J. Biomed. Health Inform.* 2021. 20. 20.

99. Harrou, F.; Zerrouki, N.; Sun, Y.; Houacine, A. An integrated vision based approach for efcient human fall detection in a home environment. *IEEE Access.* 2019, 7, 114966–114974.

100. Sarker Iqbal, H. Cyberlearning: Effectiveness analysis of machine learning security modeling to detect cyber-anomalies and multiattacks. *Internet Things.* 2021, 100, 393.

101. Blumenstock, J. Machine learning can help get covid-19 aid to those who need it most. *Nature.* 2020, 20, 20.

102. Saharan, S.; Kumar, N.; Bawa, S. An efcient smart parking pricing system for smart city environment: a machine-learning based approach. *Future Gener. Comput. Syst.* 2020, 106, 622–40.

103. Kim, J.-Y.; Seok-Jun, B.; Cho, S.-B. Zero-day malware detection using transferred generative adversarial networks based on deep autoencoders. *Inf Sci.* 2018, 460, 83–102.

104. Islam, M.Z.; Islam, M.M.; Asraf, A. A combined deep CNN-LSTM network for the detection of novel coronavirus (covid-19) using x-ray images. *Inform Med Unlocked.* 2020, 20, 100412.

105. Ale, L.; Sheta, A.; Li, L.; Wang, Y.; Zhang, N. Deep learning-based plant disease detection for smart agriculture. In *2019 IEEE Globecom Workshops (GC Wkshps)*, IEEE, 2019. p. 1–6.

106. Li, T.-H.S.; Kuo, P.-H.; Tsai, T.-N.; Luan, P.-C. CNN and LSTM based facial expression analysis model for a humanoid robot. *IEEE Access.* 2019, 7, 93998–4011.

107. Sarker Iqbal, H., Kayes, A.S.M. Abc-ruleminer: user behavioral rule-based machine learning method for context-aware intelligent services. *J. Netw. Comput. Appl.* 2020, 10, 2762.

108. Borah, A.; Nath, B. Identifying risk factors for adverse diseases using dynamic rare association rule mining. *Expert Syst. Appl.* 2018, 113, 233–263.

109. Bhavithra, J.; Saradha, A. Personalized web page recommendation using case-based clustering and weighted association rule mining. *Cluster Comput.* 2019, 22(3), 6991–7002.

110. Ruihua, X.; Luo, F. Risk prediction and early warning for air traffic controllers' unsafe acts using association rule mining and random forest. *Saf. Sci.* 2021, 135, 105125.

111. Elakkiya, R.; Subramaniyaswamy, V.; Vijayakumar, V.; Aniket, M. Cervical cancer diagnostics healthcare system using hybrid object detection adversarial networks. *IEEE J. Biomed. Health Inform.* 2021; 20:20.

112. Harrou, F.; Zerrouki, N.; Sun, Y.; Houacine, A. An integrated visionbased approach for efcient human fall detection in a home environment. *IEEE Access.* 2019, 7, 114966–114974.

113. Pan, M.; Liu, Y.; Jiayi Cao, Y.; Li, C.L.; Chen, C.-H. Visual recognition based on deep learning for navigation mark classification. *IEEE Access.* 2020, 8, 32767–32775.

Sustainable development using renewable energy sources

Artificial intelligence and machine learning contribution

Rishabh Kumar, Anuj Kumar, and Anurag Chauhan

2.1 INTRODUCTION

Sustainable development strives to harmonize economic growth with the long-term health of natural systems, guaranteeing the continued provision of critical resources and ecosystem services for the benefit of humanity. The present version of sustainable development emphasizes on environmental preservation for generations to come. Environmental sustainability is the preservation and improvement of the natural environment, with a focus on its diversity and productivity, particularly in air, water, and climate. Environmental sustainability requires society to engage in activities that meet human demands while protecting the planet's life support systems. Sustainable development hinges on the critical need to harness renewable resources at a rate that does not outpace their natural replenishment [1]. The world is grappling with a significant challenge in addressing the growing energy demand and the issues linked to fossil fuel resources. The global population growth and economic expansion are increasing energy demand, leading to increased use of fossil fuels, which are environmentally harmful, expensive, and pose a threat to future power generation. In contrast, conventional fuel sources are limited and rapidly diminishing, demanding close monitoring and long-term actions to avoid future energy crisis. Furthermore, hazardous emissions from fossil fuels, which include greenhouse gases, contribute to global warming.

However, non-conventional energy resources (renewable energy) are utilized to satisfy primary energy requirements alongside non-renewable sources, and it is especially employed to optimize efficiency in energy use. To prevent the depletion of fossil fuels, renewable energy sources have been utilized to produce electricity, tackle climate change, heat and cool spaces, heat water, and facilitate transportation [2]. The study revealed that renewable energy technologies typically produce less CO_2 throughout their lifespan, compared to coal's 1,000 g CO_2/kWh and natural gas's 475 g CO_2/kWh [3]. Because of their intermittent nature and high maintenance costs, renewable energy technologies aren't more viable than standard electric energy conversion technologies. There are many advantages of utilizing renewable energy

DOI: 10.1201/9781003581246-2

sources, such as less reliance on fossil fuels and less emissions of pollutants like NO_x, SO_x, CO_x, etc. into the atmosphere. Hence, the renewable energy becomes a more desirable option from a social perspective [4]. According to the agency, India will cut its greenhouse gas emissions by 20–30% by 2030 compared to 2020 levels [5]. Figure 2.1 shows the amount of energy delivered by renewable energy sources.

In sustainability studies, renewable energies and energy efficiency have emerged as areas of study because of the optimistic worldwide goals for a green energy by 2030 [7]. Sustainable energies have undergone a revolution thanks to advances in artificial intelligence (AI) and machine learning (ML). These algorithms enhance predictions of availability of renewable resources, site energy infrastructure optimally, operate power grids effectively, improve building and industry energy consumption, enhance energy storage options, and support decision-making by policymakers. The integration of AI and ML with sustainable energies shows potential in addressing global energy sustainability and climate change issues. Renewable energy system (RES) management and optimization have significantly improved with the use of ML tools. These technologies facilitate the development of complex management strategies for sustainable energy systems by managing difficult data-driven processes that uncover hidden relationships. In the end, ML enables the creation of intricate strategies for effectively designing, tracking, and managing renewable energy resources. In sustainable energy models, ML algorithms effectively manage intricate issues by substituting empirical

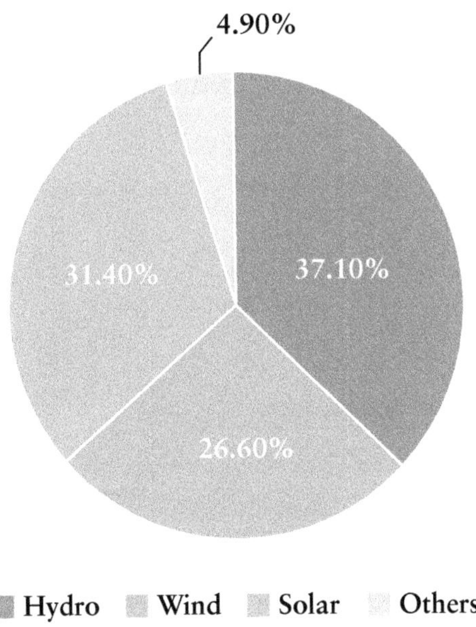

Figure 2.1 Energy capacity of different renewable energy sources [6].

programs and cutting down on processing expenses. New ML algorithms specifically designed for SE issues have been developed as a result of this intersection. This partnership between ML and sustainability has accelerated progress and will shape future civilizations that rely on advanced ML techniques in addition to renewable energy sources. In recent years, a lot of work has gone into using ML approaches with renewable energies and numerous studies have been published in tandem with the advancement of this field's research. The following article constitutes a comprehensive assessment of the most recent research on the use of AI and ML in solar, wind, hydropower, and biomass energy.

2.2 APPLICATION OF ML APPROACH WITH SOLAR ENERGY

Solar technology harvests solar energy from photovoltaic (PV) systems and solar thermal collectors (STC). Solar energy supports industrial processes, produces hot water, provides space heating and cooling, and can even produce electricity when combined with concentrated solar power (CSP) systems. With the help of this technology, solar energy may be used for a wide range of industrial, commercial, and residential purposes in a flexible and sustainable manner [8, 9]. Solar photovoltaic systems can be installed at homes or businesses, generating clean electricity to run your appliances and equipment. Solar energy systems are used to supply power for household purposes, as shown in Figure 2.2. Meanwhile, PV technologies produce electricity for a variety of purposes, including heating and cooling indoor spaces, powering household appliances, and producing hot water for the home [10].

In order to evaluate solar tower (ST) and photovoltaic technologies, authors in Ref. [12] designed a 100 MW plant in an area with abundant

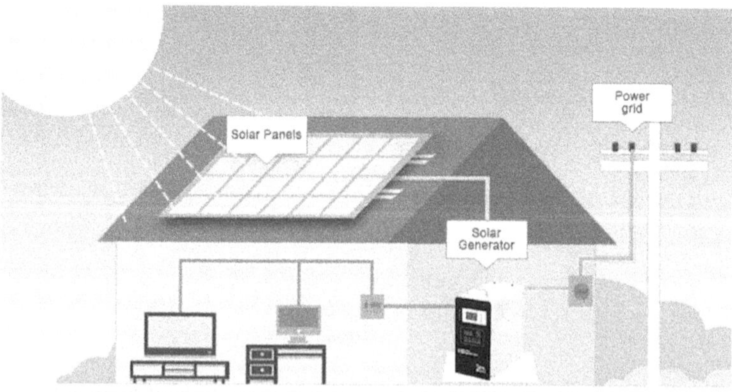

Figure 2.2 Solar energy system for residential purposes [11].

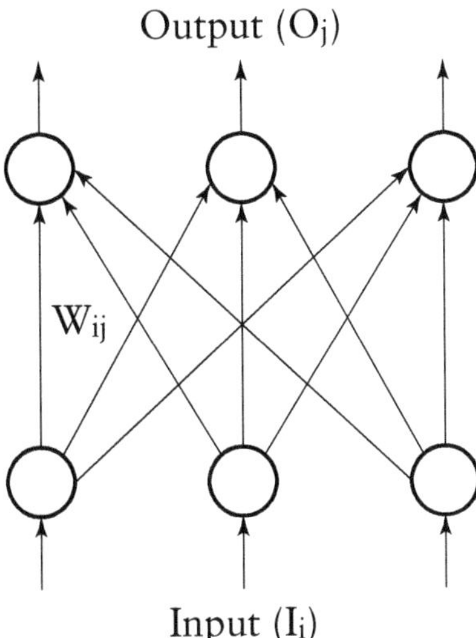

Figure 2.3 Schematic diagram of two-layer network [12].

sunshine. The objective was to equitably assess their performance in similar conditions, taking into account the particular irradiance levels. Using ANN approach, the PV plant was optimized for the ideal spacing among paralleled PV arrays, and the ST plant was optimized for solar multiple and thermal energy storage hours. A two-layer network, as shown in Figure 2.3, has been used in order to analyse the performance of ST and PV technologies. In terms of annual energy generation and capacity utilization, the ST plant performed better than the PV plant, with 58.6% compared to 30.9%. Nonetheless, the PV facility excels in terms of solar-to-electric efficiency and land usage. Despite its improved performance, the ST plant had 2.83 times better levelized energy cost than the PV plant. The ST plant is an excellent option for potential global horizontal irradiance and direct normal irradiance locations due to its greater technical capabilities; it operates more economically than the PV plant.

Authors in Ref. [13] analyzed the electrical performance of PV/T systems cooled by water-based nanofluids using ML techniques. The best approach out of ANFIS, LSSVM, and ANNs were determined by trial-and-error methods combined with statistical analysis. The ANFIS has been proven to be the best approach for simulating the performance of the solar energy system in consideration. As per the results, a different dataset with an average absolute relative deviation (AARD) of 15.21% was correctly forecasted by the ANFIS model. After optimization, it was determined that using 30 litres per hour of water-silica nano-coolant (3 wt%, 12.5 nm) under 788.285 W/m² of radiation was the most efficient configuration for reaching peak electrical efficiency in a PV/T system. The system's electrical efficiency was significantly increased by 27.7% as a result of this optimized arrangement.

Authors in Ref. [14] introduced a novel approach to optimize the porous volumetric solar receiver by combining heat transfer studies and a genetic algorithm. The study assessed fluid flow and heat transmission within the receiver using a volume-averaging simulation approach that utilized the local thermal non-equilibrium model. This approach aided in identifying a solar receiver design that minimized flow resistance and maximized thermal efficiency when combined with the genetic algorithm. The results of single-objective optimization indicated that larger porosity and higher intake velocities were preferred in order to increase thermal efficiency in the porous volumetric solar receiver. A reduction in input velocity appeared to be associated with a decrease in the optimal pore size. In contrast, broader receiver dimensions seemed to promote an increase in optimal pore size. Furthermore, both porosity and pore size were simultaneously optimized using a multi-objective optimization, which produced a Pareto front that identified receiver configurations with reduced flow resistance and higher relative thermal efficiency.

Using a multi-objective approach authors in Ref. [15] aimed to optimize the solar receiver fitted using DWVGs and dimple. Thermal efficiency enhancement and reducing TKE (turbulent kinetic energy) were the objectives of the optimization. The important factors taken into account for the optimization process were the dimple properties and configuration, in addition to the DWVG dimensions, length, height, angle, and spacing. CFD and the NSGA II optimization algorithm were used. In the process of optimization, the Reynolds number was fixed at 15,000. The study's findings indicated that optimization might establish a Pareto front and the optimized configurations significantly improved mixing and thermal efficiency. In particular, the solar receiver having DWVGs and dimples arranged inline performed better than the staggered design in terms of mixing and heat transfer. The flow impingement at the protrusion's front edge and downstream dimple regions, along with re-attachment farther downstream, resulted in improved thermal performance. They added to the enhanced mixing, as do the vortices created by protrusions and dimples. This led to a 5.2-fold increase in turbulent kinetic energy (TKE), a 166.57% increase in friction factor, and a 75.78% improvement in heat transmission. The investigation showed that overall optimization improved the system's thermal performance by 72%.

Authors in Ref. [16] proposed a weather categorization model based on convolutional neural networks and generative adversarial networks (GAN). The 33 various kinds of meteorological weather were first combined into ten groups. Subsequently, a model of GANs was adopted to enrich training data across every category of weather. Ultimately, the improved dataset having both developed and actual solar irradiance data was used to train a CNN-based weather categorization model. The structure of CNN network has been shown in Figure 2.4. The study contrasted CNN classification

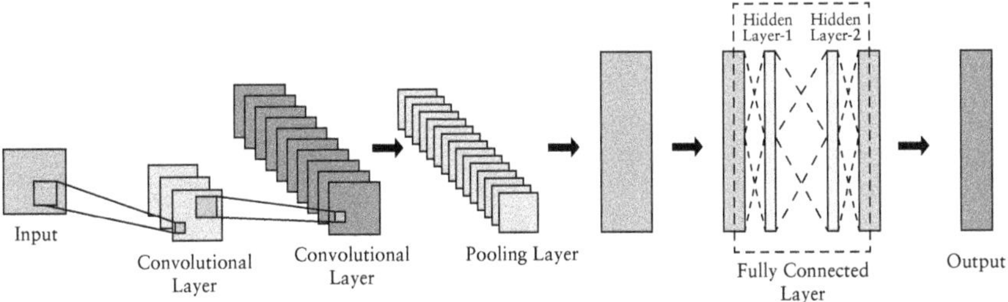

Figure 2.4 Structure of CNN model.

models with standard ML models and assessed the reliability of data created by GANs, and used them to forecast solar irradiance. The results of the simulation highlighted how well GANs function to produce fresh data that closely resembled original features without requiring training set memorization. In classification tasks, CNNs performed better than conventional ML models. By using GAN-based data augmentation, these classification models performed noticeably better. Additionally, the selection of précised day-ahead photovoltaic power forecasting models was largely dependent on the weather categorization model, which had the potential for improving total efficiency.

Authors in Ref. [17] emphasized how average forecasting outcomes can be achieved with reliable input weather data. The aim of this research was to devise a rating system to assess the effectiveness of particular ML models. This study examined four models: DT, LR, GPR, and SVM. According to the results, the RMSE values of forecasting techniques GPR, LR, SVM, and DT were 7.91%, 8.01%, 7.56%, and 7.70%, respectively. Although there was a small difference in their accuracy, their computation times differed significantly. GPR takes more than two hours to complete the forecasting process, but LR completes it in six seconds, DT completes in nearly 18 seconds, and SVM takes more than 1 hour. This significant time disparity emphasized how crucial it was to take calculation time into account when forecasting, particularly when using ML techniques to anticipate PV power output. Further, the summary of studies above has been discussed in Table 2.1.

2.3 APPLICATION OF ML APPROACH WITH WIND ENERGY

Wind is the flow of air caused by a pressure gradient. Wind power systems (as shown in Figure 2.5) can effectively utilize wind flow for power generation. As per World Economic Forum data, wind turbines generate 14% of the world's energy demands [18]. Wind turbines are classified based on size (micro and macro), rotation axis (horizontal and vertical), and turbine

Table 2.1 Summary of application of ML with solar energy

Reference	Model used	ML parameters	Result	Accuracy (R^2 value)	Remarks
[12]	ANN	–	The ST plant is an excellent option for potential global horizontal irradiance and direct normal irradiance locations due to its greater technical capabilities; it operates more economically than the PV plant	–	ST plants can be established at the location having large area of land and high availability of water source due the higher land and water requirement for the ST plant compared to PV plant
[13]	CFF, MLP, GR, RBF, ANFIS, and LS-SVR	–	• The ANFIS has been proven to be the best approach for simulating the performance of the solar energy system in consideration • The system's electrical efficiency was significantly increased by 27.7% as a result of this optimized arrangement	0.9534 (ANFIS)	The only drawback is do not use for extrapolation purposes, as recommended for the developed ANFIS model
[14]	GA	–	Furthermore, both porosity and pore size were simultaneously optimized using a multi-objective optimization, which produced a Pareto front that identified receiver configurations with reduced flow resistance and higher relative thermal efficiency	–	The authors could have used different algorithms other than GA for the optimization like NSGA, NSGA II with grid diversity enhancement, etc.
[15]	NSGA II	–	The solar receiver having DWVGs and dimples arranged inline performed better than the staggered design in terms of mixing and heat transfer	–	The comparison could have been done with the application of other algorithms like neural networks
[16]	GAN and CNN	Hidden layer = 2 with 124 neurons	The proposes model addresses the challenge of establishing classification models with minimal sample sizes for specific categories	–	Other ML models such as gradient boosting machines, decision tree, etc. can be used to develop the novel model and optimize the PV power forecasting model
[17]	LR, DT, GPR, and SVM	–	The interaction LR has outperformed the other models based on the time to complete the forecast process. However, the difference in accuracy of all models have been very less	0.9279	Authors should investigate the various possible data durations to tune the ML model

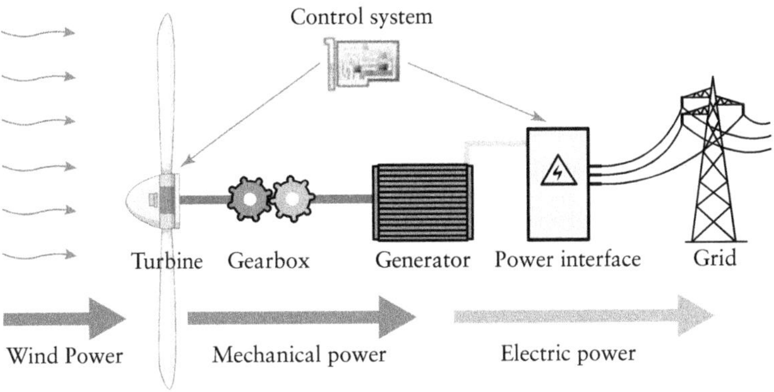

Figure 2.5 Wind power systems [19].

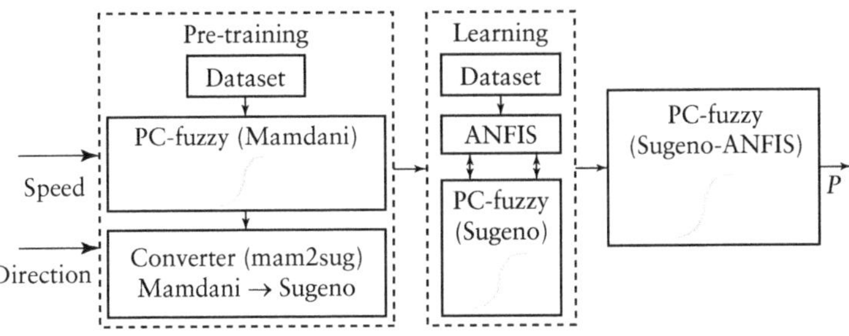

Figure 2.6 Developed Sugeno-ANFIS model through learning approach [20].

placement (on-site and off-site), and the turbine gets influenced by other factors such as placement of WT, wind speed, and turbine design.

Authors in Ref. [20] proposed a novel technique for enhancing the power curve approximation model of wind turbine of wind farms by transforming a Mamdani FIS into a Sugeno FIS. The techniques to improve the power curve approximation model were ANN, Mamdani FIS, Sugeno FIS, and Sugeno-ANFIS. The analysis revealed that the proposed fuzzy model performed better than the other models. The architecture of Sugeno-ANFIS model has been illustrated in Figure 2.6. The present study described fuzzy models that performed significantly better than comparable models when pre-inference conditions were used. The new models significantly outperformed the reference model's average power curve, particularly in wind farm 1. The models with the highest pre-established inference outperformed the rest, achieving significant accuracy gains (MAE, NMAE, and RMSE improved by 50.65%, 50.65%, and 46.68%, respectively). While enhanced wind data led to modest gains in MAE, NMAE, and RMSE (5.26%, 6.78%, and 4.70%) at wind farm 2, these pale in comparison to the significant advancements observed at wind farm 1.

Authors in Ref. [21] presented a novel numerical method for optimizing the blade design for Savonius wind turbine at low wind speeds. The blade design was optimized using the ANN approach and optimized the power coefficient of turbine. With the help of ANN, the coordinates of blade profile were altered to obtain the optimum shape. During the initial code run, the error between the actual and optimum C_p was found to be $\pm 2.5\%$. With a maximum C_p of 0.35 at a tip speed ratio of 1.1, the optimized blade design demonstrated a significant 55% improvement in power coefficient compared to the semi-circular baseline.

In order to enhance wind time-series data extraction of features, authors in Ref. [22] presented a novel method called T.S.B. K-means, which is designed for time-series data. It also considered a related cluster selection approach. Furthermore, the study combines T.S.B. K-means via MLPNN, HANTS, and DWT to establish a comprehensive model for wind power estimation. A T.S.B. K-means clustering technique provided a comprehensive cluster for input into an MLPNN. To train hourly wind power predictions, a cluster selection approach selected the cluster that was significantly correlated with testing data. To prepare MLPNN inputs, pre-processing used HANTS and DWT, and time series analysis. Several wind datasets analyzed the reliability associated with the hybrid forecasting model, as represented in Figure 2.7 and confirmed the effectiveness of a clustering. The suggested hybrid forecasting method's better accuracy was confirmed by comparison with existing models.

To maximize the power coefficient, the authors in Ref. [23] numerically analyzed an optimized design of the conventional Savonius turbines using a GA. Further, the experimental study was performed with two-bladed, three-bladed, and optimized-blade turbine and compared the results with numerical study. Based on the numerical results, the optimized blade shape yielded an improvement of 52% corresponding to conventional Savonius turbines, which leads to the maximum C_p of 0.29 for the optimized Savonius turbine. Moreover, experimental study was also carried out for the optimized blade two-blade and three-blade designs. An optimized blade design delivered a significantly higher coefficient of power (C_p) of 29% at a wind speed of 9 m/s, compared to just 14% and 11% achieved by two-blade and three-blade designs, respectively, under the same wind conditions. Moreover, at slower wind speeds of 5.5 m/s, the optimized blade performed more proficiently and produced a maximum C_p of 28%. This low speed caused minimum flow separation, helping the turbine to maximize wind energy usage while minimizing losses.

Authors in Ref. [24] determined advanced ML techniques for gap filling that can lower uncertainties and produced a more dependable time series that satisfies all of the prerequisites for such types of measurements. The study aimed to identify the most effective ML models, revealing that even minor improvements in uncertainty could lower project finance costs, enhancing the viability of VMMs in real-world applications. Nine models

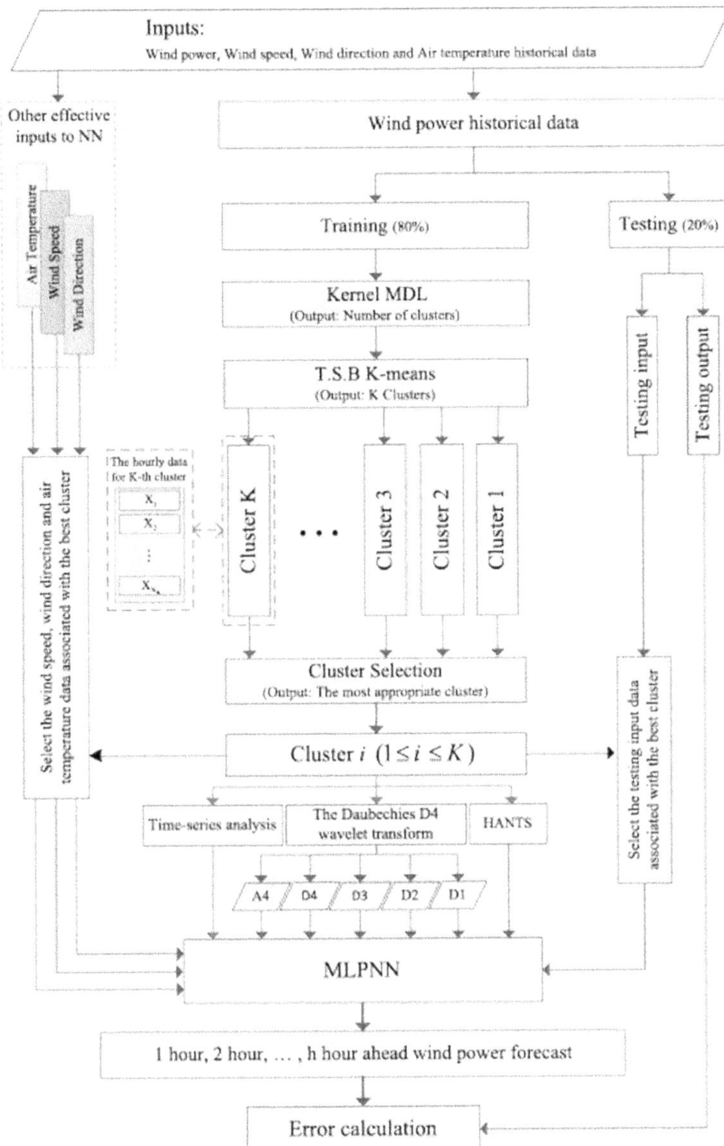

Figure 2.7 Schematic diagram of the hybrid model proposed in this study [22].

were employed, such as linear regression, lasso regression, ridge regression, KNN, DT, RF, AdaBoost, GP, MOMoGP, and MONN. On comparing to nine ML methods, the MCP approach outperformed the linear regression model in improving association of Meteorological Masts while employed with advance methods. Based on results, the AdaBoost regression offered the lowest errors in air temperature predictions, while MOMoGP, weather research and forecasting (WRF), KNN, and ERA5 were the most effective for wind speed and direction predictions. Models AdaBoost and KNN were recommended for further research and real-world applications, MOMoGP excelled in certain conditions, whereas appearing unstable in other conditions. The suggested regression models may be adopted as an initial point to develop end-user services and apps.

Authors in Ref. [25] proposed a novel blade shape for Savonius wind turbine using a semi-elliptical shape. The height of concave and convex sides of blades (a1 and a2) were used as design points to describe the shape of the blade. The range of a_1 and a_2 was chosen as 0.275–0.55 and 0.2–0.55, respectively. Based on these two variables, different 68 blade shapes were obtained and analyzed using CFD. An optimization study using KRG surrogate model and the PSO was performed to find the optimal design. The framework of PSO is illustrated in Figure 2.8. The blade design with value of a_1 and a_2 as 0.3936 and 0.2743 was identified as optimal blade design which achieved a 4.41% improvement in C_p compared to the conventional design, reaching a maximum C_p of 0.2580.

Authors in Ref. [26] proposed an integrated approach with a combination of FCBF algorithm and VMD algorithms for extracting features from wind resources data. Further, the LSSVM was adopted to forecast short-term wind speeds. In addition to this, the outcomes of point predictions and the

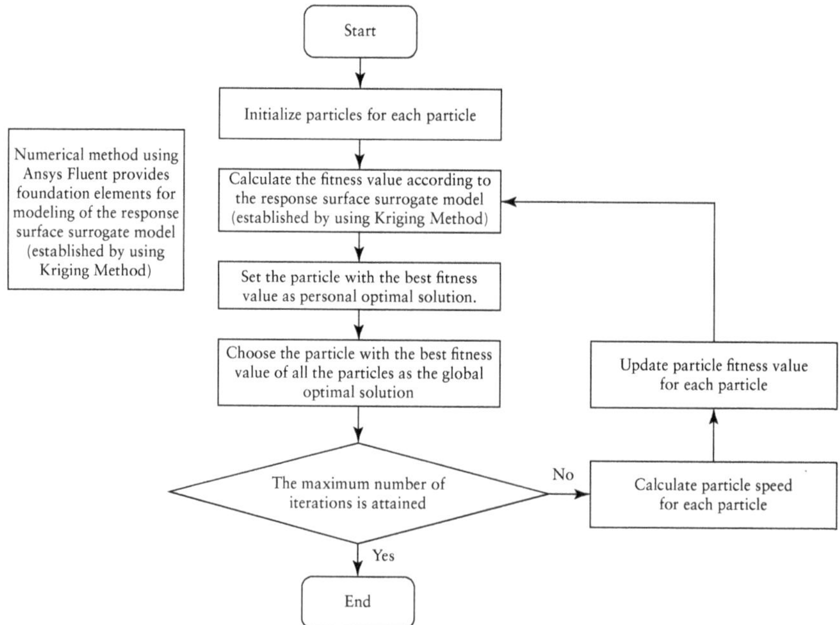

Figure 2.8 The methodology adopted for PSO technique [25].

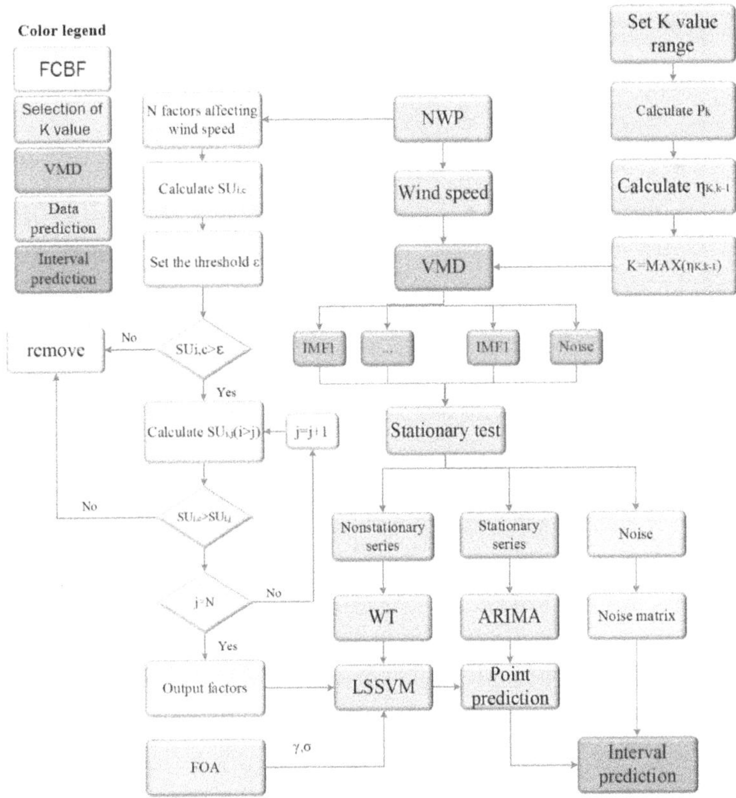

Figure 2.9 Framework for the model proposed in this study [26].

discrepancies detected in the training dataset were combined to manage the prediction error. This combination technique was used to forecast wind speed intervals. The framework of the technique used is illustrated in Figure 2.9. To validate the efficiency of this model, the simulations were utilizing numerical weather prediction (NWP) data from Changma Wind Farm and Sotavento Wind Farm. The integrated model competently combined different approaches, demonstrating accurate data preparation and precise numerical wind speed forecasts. It resulted in significant improvement of wind speed handling with minimum forecast error in the forecasting system. Finally, the FCBF-VMD-FOA-LSSVM-ARIMA model delivered reliable and anticipated results. Table 2.2 summarizes the studies for wind energy with application of ML.

2.4 APPLICATION OF ML APPROACH WITH HYDROPOWER

Hydropower is the process of converting a water source's gravitational potential or kinetic energy into power that can be used to generate electricity

Table 2.2 Summary of application of ML with wind energy

Reference	Model used	ML parameters	Result	Accuracy (R^2 value)	Remarks
[20]	ANN, ANFIS	–	The fuzzy models significantly outperformed the reference model's average power curve, particularly in wind farm I	MAE improved by 50.65%	Neural networks could have been used for comparison with ANFIS algorithm
[21]	ANN	–	The optimized blade design demonstrated a significant 55% improvement in power coefficient compared to the semi-circular baseline	–	Comparison of blade design obtained using different neural networks could have been carried out
[22]	T.S.B. K-means via MLPNN, HANTS, and DWT	Cluster selection method for training MLPNN	The suggested hybrid forecasting method's better accuracy was confirmed by comparison with existing neural network models	0.523 (avg. MAE)	The study can be performed for the complex and realistic data without any simplifications to tune the model and improve its accuracy
[23]	GA	–	The optimized blade shape yielded an improvement of 52% corresponds to conventional Savonius turbines, which leads to the maximum C_p of 0.29 for the optimized Savonius turbine	–	The performance of turbine can be tested for different design parameters

(Continued)

Table 2.2 (Continued) Summary of application of ML with wind energy

Reference	Model used	ML parameters	Result	Accuracy (R^2 value)	Remarks
[24]	KNN, DT, RF, AdaBoost, GP, MOMoGP, and MONN	Hidden layer = 1 with 1,000 hidden units for MOMoGP and MONN	The AdaBoost regression offered the lowest errors in air temperature predictions, while MOMoGP, weather research and forecasting (WRF), KNN, and ERA5 were the most effective for wind speed and direction predictions	0.983 for AdaBoost, 0.873 for KNN	The advanced regression models must be used to reduce the uncertainty
[25]	KRG and PSO	–	The blade design with value of a_1 and a_2 as 0.3936 and 0.2743, respectively, was identified as optimal blade design, which achieved a 4.41% improvement in C_p compared to the conventional design, reaching a maximum C_p of 0.2580	–	The stress analysis can be performed on the new blade design because both the blades are different in shapes
[26]	LSSVM	–	The hybrid model resulted in significant improvement of wind speed handling with minimum forecast error in the forecasting system	MAE = 0.1654 and 0.1451 for Changma and Sotavento farm, VMD-FOA-LSSVM-ARIMA model	The different optimization algorithms and different hybrid models can be used to enhance the performance of the model

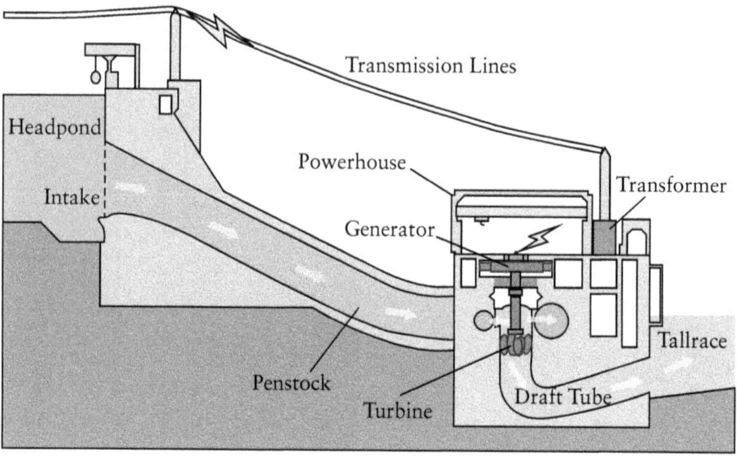

Figure 2.10 Working of hydropower system [28].

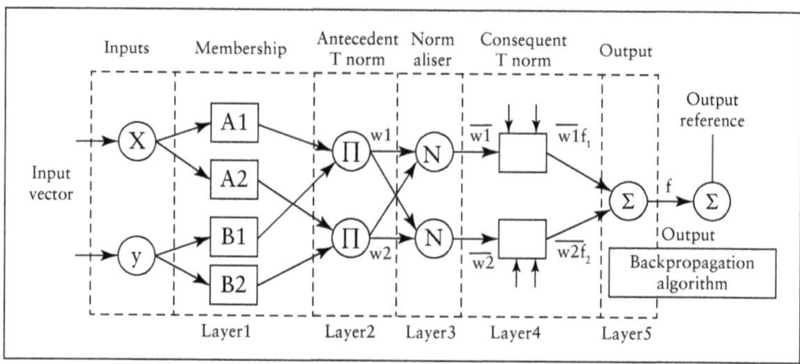

Figure 2.11 Structure of the ANFIS model [29].

as illustrated in Figure 2.10. Hydropower is a sustainable and reliable alternative to fossil fuels as it does not emit carbon dioxide or other pollutants. Hydro power plants pose significant global social and economic challenges and require a sufficient energy source like a river or elevated lake. Hydroelectricity accounts for 15% of worldwide electricity generated and at least 50% of total power supply in over 35 countries. As per report on hydropower [27], hydropower contributed a significant role in generating renewable energy and has generated 4,400 TWh of electricity in 2021.

Authors in Ref. [29] developed an effective load shedding strategy for microgrids by combining ANN and ANFIS. This method effectively predicted the precise load decline required, completing it faster than traditional approaches. An ANFIS was formed by combining a neural network and fuzzy logic to improve load shedding efficiency across varied input data ranges. The framework for the model proposed in this study is illustrated in Figure 2.11.

This algorithm estimated the most appropriate load shedding values for the data input throughout the trained dataset, considerably reducing errors. The developed method accomplished load shedding efficiently and quickly with the required amount to be shed. Simulations showed that this strategy allowed for the rapid removal of the intended load, hence improving system stability.

Growing computational complexity hinders the optimization of large hydropower systems with traditional techniques like POA. In order to address this problem, authors in Ref. [30] proposed a new approach known as the UPOA. The model comparison of POA and UPOA is shown in Figure 2.12. This approach used a uniform design to evaluate a subset of possible solutions by breaking the large problem down into smaller two-stage optimization sub-problems. Subsequently, it employed successive approximation to gradually improve the quality of the result. The outcomes of China's Lancang hydropower system were used to assess the performance of POA and UPOA. UPOA outperformed POA in both execution time and convergence speed. These three techniques (uniform design, two-stage optimization, and successive approximation), strategically used by UPOA, are the key to this advantage. The proposed UPOA approach had an excellent global search capability and can identify suitable solutions in an appropriate period of time.

Reference [31] proposed a ML-based model for predicting hydropower generation with different time increments. This tool, with its flexibility in

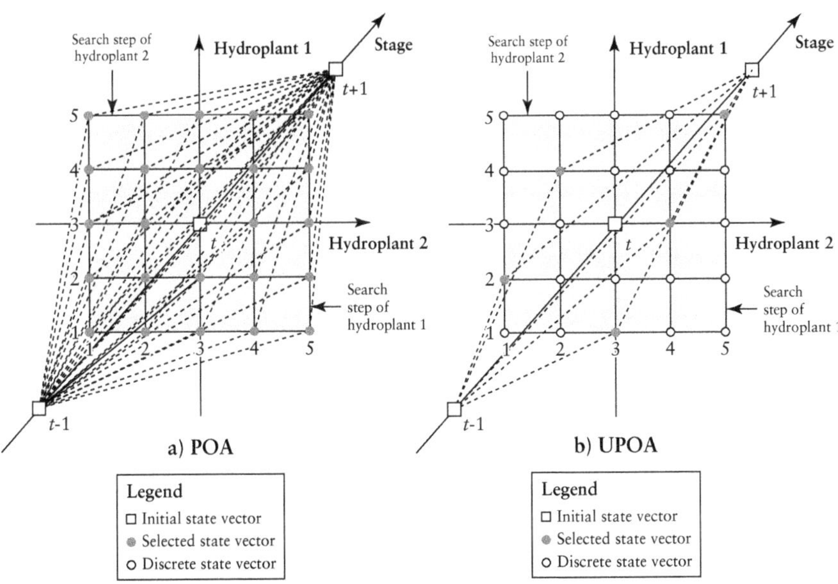

Figure 2.12 Model comparison of POA and UPOA for different hydropower systems [30].

prediction range, promises to be a valuable asset for hydropower system operators in making informed decisions. Authors in Ref. [31] also investigated various input/output model designs to improve the accuracy of predictions by employing various lag-time input techniques and determined the most significant input variables influencing the model's predicted accuracy. The models adopted for the study were ANN, SVM, and ARIMA. A schematic diagram of SVM model is illustrated in Figure 2.13. The study's findings demonstrated that ML techniques were found to be superior to traditional methods for modelling hydropower generation time series. This is likely due to the unique characteristics of hydropower data, the relative ease of implementing ML models, and their inherent ability to handle complex problems. The influence of input configurations on model behaviour was assessed through Taylor diagram–based sensitivity analysis for both ANN and SVM approaches. The effectiveness of the proposed models was rigorously assessed through uncertainty evaluations (95% predicted uncertainty and d-factor analysis). The results were highly positive, indicating all models effectively minimize uncertainty. The authors also compared the performance of the proposed ANN to SVM under different conditions to assess accuracy. In all cases, the ANN achieved significantly higher correlation values, ranging from 0.94 to 0.96. This demonstrated that the ANN was consistently more accurate than SVM in all the conditions. Authors reported that results were encouraging, suggesting that the ANN algorithm had a substantial potential for long-term as well as short-term forecasting of hydropower generation time series.

Authors in Ref. [32] utilized different ANN models in order to predict the hydroelectric production of Ecuador. The models used for the prediction were MLP, LSTM, and seq2seq. Two situations were considered based

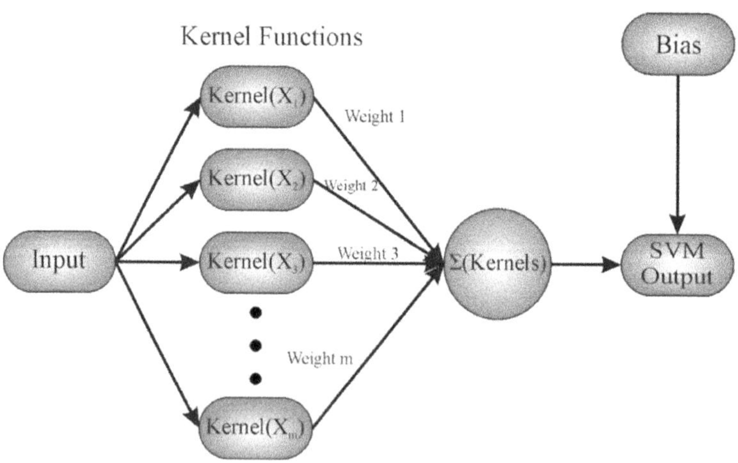

Figure 2.13 Framework for the SVM model [31].

on the projected horizon: 1-month (one-step) and 12 months (multi-step). MLPs with differentiation achieved the best results for both univariate and bivariate series. This demonstrated the advantage of ANN models in time series analysis over conventional techniques for statistical analysis. An analysis of model execution time revealed an average duration of 1.48 minutes per model for single-step situations and 1.37 minutes per model for multi-step situations. Training these models required significantly less computational power when the data was organized in sequences, compared to one-step forecasting problems. Differentiating as a pre-processing step for the monthly gross production series improved prediction errors for all ANN configurations tested in both one-step and multi-step scenarios. Similarly, adding a second predictive variable (precipitation) to the bivariate analysis improved the models' capability for learning and adaption in both cases.

Authors in Ref. [33] proposed a novel QP algorithm to deal with power shortages in hydropower plants. In QP algorithm, domain knowledge was utilized to predict the initial operational conditions from upstream to downstream. It managed the non-linear generating features of hydro plants by dealing with iterative sub-problems and dynamically modifying water head to gradually optimize and improve the solution's quality. Two case studies were investigated: a mature hydropower system and a real-world operating under seasonal load variations (summer and winter). Instead of using separate curves for each water head, the QP algorithm was the perfect choice because it could iteratively calculate and account for changes in water head throughout the scheduling period. These two hydropower case studies demonstrated clear performance advantages compared to the original load demand curve. Moreover, the summary of hydropower studies having application of ML has been discussed in this study (Table 2.3).

2.5 APPLICATION OF ML APPROACH WITH BIOMASS ENERGY

Biomass, a renewable energy source, includes a wide range of organic substances and wastes. It consists of specific energy crops (such as castor seed, copra, sesame, jatropha, and so on), residual of agricultural crops (such as wheat straw, rice straw, corn stover and so on), and residues from forests, all of these are used to generate biomass energy [34]. Figure 2.14 describes the ways to generate the biomass energy. The consumption of modern biomass energy has a number of advantages, including environmental advantages such as reduced utilization of organic materials, landfill waste, and the carbon neutrality. It also brings economic prospects, as biomass products have the potential to be less expensive and more lucrative than fossil fuels. Furthermore, the societal benefits of using biomass energy involves cleaner air and better public health due to lower toxic emissions [35–37].

Table 2.3 Summary of application of ML with hydropower

Reference	Model used	ML parameters	Result	Accuracy (R^2 value)	Remarks
[29]	ANN and ANFIS	Optimizer = Stochastic gradient descent hidden layer = 4	The combination of ANN and ANFIS algorithm estimated the most appropriate load shedding values for the data input throughout the trained dataset, considerably reducing errors. The developed method accomplished load shedding efficiently and quickly with the required amount to be shed	–	The other combination of algorithms must be used for optimizing the shedding of load and compared with the developed model
[30]	POA and UPOA	–	UPOA outperformed POA in both execution time and convergence speed. These three techniques (uniform design, two-stage optimization, and successive approximation), strategically used by UPOA, are the key to this advantage	–	The hybrid models should be used to enhance the performance of hydropower system
[31]	ANN, ARIMA, and SVM	Hidden layer = 1 with number of node = 3, 5, 7, 10, and 15	The study's findings demonstrated that ANN was found to be superior to traditional methods for modelling hydropower generation time series	0.94–0.96	The different parameters such as temperature, degree of inflow, climate change and drought variability can be introduced to predict the time series for hydropower generation

(Continued)

Table 2.3 (Continued) Summary of application of ML with hydropower

Reference	Model used	ML parameters	Result	Accuracy (R^2 value)	Remarks
[32]	LP, LSTM, and seq2seq	Optimizer = stochastic gradient descent For one-step scenario: number of nodes = 64, act_hid = sigmoid functions and act_out = sigmoid functions For multi-step scenario: Number of nodes = 84, act_hid = tanh functions and act_out = linear	MLPs with differentiation achieved the best results for both univariate and bivariate series. This demonstrated the advantage of ANN models in time series analysis over conventional techniques for statistical analysis	–	The proposed model should be validated with the dataset of hydropower system within or outside the Ecuador to generalize the findings
[33]	QP	–	Instead of using separate curves for each water head, the QP algorithm was the perfect choice because it could iteratively calculate and account for changes in water head throughout the scheduling period	–	The sensitivity of a QP technique to parameters such as load demand, forecast errors in inflows, and other parameters for developing a robust optimization technique

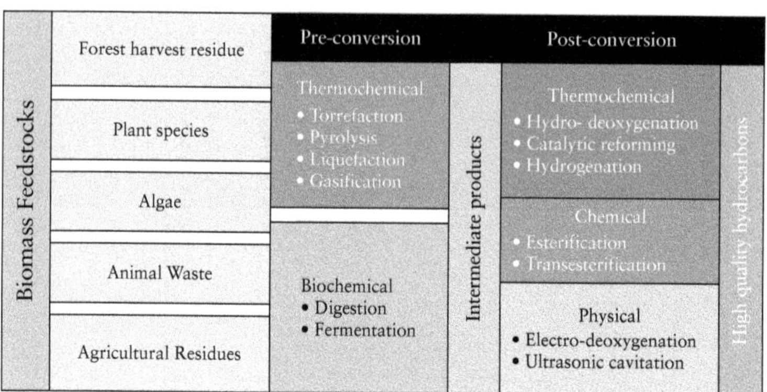

Figure 2.14 Various process techniques for the conversion of bioenergy [38].

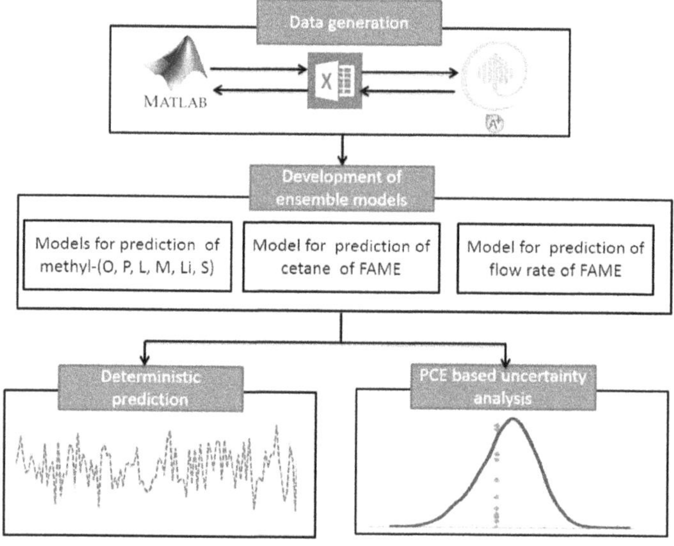

Figure 2.15 Framework for proposed model [39].

Authors in Ref. [39] developed a data-driven soft sensing system to esti-mate the flow velocity, cetane number (CN), and composition of FAME. A non-intrusive PCE technique has been incorporated with the soft sensor design to handle uncertainties. By incorporating PCE, the prediction model was transformed from deterministic to stochastic approach. Figure 2.15 illustrated the structure of proposed model used for the study. This modifica-tion enabled to visualize the impact of process uncertainties using predictive distributions, resulting in a clearer picture of the way uncertainties influence the model's predictions. Individual models (soft sensors) were developed to address each of the component (methyl-Li, methyl-M, methyl-O, methyl-S, and methyl-P), flow rate, and cetane number. The models were aimed to precisely anticipate FAME's specific characteristics. As per results reported, the prediction accuracies for methyl-P, methyl-S, methyl-O, methyl-M, methyl-Li, CN, and FAME flow rates were 0.9939, 0.9833, 0.9915, 0.9890, 0.9877, 0.9953, and 0.9952, respectively. Considering 1% uncertainty with input variables of soft sensor, the MADP (mean absolute deviation percent) values for methyl-P, methyl-S, methyl-O, methyl-M, methyl-Li, CN, and FAME flow rates were estimated as 0.82135, 0.1651, 0.41208, 0.32227, 0.27479, 0.97013, and 0.96546, respectively. These sensors excel at both precise predictions and estimating their own margins of error, making them ideal for real-time use.

Authors in Ref. [40] introduced three prediction approaches: SVM model, ANFIS model, and FNN model optimized using the GA, SA, and LM algorithms. These models were designed for accurately predicting the viscosity of biodiesels. The goal of this study was to compare the predicted accuracy of these models by using large experimental biodiesel viscosity models, while accounting for environmental impacts and the renewability of diesel fuels. The viscosity data considered for the study was the experimental data produced by Knothe and Steidley [41], Freitas et al. [42], Yuan et al. [43, 44], Krisnangkura et al. [45], Fietosa et al. [46], and Noguiera et al. [47] at 40° C. Based on the results, the FNN trained by GA and SA showed superior accuracy compared to SVM, ANFIS, and LM on new datasets due to efficient stochastic optimization techniques and improved generalization capabilities.

Authors in Ref. [48] utilized ANN model combined with PSO and TLBO in order to generate CN of bio fuel and other fuels from its FAMEs profile. The architecture of TLBO method has been represented in Figure 2.16. Based on available literature [49–56], 232 samples were collected and fed to optimization models. The input variables were double bond, carbon number, and molar weight, while output variable was CN. All proposed models performed outstandingly well but the TLBO-ANN model proved better as the mean square error was 3.538% and R^2 values for TLBO-ANN model was 0.973. The sensitivity study revealed that the double bond had greatest influence on CN of biodiesel.

Authors in Ref. [57] aimed to predict the biofuel properties from fatty acids profile using ANN. The biodiesel properties were considered as density, flash point (FP), CN, and kinematic viscosity (KV). The five fatty acid profiles (FAs) out of 55 distinct feedstocks collected from the literature were used as input parameters for the networks. But for density network, authors considered an additional parameter which is temperature. The ANN model had five input layers along with two hidden layers of logsig and purelin transfer functions as shown in Figure 2.17. The regression coefficients of FP, KV, density, and CN were 0.991, 0.958, 0.994, and 0.967, respectively. Every network had the ability to predict values that were almost identical to the experimental values.

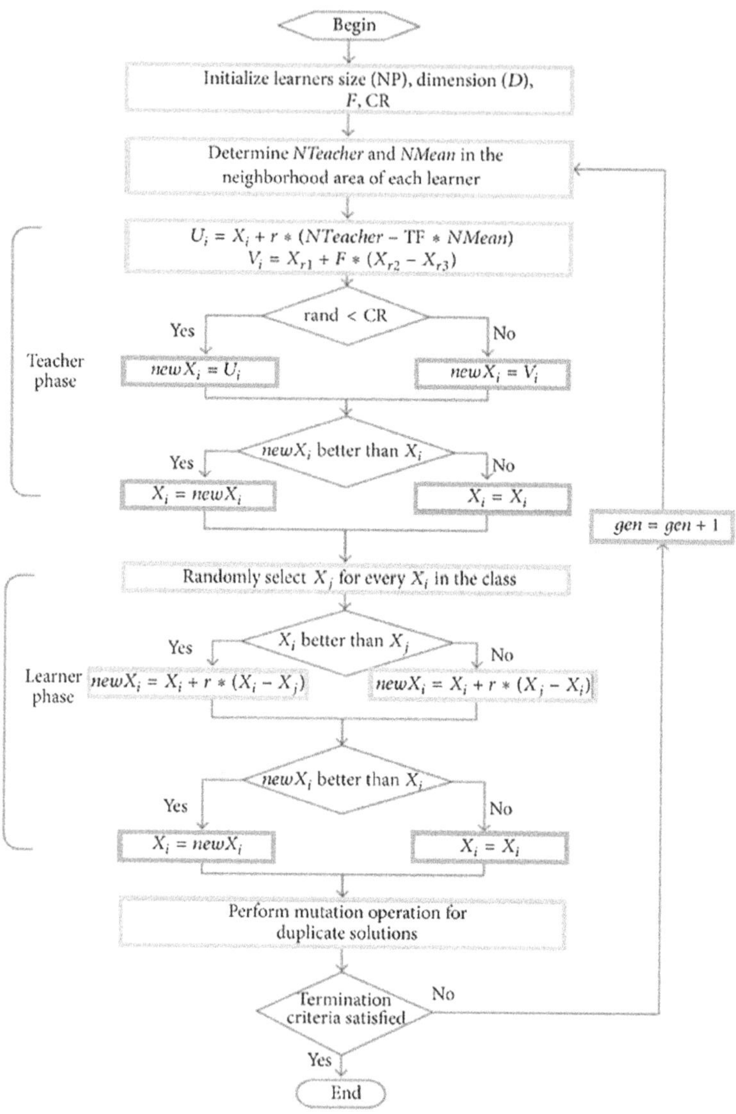

Figure 2.16 TLBO model structure with ANN [48].

Authors in Ref. [58] introduced ANN in order to predict the CN of biofuel consisting of furanic chemical compounds. The DCN was an equivalent data obtained using an ignition quality tester (IQT). By using a theoretical inverse correlation based on an ignition delay recorded in the fixed volume of combustion chamber device, it acted as a direct measurement alternative for the CN. The data points of DCN captured using an IQT have been used to validate the model and enrich the experimental dataset. The architecture for

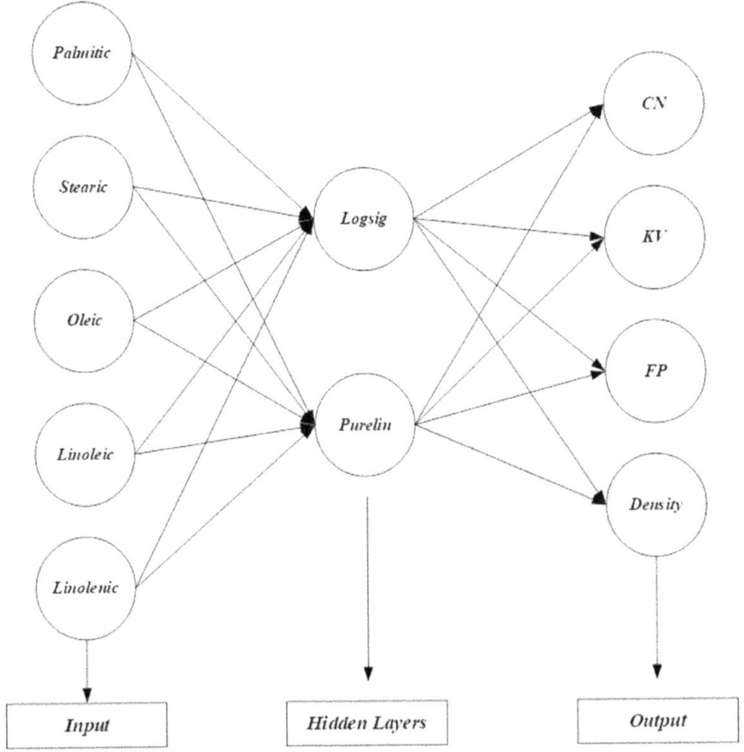

Figure 2.17 ANN structure considered in the study [57].

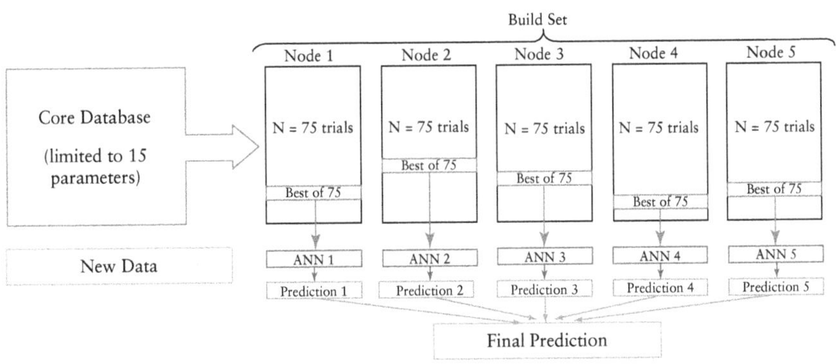

Figure 2.18 Architecture of the study proposed [58].

the proposed study has been represented in Figure 2.18. The current study improved an existing model by optimizing it and its major learning variables via ANN. Furthermore, it anticipated to improve the model's adaptability to a higher range of probable types of fuel, especially by incorporating furans and their derivatives into the model to assure the integration of these specific molecules. As per results, two biofuel additive alternatives had CNs that were suitable for application in conventional diesel engines. A furan segment within the carbon compounds was hydrogenated in order to yield its tetrahydrofuranyl analogy, exhibiting CN values of 60.4 and 59.8, respectively. This new approach yielded a 5.54% improvement (0.35 CN units) over previous attempts. Without eliminating any measurements from the database, the total root mean square error for the core dataset was 5.97. CNs were effectively predicted for a variety of molecules, including biomass-derived hydrocarbons, furanic compounds, and FAME. The predictions were around 95% confidence interval when using trustworthy published values as references for various biofuel categories.

Authors in Ref. [59] used the ANN model and a non-linear regression model in order to forecast the rheological characteristics of waste vegetable oil, waste vegetable oil biodiesel modified water-based mud and water-based mud. The rheological characteristics of oils were defined by yield point apparent and plastic viscosity. The LM-BP method was used with a three-layer FFBPN for ANN predictions, and the MLNR model was considered for non-linear predictions of rheological characteristics of oils. It was noticed that the mean absolute percent error of the ANN model for AV, PV, and YP was found to be minor, such as 0.1432, 0.3740, and 0.0506, respectively. Also, the PV, AV, and YP had R^2 values of 0.996, 0.997, and 0.997. The MLNR model had a very high mean absolute percent error and the values of R^2 were quite low too, compared to ANN model. It was concluded that the ANN model had superior performance and consistency in their predictions across all predicted outcomes. Furthermore, Table 2.4 discusses the summary of the studies for biomass energy with ML application.

2.6 CASE STUDY: APPLICATION OF DIFFERENT ML MODELS FOR SAVONIUS HYDROKINETIC TURBINE

A regression analysis is carried out in order to optimize the slot parameters and predict the power coefficient (C_p) of the Savonius hydrokinetic turbine. The regression analysis is carried out using the different regression models such as LR, PR, DT, and SVR and RF models. The Savonius turbine with slotted blade has been optimized using the slot parameters such as slot shape, slot position, slot gap, and divergent ratio. The slot shape, slot position, slot gap, divergent ratio, and TSR are considered as input parameters, and on the other hand C_p is taken as an output parameter. The range of slot parameters and turbine design parameters selected for this study are mentioned in Table 2.5.

The datasets available for training and testing has been generated using the commercial software Ansys (v.2022). The available dataset is split in such a way that 80% of the data is available for training, and the remaining 20%

Table 2.4 Summary of application of ML with biomass energy

Reference	Model used	ML parameters	Result	Accuracy (R^2 value)	Remarks
[39]	Non-intrusive PCE	–	These sensors excel at both precise predictions and estimating their own margins of error, making them ideal for real-time use	As per results, the prediction accuracies for methyl-P, methyl-S, methyl-O, methyl-M, methyl-Li, CN, and FAME flow rate were 0.9939, 0.9833, 0.9915, 0.9890, 0.9877, 0.9953, and 0.9952, respectively	The multi-objective optimization with soft sensing strategy to maximize quality, yield, and minimize cost can be adopted
[40]	SVM, ANFIS, and FNN	Three-layer FNN with nine hidden nodes	Based on the results, the FNN trained by GA and SA showed superior accuracy compared to SVM, ANFIS, and LM on new datasets due to efficient stochastic optimization techniques and improved generalization capabilities	0.9751 and 0.9012 for FNN-GA and FNN-SA, respectively	For predicting the application and reliability, the experimental validation should be carried out for the different types of blends and biodiesels
[48]	ANN model combines with PSO and TLBO	Hidden layer = 1 with ten nodes (ANN–TLBO)	All proposed models performed outstandingly well but the TLBO–ANN model proved better as the mean square error was 5.54% and R^2 values for TLBO–ANN model was 0.973	0.973 and 0.951 for TLBO–ANN and PSO–ANN, respectively	Comparison with the other optimization algorithm and machine learning methods for enhancing the accuracy of predictions of CN
[57]	ANN	Hidden layers = 1 to 4 Number of neurons = 5 (density), 6 (for FP and CN), and 7 (KV)	Every network had the ability to predict values that were almost identical to the experimental values	0.991, 0.958, 0.994, and 0.967 for FP, KV, density, and CN, respectively	To make the model more comprehensive, the different parameters like pour point higher heating value, and cloud point could be considered
[59]	ANN and multiple non-linear regression model	Hidden layer = 1 Number of neurons = 6 (tansig function) and 9 (logsig function)	The ANN model had superior performance and consistency in their predictions across all predicted outcomes	0.996, 0.997, and 0.997 for PV, AV, and YP, respectively	The real-time monitoring and optimization strategy could be adopted by integrating the other accurate predictive models for enhancing the performance

Table 2.5 Design parameters for present study

Parameters	Values
Overlap ratio (e/d)	0.1
No. of blades	2
Blade diameter (d)	80 mm
Turbine diameter (D)	160 mm
Slot shapes	Convergent and divergent
Slot position (f)	(1–9%) of blade diameter (d)
Slot gap (b)	3–7 mm
Tip speed ratio (λ)	0.6–1.2

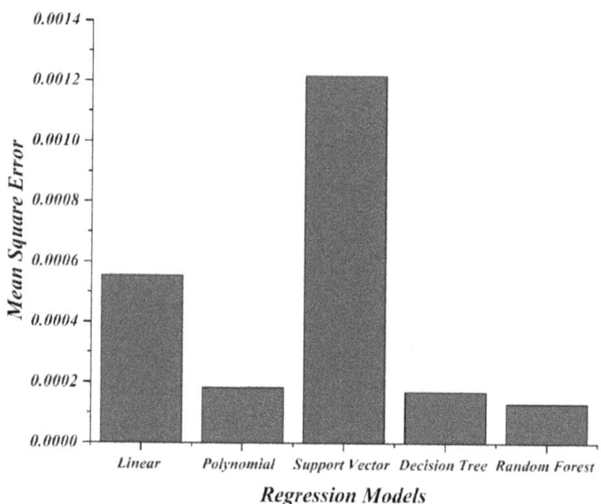

Figure 2.19 Variation of mean square error based on different regression models.

for testing. There are five different models which are used to predict the output parameter values based on data available of input parameters and output parameters. For determining the performance of the regression models, the mean square error (MSE) should be less when compared with other regression models, as shown in Figure 2.18, and the correlation coefficient (R) must be close to 1 on comparing with other regression models as shown in Figure 2.19. However, the value of MSE and R^2 was found to be 1.82×10^{-4} and 0.934, respectively. However, the value of R^2 was very much close to 1 for polynomial regression model, and MSE is quite low, which made the polynomial regression model as best suited for this study compared to other regression models.

According to the polynomial regression model, the C_p is predicted for different slot parameters and λ. The maximum predicted C_p was 0.2903

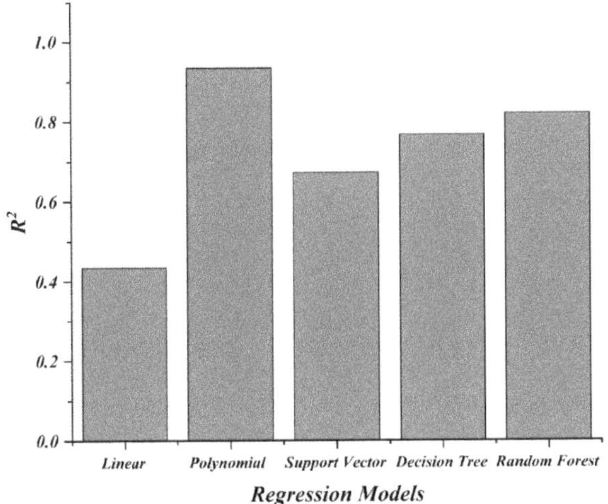

Figure 2.20 Variation of R^2 based on different regression models.

for the divergent slot shape at a slot position of 5% of the blade, with a slot gap of 3 mm and a divergent ratio of 0.2 corresponding to 0.9 value of λ, as illustrated in Figure 2.20. The numerical study determined that the maximum C_p achieved a value of 0.2942 at a λ of 0.9 for the same slot parameters. The error between the maximum C_p from both the studies is about 1.34%. Hence, the maximum C_p obtained from computational study and ML analysis are in good agreement with each other.

Based on regression analysis, the correlation has been developed in order to predict the value of power coefficient based on slot parameters and TSR. The correlation equation is expressed as follows:

$$y = (-0.190095349735671 + (0.0229601271607683 * S_p) + (-0.0164242972329961 * S_g) + (-0.00775468071099217 * D_r) + (1.08048754061739 * \lambda) + (-0.00180905869059508 * S_p^2) + (0.00318319255034801 * S_p * S_g) + (-0.0195386592635849 * S_p * D_r) + (0.0121326537898164 * S_p * \lambda) + (0.000308131889375975 * S_g^2) + (-0.0314761907494742 * S_g * D_r) + (0.0133608466770287 * S_g * \lambda) + (0.0586948394524399 * D_r^2) + (0.00648261846834469 * D_r * \lambda) + (-0.625262739607052 * \lambda^2)),$$

Here S_p, S_g, D_r, and λ are slot position, slot gap, divergent ratio, and tip speed ratio, respectively.

2.6.1 Validation of correlation

Figure 2.22 shows the predicted value of C_p for divergent ratio of 0.8 at an optimum slot position, slot gap, and at λ value of 0.9 based on the correlation

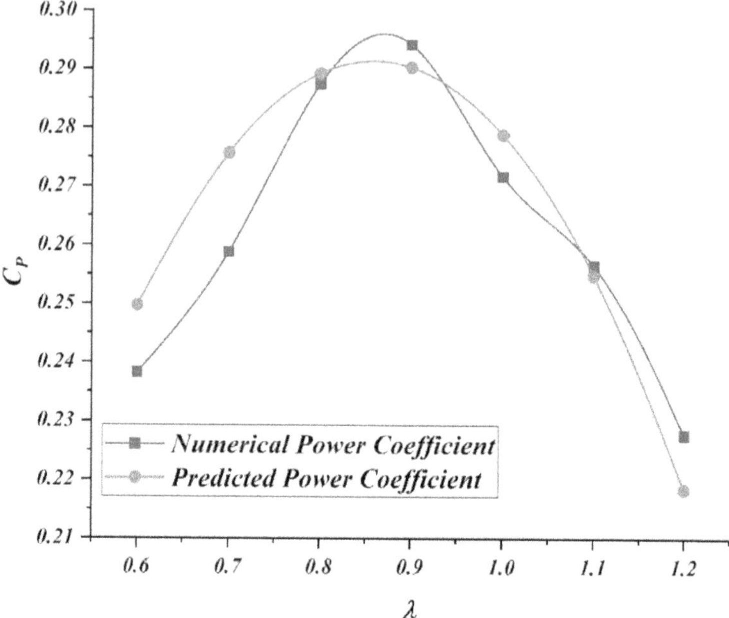

Figure 2.21 Variation of C_p obtained by numerical study and ML analysis.

```
In [34]:   x1= 4
           x2= 3
           x3= 0.8
           x4= 0.9
           y = -0.190095349735671 + 0.0229601271607683*x1 + -0.0164242972329961*x2 + -0.00775468071099217*x3 +
           1.08048754061739*x4 + -0.00180905869059508*x1^2 + 0.00318319255034801*x1*x2 + -0.0195386592635849*x1*x3 +
           0.0121326537898164*x1*x4 + 0.000308131889375975*x2^2 + -0.0314761907494742*x2*x3 + 0.0133608466770287*x2*x4 +
           0.0586948394524399*x3^2 + 0.00648261846834469*x3*x4 + -0.625262739607052*x4^2

Out [34]:   0.220833396
```

Figure 2.22 Predicted value of C_p for the ε of 0.8 at an optimum slot position, slot gap, and at λ value of 0.9.

developed through polynomial regression. With respect to λ value of 0.9, the value of power coefficient obtained was 0.2208. This value of power coefficient has been validated with the numerical study as shown in Figure 2.20.

2.7 CONCLUSION

This study represented a utilization of AI and ML in sustainable development via RE sources. The studies subjected to solar energy, wind energy, hydropower energy, and biomass energy with AI and ML predictions were utilized to represent the application of RE to achieve sustainability. A co-optimization for policy-based AI and ML technologies was expected to substitute standard optimization methods for achieving optimal results in power system difficulties. Strategic plans, extrapolations, and complicated models were among the current AI- and ML-driven renewable energy explorations. Also, AI and ML were able to predict the swift operation of the power systems and optimum power productions and improve the energy infrastructure and will help authorities in making decisions. To achieve realistic responses, multi-objectives, including social, environmental, and uncertainty elements must be considered concurrently, as single aims alone are ineffective. Moreover, integrating multiple algorithms can improve problem-solving speed, reliability, and effectiveness. Combining AI approaches creates a hybrid approach to system design that prioritizes speed, quick convergence, and resilience. However, this approach demands greater complexity and can be harder to develop.

The case study was also carried out to optimize the slot parameters of Savonius hydrokinetic turbine and predict the power coefficient. The regression analysis was performed using the five models: LR, PR, SVR, DT, and RF. Based on the analysis, it was found that polynomial regression performed better than other models and achieved a 0.934 R^2 value. The maximum C_p obtained using numerical study was 0.2942, whereas the maximum C_p predicted using PR model was 0.2903 for $\lambda = 0.9$. The error between numerical and predicted maximum C_p was 1.34%, suggesting that polynomial regression can be used to optimize and predict the performance of Savonius turbine. However, researchers have concentrated on AI and ML approaches, which are deemed more acceptable than traditional methods because of their capacity to analyse for global optimal, high accuracy, and quick convergence; however, AI and ML still encounter difficulty owing to the larger number of desired inputs. Therefore, in future research will focus on building and hybridizing best AI and ML approaches for varied goal objectives. Moreover, there is also need of developing sophisticated multidimensional dynamic models that go beyond physical and time-based dimensions. This demands employing innovative graph neural network adaptations to comprehend sophisticated relationships and interactions among various data formats.

NOMENCLATURE

AI	artificial intelligence
ANFIS	adaptive neuro-fuzzy inference systems
ANN	artificial neural network
ARM	association rule mining
ARIMA	autoregressive integrated moving average
BFGS	Broyden–Fletcher–Goldfarb–Shanno
CN	cetane number
CNN	convolutional neural network

C_p	power coefficient
DCN	derived cetane number
DT	decision tree
DWT	discrete wavelet transforms
DWVG	delta winglet vortex generators
ERA5	European Centre for Medium-Range Weather Forecasts Reanalysis 5
FAME	fatty acid methyl esters
FCBF	fast correlation-based filtering
FFBPN	feedforward backprop network
FIS	fuzzy inference system
FNN	feedforward neural network
GA	genetic algorithm
GAN	generative adversarial networks
GP	Gaussian process
GPR	Gaussian process regression
HANTS	harmonic analysis time series
IQT	ignition quality tester
KNN	k-nearest neighbours
KRG	Kriging model
LM	Levenberg–Marquardt
LM-BP	Levenberg–Marquardt backpropagation
LR	linear regression
LSSVM	least squares support vector model
LSTM	long short-term memory
MAE	mean absolute error
MCP	measure–correlate–predict
ML	machine learning
MLNR	multiple linear regression
MLP	multilayer perceptron
MLPNN	multilayer perceptron neural network
MOMoGP	multi-output mixture of Gaussian process
MONN	multi-output neural network
NMAE	normalized mean absolute error
NSGA II	non-dominated sorting genetic algorithm II
PCE	polynomial chaos expansion
POA	progressive optimality algorithm
PR	polynomial regression
PSO	particle swarm optimization
QP	quadratic programming
RF	random forest
RMSE	root means square error
SA	simulated annealing
seq2seq	sequence-to-sequence
SVM	support vector machine

SVR	support vector regression
SWT	Savonius wind turbine
TLBO	teaching–learning-based optimization
UPOA	uniform progressive optimality algorithm
VMD	variational mode decomposition
VMM	virtual metrology models

REFERENCES

1. H.E. Daly, Toward some operational principles of sustainable development, *Ecol. Econ.* 2 (1990) 1–6. https://doi.org/10.1016/0921-8009(90)90010-R.
2. S. Ryan, Electricity generation from renewable energy — national geographic society [WWW Document], *Natl. Geogr. Mag.*, 2015.
3. B. Tierney, S., Setting the record straight about renewable energy | World Resources Institute, 2020.
4. A. Skoglund, M. Leijon, A. Rehn, M. Lindahl, R. Waters, On the physics of power, energy and economics of renewable electric energy sources — Part II, *Renew. Energy.* 35 (2010) 1735–1740. https://doi.org/10.1016/j.renene.2009.08.031.
5. MNRE, Current status | Ministry of new and renewable energy, government of India *Sol. Energy.*, 2020. https://mnre.gov.in/.
6. IRENA, Technologies, (2022). https://www.irena.org/Data/View-data-by-topic /Capacity-and-Generation/Technologies.
7. A. Jäger-Waldau, I. Kougias, N. Taylor, C. Thiel, How photovoltaics can contribute to GHG emission reductions of 55% in the EU by 2030, *Renew. Sustain. Energy Rev.* 126 (2020), 109836. https://doi.org/10.1016/j.rser.2020.109836.
8. M.H. Khoshgoftar Manesh, S.A. Mousavi Rabeti, M. Nourpour, Z. Said, Energy, exergy, exergoeconomic, and exergoenvironmental analysis of an innovative solar-geothermal-gas driven polygeneration system for combined power, hydrogen, hot water, and freshwater production, *Sustain. Energy Technol. Assessments.* 51 (2022), 101861. https://doi.org/10.1016/j.seta.2021.101861.
9. Z. Said, M. Ghodbane, A.K. Tiwari, H.M. Ali, B. Boumeddane, Z.M. Ali, 4E (Energy, Exergy, Economic, and Environment) examination of a small LFR solar water heater: An experimental and numerical study, *Case Stud. Therm. Eng.* 27 (2021), 101277. https://doi.org/10.1016/j.csite.2021.101277.
10. A. Allouhi, S. Rehman, M.S. Buker, Z. Said, Up-to-date literature review on Solar PV systems: Technology progress, market status and R&D, *J. Clean. Prod.* 362 (2022), 132339. https://doi.org/10.1016/j.jclepro.2022.132339.
11. X. Power, 1KW/1KVA Off Grid Solar System 12V Factory Wholesale, (n.d.). https://www.xindunpower.com/product/1kw-1kva-off-grid-solar-system-12v -factory/.
12. A.B. Awan, M. Zubair, K.V.V. Chandra Mouli, Design, optimization and performance comparison of solar tower and photovoltaic power plants, *Energy.* 199 (2020), 117450. https://doi.org/10.1016/j.energy.2020.117450.
13. Y. Cao, E. Kamrani, S. Mirzaei, A. Khandakar, B. Vaferi, Electrical efficiency of the photovoltaic/thermal collectors cooled by nanofluids: Machine learning simulation and optimization by evolutionary algorithm, *Energy Rep.* 8 (2022) 24–36. https://doi.org/10.1016/j.egyr.2021.11.252.

14. S. Du, Y.L. He, W.W. Yang, Z. Bin Liu, Optimization method for the porous volumetric solar receiver coupling genetic algorithm and heat transfer analysis, *Int. J. Heat Mass Transf.* 122 (2018) 383–390. https://doi.org/10.1016/j.ijheat-masstransfer.2018.01.120.

15. L. Luo, W. Du, S. Wang, L. Wang, B. Sundén, X. Zhang, Multi-objective optimization of a solar receiver considering both the dimple/protrusion depth and delta-winglet vortex generators, *Energy.* 137 (2017) 1–19. https://doi.org/10.1016/j.energy.2017.07.001.

16. F. Wang, Z. Zhang, C. Liu, Y. Yu, S. Pang, N. Duić, M. Shafie-khah, J.P.S. Catalão, Generative adversarial networks and convolutional neural networks based weather classification model for day ahead short-term photovoltaic power forecasting, *Energy Convers. Manag.* 181 (2019) 443–462. https://doi.org/10.1016/j.enconman.2018.11.074.

17. C.K.G. Zaim Zulkifly, Kyairul Azmi Baharin, Improved machine learning model selection techniques for solar energy forecasting applications, *Int. J. Renew. Energy Res.* 11 (2021), 308–319.

18. V. Masterson, Wind and solar generated 10% of global electricity in 2021, 2022. https://www.weforum.org/agenda/2022/04/wi%0Aand-solar-electricity -global-energy/.

19. Wind Turbine Power Brushes, Graphitestore. (n.d.). https://www.graphitestore .com/wind-turbine-power-brushes.

20. M.M.S.L. Jonata C. de Albuquerque, Ronaldo R. B. de Aquino, Otoni Nóbrega Neto, M.A. de C.J. Aida A. Ferreira, Power curve modelling for wind turbine using artificial intelligence tools and pre-established inference criteria, *J. Mod. Power Syst. Clean Energy.* 9 (2021), 526–533.

21. S.A. Al-Shammari, A.H. Karamallah, S. Aljabair, Blade shape optimization of savonius wind turbine at low wind energy by artificial neural network, *IOP Conf. Ser. Mater. Sci. Eng.* 881 (2020), 012154. https://doi.org/10.1088/1757 -899X/881/1/012154.

22. R. Azimi, M. Ghofrani, M. Ghayekhloo, A hybrid wind power forecasting model based on data mining and wavelets analysis, *Energy Convers. Manag.* 127 (2016) 208–225. https://doi.org/10.1016/j.enconman.2016.09.002.

23. A. Ramadan, K. Yousef, M. Said, M.H. Mohamed, Shape optimization and experimental validation of a drag vertical axis wind turbine, *Energy.* 151 (2018) 839–853. https://doi.org/10.1016/j.energy.2018.03.117.

24. S. Schwegmann, J. Faulhaber, S. Pfaffel, Z. Yu, M. Dörenkämper, K. Kersting, J. Gottschall, Enabling virtual met masts for wind energy applications through machine learning-methods, *Energy AI.* 11 (2023), 100209. https://doi.org/10.1016/j.egyai.2022.100209.

25. W. Tian, Z. Mao, B. Zhang, Y. Li, Shape optimization of a Savonius wind rotor with different convex and concave sides, *Renew. Energy.* 117 (2018) 287–299. https://doi.org/10.1016/j.renene.2017.10.067.

26. Y. Zhang, R. Li, Short term wind energy prediction model based on data decomposition and optimized LSSVM, *Sustain. Energy Technol. Assessments.* 52 (2022), 102025. https://doi.org/10.1016/j.seta.2022.102025.

27. G. hydro Power, Installed global hydropower capacity could reach 1,200 GW in 2022, report says, (2022). https://www.hydroreview.com/h%0Aydro-indus-try-news/installed-global-hydropower-capacity-could-reach%0A-1200-gw-in -2022-report-says/#gref.

28. Types Of Water Turbines - Waterturbines, Pinterest. (n.d.). https://www.pinterest.com/pin/202239839507138881/.

29. F. Conteh, S. Tobaru, M.E. Lotfy, A. Yona, T. Senjyu, An effective Load shedding technique for micro-grids using artificial neural network and adaptive neuro-fuzzy inference system, *AIMS Energy*. 5 (2017) 814–837. https://doi.org/10.3934/energy.2017.5.814.

30. Z. Kai Feng, W. Jing Niu, C. Tian Cheng, Optimizing electrical power production of hydropower system by uniform progressive optimality algorithm based on two-stage search mechanism and uniform design, *J. Clean. Prod.* 190 (2018) 432–442. https://doi.org/10.1016/j.jclepro.2018.04.134.

31. M. Sattar Hanoon, A. Najah Ahmed, A. Razzaq, A.Y. Oudah, A. Alkhayyat, Y. Feng Huang, P. Kumar, A. El-Shafie, Prediction of hydropower generation via machine learning algorithms at three Gorges Dam, China, *Ain Shams Eng. J.* 14 (2023), 101919. https://doi.org/10.1016/j.asej.2022.101919.

32. J. Barzola-Monteses, J. Gómez-Romero, M. Espinoza-Andaluz, W. Fajardo, Hydropower production prediction using artificial neural networks: an Ecuadorian application case, *Neural Comput. Appl.* 34 (2022) 13253–13266. https://doi.org/10.1007/s00521-021-06746-5.

33. W. Jing Niu, Z. Kai Feng, C. Tian Cheng, Optimization of variable-head hydropower system operation considering power shortage aspect with quadratic programming and successive approximation, *Energy*. 143 (2018) 1020–1028. https://doi.org/10.1016/j.energy.2017.11.042.

34. W. Yang, D. Pudasainee, R. Gupta, W. Li, B. Wang, L. Sun, An overview of inorganic particulate matter emission from coal/biomass/MSW combustion: Sampling and measurement, formation, distribution, inorganic composition and influencing factors, *Fuel Process. Technol.* 213 (2021), 106657. https://doi.org/10.1016/j.fuproc.2020.106657.

35. M. Aydin, The effect of biomass energy consumption on economic growth in BRICS countries: A country-specific panel data analysis, *Renew. Energy*. 138 (2019) 620–627. https://doi.org/10.1016/j.renene.2019.02.001.

36. F. Bilgili, E. Koçak, Ü. Bulut, S. Kuşkaya, Can biomass energy be an efficient policy tool for sustainable development?, *Renew. Sustain. Energy Rev.* 71 (2017) 830–845. https://doi.org/10.1016/j.rser.2016.12.109.

37. Z. Wang, Q. Bui, B. Zhang, The relationship between biomass energy consumption and human development: Empirical evidence from BRICS countries, *Energy*. 194 (2020), 116906. https://doi.org/10.1016/j.energy.2020.116906.

38. M. Meena, S. Shubham, K. Paritosh, N. Pareek, V. Vivekanand, Production of biofuels from biomass: Predicting the energy employing artificial intelligence modelling, *Bioresour. Technol.* 340 (2021), 125642. https://doi.org/10.1016/j.biortech.2021.125642.

39. I. Ahmad, A. Ayub, U. Ibrahim, M.K. Khattak, M. Kano, Data-based sensing and stochastic analysis of biodiesel production process, *Energies*. 12 (2019), 12010063. https://doi.org/10.3390/en12010063.

40. A. Aminian, B. ZareNezhad, Accurate predicting the viscosity of biodiesels and blends using soft computing models, *Renew. Energy*. 120 (2018) 488–500. https://doi.org/10.1016/j.renene.2017.12.038.

41. G. Knothe, K.R. Steidley, Kinematic viscosity of biodiesel components (fatty acid alkyl esters) and related compounds at low temperatures, *Fuel*. 86 (2007) 2560–2567. https://doi.org/10.1016/j.fuel.2007.02.006.

42. S. V.D. Freitas, M.J. Pratas, R. Ceriani, Á.S. Lima, J.A.P. Coutinho, Evaluation of predictive models for the viscosity of biodiesel, *Energy Fuels*. 25 (2011) 352–358. https://doi.org/10.1021/ef101299d.

43. W. Yuan, A.C. Hansen, Q. Zhang, Predicting the temperature dependent viscosity of biodiesel fuels, *Fuel*. 88 (2009) 1120–1126. https://doi.org/10.1016/j.fuel .2008.11.011.

44. W. Yuan, A.C. Hansen, Q. Zhang, Z. Tan, Temperature-dependent kinematic viscosity of selected biodiesel fuels and blends with diesel fuel, *JAOCS, J. Am. Oil Chem. Soc*. 82 (2005) 195–199. https://doi.org/10.1007/s11746-005-5172 -6.

45. K. Krisnangkura, T. Yimsuwan, R. Pairintra, An empirical approach in predicting biodiesel viscosity at various temperatures, *Fuel*. 85 (2006) 107–113. https://doi.org/10.1016/j.fuel.2005.05.010.

46. F.X. Feitosa, M.D.L. Rodrigues, C.B. Veloso, C.L. Cavalcante, M.C.G. Albuquerque, H.B. De Santana, Viscosities and densities of binary mixtures of coconut + colza and coconut + soybean biodiesel at various temperatures, *J. Chem. Eng. Data*. 55 (2010) 3909–3914. https://doi.org/10.1021/je901060j.

47. C.A. Nogueira, F.X. Feitosa, F.A.N. Fernandes, R.S. Santiago, H.B. De Sant'ana, Densities and viscosities of binary mixtures of babassu biodiesel + cotton seed or soybean biodiesel at different temperatures, *J. Chem. Eng. Data*. 55 (2010) 5305–5310. https://doi.org/10.1021/je1003862.

48. A. Baghban, M.N. Kardani, A.H. Mohammadi, Improved estimation of Cetane number of fatty acid methyl esters (FAMEs) based biodiesels using TLBO-NN and PSO-NN models, *Fuel*. 232 (2018) 620–631. https://doi.org/10.1016/j.fuel .2018.05.166.

49. G. Knothe, A comprehensive evaluation of the cetane numbers of fatty acid methyl esters, *Fuel*. 119 (2014) 6–13. https://doi.org/10.1016/j.fuel.2013.11 .020.

50. R. Piloto-Rodríguez, Y. Sánchez-Borroto, M. Lapuerta, L. Goyos-Pérez, S. Verhelst, Prediction of the cetane number of biodiesel using artificial neural networks and multiple linear regression, *Energy Convers. Manag*. 65 (2013) 255–261. https://doi.org/10.1016/j.enconman.2012.07.023.

51. B.R. Moser, R.L. Evangelista, G. Jham, Fuel properties of Brassica juncea oil methyl esters blended with ultra-low sulfur diesel fuel, *Renew. Energy*. 78 (2015) 82–88. https://doi.org/10.1016/j.renene.2015.01.016.

52. T.T. Kivevele, Z. Huan, An analysis of fuel properties of fatty acid methyl ester from Manketti seeds oil, *Int. J. Green Energy*. 12 (2015) 291–296. https://doi .org/10.1080/15435075.2014.886579.

53. K.G. Georgogianni, M.G. Kontominas, E. Tegou, D. Avlonitis, V. Gergis, Biodiesel production: Reaction and process parameters of alkali-catalyzed transesterification of waste frying oils, *Energy and Fuels*. 21 (2007) 3023–3027. https://doi.org/10.1021/ef070102b.

54. A.O.E. Solomon olasunkanmi odeyale, prediction of fuel properties of biodiesel using two-layer artificial neural network, *Walailak J. Sci. Technol*. 12 (2014) 325–341.

55. L.F. Ramírez-Verduzco, J.E. Rodríguez-Rodríguez, A.D.R. Jaramillo-Jacob, Predicting cetane number, kinematic viscosity, density and higher heating value of biodiesel from its fatty acid methyl ester composition, *Fuel*. 91 (2012) 102–111. https://doi.org/10.1016/j.fuel.2011.06.070.

56. M.M. Azam, A. Waris, N.M. Nahar, Prospects and potential of fatty acid methyl esters of some non-traditional seed oils for use as biodiesel in India, *Biomass Bioenergy.* 29 (2005) 293–302. https://doi.org/10.1016/j.biombioe.2005.05.001.

57. S.O. Giwa, S.O. Adekomaya, K.O. Adama, M.O. Mukaila, Prediction of selected biodiesel fuel properties using artificial neural network, *Front. Energy.* 9 (2015) 433–445. https://doi.org/10.1007/s11708-015-0383-5.

58. T. Kessler, E.R. Sacia, A.T. Bell, J.H. Mack, Artificial neural network based predictions of cetane number for furanic biofuel additives, *Fuel.* 206 (2017) 171–179. https://doi.org/10.1016/j.fuel.2017.06.015.

59. A.P. Tchameni, L. Zhao, J.X.F. Ribeiro, T. Li, Predicting the rheological properties of waste vegetable oil biodiesel-modified water-based mud using artificial neural network, *Geosystem Eng.* 22 (2019) 101–111. https://doi.org/10.1080/12269328.2018.1490209.

Chapter 3

Artificial intelligence as a tool for building more resilient cities in the climate change era

A systematic literature review

Mahdi Suleimany, Mohammad Riahi SeGonbad, Sara Naghibizadeh, and Sahar Dehghan Niri

3.1 INTRODUCTION

Climate change and its consequences, such as global warming, prolonged droughts, and destructive floods, have been posing significant threats to cities and urban communities. These threats include the destruction of natural habitats and the reduction of biodiversity [1], the exacerbating burden on the economic and energy systems [2], the growing number of heat mortalities [3], and even social anomalies such as segregation [4]. As a specific example, reports indicate that extreme heat events in Europe in 2003 were responsible for the death of more than 72,000 inhabitants [5]. Moreover, estimations show that climate change and its consequences will be more frequent, severe, and intense, along with urbanization and industrial growth [6]. These threats have induced scholars and policymakers to develop and adopt strategies and policies to build more resilient cities and urban communities to climate change and its consequences [7].

Urban climate resilience is the ability of cities and urban systems to cope with and recover from climate change consequences while maintaining their primary functions and services [8]. Urban climate resilience is a critical aspect of sustainable urban development as cities are more vulnerable to the mentioned consequences and threats due to high population density, complex infrastructure systems, and often location in climate-sensitive areas such as coastlines or river deltas. On the other hand, metropolitan areas often contribute significantly to climate change through high levels of energy consumption and greenhouse gas emissions [9, 10]. Building urban climate resilience involves a range of sophisticated strategies, policies, and approaches, promoting urban climate mitigation and adaptation capacities:

- First, the imperative for addressing climate change is to curtail greenhouse gas emissions, which can be attained by adopting strategic measures such as optimizing energy efficiency, augmenting the utilization of renewable energy resources, and propagating sustainable transportation practices [10].

DOI: 10.1201/9781003581246-3

- The second category of policies focuses on adapting cities to cope with the consequences of climate change. This involves implementing measures to improve the structural and architectural resilience of buildings and infrastructure against severe climate events. Additionally, integrating green infrastructure solutions, including urban forests and green roofs, can play a crucial role in reducing heat and stormwater [10].
- The third category of urban climate resilience policies pertains to mitigating social vulnerabilities. The adverse effects of climate change tend to disproportionately affect marginalized communities, necessitating measures to decrease urban climate injustice and promote adaptation to climate change for all inhabitants, particularly sensitive socio-economic groups [10].
- Creating urban climate resilience necessitates implementing comprehensive and inclusive planning procedures. Since climate change repercussions transcend sectoral boundaries, coordinated responses across diverse sectors and government levels are imperative. Additionally, involving residents and other relevant stakeholders in planning is crucial to ensure measures are tailored to meet local requirements and circumstances [10].

Building urban climate resilience requires integrated urban governance, planning, and design. This procedure necessitates a comprehensive approach to tackling both the physical repercussions of climate change and the social disparities that could further aggravate these consequences. Therefore, up-to-date systems and tools, such as artificial intelligence (AI) and its subsets, must be utilized to inform policymakers of probable climatic issues, changing vulnerability patterns, and the required measures [11].

AI is a discipline within computer science that aims to develop systems capable of performing tasks that would typically necessitate human intelligence. These tasks comprise, but are not limited to, learning from experience, comprehending natural language, identifying patterns, resolving problems, and making informed decisions. AI can be categorized into two primary types: narrow AI, specifically designed to execute a singular task, and General AI, which can comprehend, learn, and utilize knowledge across a broad spectrum of tasks [12].[1] The evolution of AI and its subsets has had a profound impact on various fields, including public policy and urban planning.

At the outset, AI was employed in public policy and planning to analyse data and develop predictive models. Predictive models produced by AI algorithms offered policymakers insightful predictions on economic trends, social issues, and environmental changes [13, 14], influencing their decision-making processes. However, AI's role in policymaking has significantly expanded with technological advancements. Presently, AI is utilized not only for data analysis, but also for decision-making and strategy implementation. AI algorithms optimize resource allocation, enhance policy enforcement, and improve service delivery. Moreover, AI is employed to forecast the impact of policy decisions and monitor and evaluate policy outcomes [15]. Furthermore, AI has revolutionized public engagement in policymaking by facilitating public consultations, gathering public opinions, and analysing public sentiments. This has made policymaking more transparent and inclusive [16].

Using AI in public policy and planning presents several challenges and considerations. One of the primary concerns is the ethical implications of AI, particularly in terms of privacy and data protection. As AI systems often rely on large amounts of data, there is a risk of infringing on individuals' privacy rights. Another challenge is ensuring transparency and accountability in AI decision-making processes. AI algorithms are often complex and opaque, making it difficult to understand how decisions are made. This lack of transparency can lead to issues of accountability, mainly when AI is used to make significant policy decisions [17]. Furthermore, ensuring fairness and avoiding bias in AI systems is highly challenging. AI systems are trained on the input data and can, therefore, inherit and perpetuate any biases present in that data. This can lead to discriminatory outcomes, which is a significant concern in the context of public policy [18].

Despite the vast body of literature on the potential capability of AI in analysing large datasets, predicting environmental patterns [19], and optimizing resource allocation [20], its application in climatically resilient urban planning and development is still in its nascent stages. Many city planners and policymakers do not have adequate knowledge about AI technologies and their potential applications in this field. There is also a need for more research and case studies demonstrating the successful implementation of AI in enhancing urban climate resilience [21]. Besides, the requirements and challenges of utilizing AI in climatically resilient urban planning are almost understudied. This knowledge gap may hinder the effective integration of AI and urban planning strategies in response to the consequences of climate change. This study aims to address this gap by developing an AI application framework for planning more climatically resilient cities based on the evidence from the relevant literature.

The following are the subsequent research questions:

(1) *What specific applications do AI and AI-based tools have in climatically resilient urban planning?*
(2) *What requirements and challenges exist in the implementation of AI in this domain?*

This study can assist decision-makers and contribute to sustainable urban development by creating an application framework for employing AI and its subsets, as evolving technologies, in building more resilient cities to climate change and its consequences. This framework can promote the knowledge of policymakers and designers of AI applications in the mentioned field and help them meet the requirements and address the existing ethical and technological challenges.

3.2 METHODOLOGY

Responding to the research questions, this study employs the systematic literature review (SLR) method to elicit the applications, requirements, and challenges of utilizing AI in planning for more climatically resilient cities. The systematic literature review is a methodical approach to identify and synthesize the available evidence on a specific topic within the literature. It involves a rigorous and detailed search of all relevant studies, a critical appraisal of the selected studies, and a synthesis of the findings. The first step in the systematic literature review is to define a clear and specific research question. The next step is to develop a detailed search strategy. This strategy applies to identifying the appropriate databases and other sources to search and the search terms to use. Once the search is complete, the reviewers will screen and refine the output studies for relevance based on predefined inclusion and exclusion criteria. The selected studies are then critically appraised for quality and risk of bias. This step is crucial as it ensures the findings are based on high-quality evidence. The final step is the synthesis of the findings, which can be accomplished via a narrative summary, a meta-analysis, or both. The results are then interpreted in the context of the overall evidence base, and the implications for practice and future research are discussed [22, 23].

In this study, we formulated the research questions in the first step based on the knowledge gap identified (in the introduction section). The second step is dedicated to the systematic search for related literature, screening, and selecting relevant and optimal topics to prepare the research input. We used the Scopus database to explore relevant literature. The literature reviewed in this research were scientific studies (including books, academic papers, reviews, and case studies) and reports published in English from 2016 to 2023. We opted for this period to diminish the volume of unrelated input studies as, prior to this time, most of the papers in the AI field were primarily technical studies, focusing on how to develop technology. Searching for the related studies on 16 January 2023, we designed a broad string (as provided in Table 3.1) to cover a vast body of relevant literature. This search brought about 1,669 pieces of research, most of which were in the computer engineering and data science fields. Therefore, we limited our search by selecting documents within social sciences, decision sciences, management, and humanities, using the subject area filtering tool of the database. This led to the reduction of outputs to 551 rows. Afterwards, four scholars screened the abstract of documents to refine research inputs by removing irrelevant or duplicate findings.

Table 3.1 Literature search criteria and description

Search criteria	Description
Document type	Scientific studies (books, papers, reviews, and case studies) and reports
Language	English
Subject area	Social sciences, decision sciences, management, and humanities
Final search string	[("artificial intelligence" OR "ai") OR ("machine learning" OR "ml") OR ("deep learning" OR "dl")] AND [(climate) AND ("resilien*" OR "mitigat*" OR "adapt*")]

Subsequently, 109 documents were selected based on the abstract screening. We also activated the database update notification to receive the relevant papers published after the initial search, which added 11 more papers to our dataset until 15 March 2024. Moreover, we found five administrative reports from other resources and added them to our dataset as SLR procedures, such as PRISMA,[2] allow for the use of inputs other than the database search [24]. Ultimately, 125 documents remained to be reviewed and deeply analysed using inductive content coding. Four scholars analysed these documents via inductive content coding using the MaxQDA software. The analysis process is as follows: first, each scholar analysed ten documents by inductive content-coding, considering the applications of AI in climatically resilient urban planning and the related requirements and challenges. Afterwards, we synchronized the codes and developed the initial coding list to avoid inconsistency among researchers [25]. Then, the rest of the documents were coded, and the codes were categorized regarding the application of AI on three levels of urban governance, planning, and design, and considering their contribution to urban climate mitigation and adaptation capacities. Finally, we combined the codes and used a narrative summary approach [21] to answer the research questions.

Figure 3.1 indicates the process of finding, screening, and reviewing/analysing documents based on the PRISMA flowchart.

3.3 RESULTS AND DISCUSSION

AI and its subsets have various applications in developing cities that are more resilient to climate change and its consequences. Some of these applications generally promote urban resilience, while others directly contribute to increasing the urban climate mitigation and adaptation capacities. By improving the climate resilience of cities in several environmental, physical, infrastructural, economic, and social dimensions, these applications can lead to sustainable urban development. In this section, we will explore these applications and discuss the requirements for implementing them, as

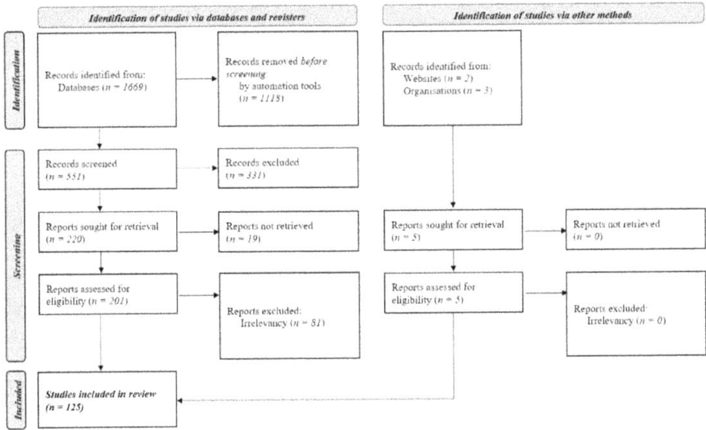

Figure 3.1 The process of systematic literature review in this study based on the PRISMA.

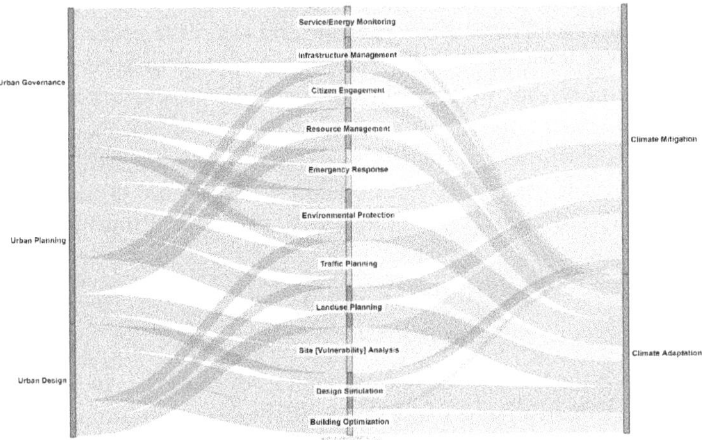

Figure 3.2 The Sankey diagram of AI applications in building more resilient cities to climate change.

well as the potential challenges based on the existing theoretical evidence. AI applications in building more resilient cities in the climate change era mainly incorporate service/energy monitoring, infrastructure management, citizens engagement, resource management, emergency response, environmental protection, traffic planning, land use planning, site analysis, design simulation, and building optimization. Although the applications of AI in building climatically resilient cities are partly shared among different levels of urban decision-making (i.e. governance, planning, and design), we will discuss these applications separately for a more specialized approach. Figure 3.2 provides the Sankey diagram [26] of the AI applications in building resilient cities to climate change and its consequences. This diagram provides an overall insight into the mentioned AI applications at different levels of urban decision-making and their contributions to urban climate mitigation and adaptation capacities.

3.3.1 Applications of AI in climatically resilient urban governance

The main AI applications in climatically resilient urban governance, as investigated in the literature, are service/energy monitoring, infrastructure management, citizens engagement, and resource management. These applications can support urban governance sectors in precisely monitoring the condition and performance of the primary services and infrastructure of cities in pre- and post-climate disasters to maintain the overall function of city life. AI-based tools can act as a real-time support system for urban decision-making, which is essential for managing cities in harsh climatic conditions by analysing and forecasting trends based on big urban data and different management datasets [27]. AI can also assist policymakers in participating with various key players and stakeholders to decarbonize cities by analysing the relevant conflicts over resources and pollution sources within metropolitan areas [28]. These applications can also optimize urban governance strategies by constantly assessing the impacts of urban policies and their interrelationship with climate change [20]. This section will detail the main applications of AI in climatically resilient urban governance.

3.3.1.1 Service/energy monitoring

Climate change and its consequences, such as extreme heatwaves, drought, and flooding, can adversely affect the performance of urban systems. These effects can cause significant disturbances in urban service delivery by suspending energy and transportation systems [9]. Since most of the public and commercial services within cities rely on urban systems and service delivery, harsh climatic conditions often lead to economic loss or crisis in cities. Besides, climatic anomalies, such as urban heat islands, can significantly exacerbate the burden on energy systems, which may cause periodic power outages and subsequent challenges for urban services [29]. AI can provide urban governments with tools to monitor service delivery in harsh climatic conditions and maintain overall service efficiency. These tools contain various analytic devices that analyse and anticipate the condition of urban services, public demands, and energy deficiencies [30]. Numerous cases have incorporated AI in their urban service delivery plans to promote the efficiency and effectiveness of public services, including from essential residential services [31] to large-scale health centres [32].

3.3.1.2 Infrastructure management

Equitable access to urban infrastructure, such as residential, transportation, energy, and natural infrastructure, is crucial to developing sustainable and climatically resilient communities [33]. These infrastructures are one of the main elements of urban service delivery and help reduce the threats caused by harsh climatic conditions. In other words, these infrastructures can support the performance of cities and mitigate potential losses. The use of optimized infrastructure, especially in the transportation and energy sectors, also significantly lowers greenhouse gas emissions and can reduce the carbon footprint of cities [34]. One of the main critical tasks of urban governance in maintaining the climate resiliency of cities is to improve the efficiency of urban infrastructure and ensure fair access to it by communities, particularly sensitive socio-economic groups [35]. AI and its subsets can be utilized as an analytical tool to help urban governors analyse the urban infrastructure system deficiencies and their capacity/service level, such as the capacity of drainage [36], water supply [37], and sewage systems [34], and consequently, assist them in adopting optimization and maintenance policies.

3.3.1.3 Citizens engagement

Building climatically resilient cities highly demands a participatory approach in adopting and implementing relevant strategies, as there are various key players and stakeholders and, subsequently, considerable conflicts in this process [38]. AI-based tools have the potential to contribute significantly to improving citizens and stakeholder engagement. By automating data collection and analysis related to climate risks, AI can provide stakeholders with information applicable to design effective mitigation and adaptation strategies [39]. AI can also facilitate real-time communication and feedback, fostering a sense of collaboration among citizens. AI-powered open data services can enhance citizen engagement processes and promote resiliency on its basic levels (i.e. individual level) by enlightening citizens about climate change and related issues [40]. Moreover, AI-based innovations can support the participation of indigenous and marginalized communities in urban governance and planning, which is essential for enhancing climate resilience and justice by analysing demands and resource conflicts and diminishing bureaucratic challenges [41].

3.3.1.4 Resource management

Managing limited environmental, energy, and financial resources is one of the main keys to sustainable urban development and also affects climate mitigation and adaptation capacities. Theoretical evidence suggests that AI has the potential to make significant contributions to resource management by optimizing energy consumption and mitigating resource conflicts. Reports suggest that smart AI-based manufacturing can reduce energy consumption, waste, and carbon emissions in various industries by 30–50%. AI-based urban transportation also requires less energy and has up to 60% lower greenhouse gas emission rate than traditional modes [42]. Moreover, AI tools can enhance energy distribution by analysing available energy sources, their efficiency, and the alternatives for clean and low-carbon developments [43]. Utilizing AI to manage environmental resources in cities can further promote sustainability by helping to monitor and preserve limited natural assets and shift to more renewable ones [34]. However, it must be noted that implementing AI-based solutions depends on careful planning, adequate infrastructure, and active participation from all stakeholders.

3.3.2 Applications of AI in climatically resilient urban planning

Integrating AI into urban planning procedure and tools to build resilient cities to climate change can aid policymakers in optimizing their plans considering the existing and projected climate conditions. The main applications of AI in climatically resilient urban planning are emergency response, environmental protection, traffic planning, and land use planning. Urban decision-makers can implement these applications to enhance climate mitigation and adaptation and prepare cities for pre- and post-crisis conditions. This section discusses the mentioned AI applications in planning climatically resilient cities and communities.

3.3.2.1 Emergency response

One of the basic steps in building resilient cities is accurately identifying hazards and threats. It is complicated to predict hazards and, accordingly, adopt preparation strategies in the climate change era, as there are various interacting variables. Hence, integrating AI into planning for climate-resilient cities may be useful as AI tools can facilitate identifying and forecasting climate crises and the subsequent threats to cities [40]. Using AI and ML algorithms, the researchers set up early warning systems to support urban governors, planners, and related public service providers, such as medical centres and fire departments, in emergency response to adverse climatic conditions [44]. AI-based devices can also assist planners in post-disaster recovery by allowing the quick estimation of the induced damage and the required sources for recovery [45]. The applications of AI in enhancing emergency response go far beyond data-driven analytics and warning systems. AI robotics, such as rescue robots and robotic dogs, enable decision-makers to maintain public safety in adverse conditions by improving the search and rescue processes, during dangerous natural disasters [46].

3.3.2.2 Environmental protection

Safeguarding the environment and natural habitats is crucial for building climatically resilient cities as they significantly influence communities' mitigation and adaptation capacities. AI can significantly assist planners in both monitoring the condition of natural habitats and assessing the environmental impact of their policies. Integrating various sensors and analytical devices, AI enables decision-makers to identify the natural habitats that undergo environmental degradation and require preservation plans [47]. These tools are capable of precisely measuring the biodiversity of natural habitats and their fluctuation due to urbanization and climate change. Moreover, they can enforce environmental protection laws by observing and estimating the damage caused by individual or industrial activities [48]. Many cases implement AI tools to analyse and protect their green areas and infrastructure, which are valuable urban assets for improving climate resiliency and adaptiveness by ensuring food security, improving thermal comfort, and absorbing runoffs [49, 50]. Planners may also employ AI to reduce climate injustice, particularly in areas with adverse climates, by selecting the optimum species and supporting their equitable distribution [51].

3.3.2.3 Traffic and land use planning

Traffic and land use planning are two main urban planning sectors that considerably impact cities' climate mitigation and adaptation capacities. AI can increase the efficiency of urban transportation systems by optimizing facilities within highly demanded areas and, consequently, reduce overall energy consumption and street-level greenhouse gas emissions [40]. Some cases utilize AI to develop integrated urban mobility systems, making their urban spaces more sustainable by supporting clean transportation modes like walking or biking [52]. Furthermore, AI-optimized land use planning can induce more walkable neighbourhoods by providing suitable accessibility to the required amenities. Urban planners use AI to analyse the land use compatibility and identify and discharge polluting centres to mitigate air pollution in cities [53, 54]. Also, AI can evaluate citizens' accessibility and satisfaction with primary land uses and services by analysing urban networks and citizens' points of interest (POI) [55]. These land uses, including healthcare centres, fire departments, and urban parks, are critical for mitigating the side effects of climate change and decreasing citizens' exposure rate.

3.3.3 Applications of AI in climatically resilient urban design

AI and its subsets can also support building climatically resilient cities and communities on the design scale. Based on the reviewed theoretical evidence, the main contributions of AI to this level are site [vulnerability] analysis, design simulation, and building optimization. AI-based tools can assist urban designers and landscape architects in promoting their designs by analysing the specific conditions and needs of different urban sites, integrating nature-based solutions [56], and optimizing the outputs by simulation under various climatic conditions. This section delves into the mentioned AI applications.

3.3.3.1 Site (vulnerability) analysis

One of the basic measures in building climatically resilient cities is the vulnerability assessment and identifying susceptible communities, which is often conducted based on the analysis of exposure rates, the population of residing sensitive groups, and the accessibility to the primary infrastructure [31]. AI-based tools can help accurately analyse the intensity and spatial patterns of climate crises, such as urban floods and heat islands, by monitoring and exploring a large amount of collected data over time [57]. This tool can also estimate the exposure rate of sensitive groups, such as older adults, people with disabilities, and low-income households, by simulating the existing environmental conditions [58]. Besides, this tool is used to evaluate the urban communities' deprivation in access to basic infrastructure, which may significantly affect their adaptive capacity [59].

3.3.3.2 Design simulation

Urban designers and architects often face the challenge of accurate design simulation, as their design depends on numerous influential variables, especially for developing climatically resilient urban spaces. AI-based computer applications can provide design experts with a suitable simulation environment to visualize and evaluate their designs under multiple scenarios, particularly in harsh climatic conditions, to reveal their robustness or deficiencies [60]. Furthermore, these applications can explore and assess the impacts of potential climate-friendly design policies, such as green roofs, permeable pavements, and reflective surfaces, to opt for the most efficient strategies and policies [61]. These AI-based simulators may aid designers in also detecting the side effects of their projects on micro-climate, biodiversity, and natural resources, which is almost infeasible by traditional design tools [62].

3.3.3.3 Building optimization

Constructing energy-efficient buildings and urban forms while reducing the citizens' heat exposure and maintaining indoor and outdoor thermal comfort is one of the keys to designing sustainable and more climatically resilient urban communities and spaces. Hence, urban designers and architects seek to develop plans to passively reduce temperature in places facing global warming. In some cases, they also adopt water-sensitive design approaches to tackle the challenges of drought and arid climate [63]. AI-based tools are valuable assets that can deeply analyse the energy efficiency and climate compatibility of designs and, consequently, select the optimum case by implementing building information modelling (BIM) algorithms. In other words, they are capable of comparing numerous designs and plans under different conditions and scenarios to support the parametric design and select the optimum case [58, 64, 65].

Table 3.2 summarizes the applications of AI in building more resilient cities in the climate change era, which can be used as a framework for integrating AI-based tools into urban governance, planning, and design procedures to promote cities' climate mitigation and adaptation capacities.

Table 3.2 The framework of using AI as a tool for building more resilient cities in the climate change era.

Decision-making level	Application of AI	Contribution to climate mitigation and adaptation
Urban governance	Service/energy monitoring	Mitigating the side effects of climate change by helping to secure the provision of urban services and energy
	Infrastructure management	Analysing and ensuring the efficiency and accessibility of primary urban services in adverse climatic conditions
	Citizens engagement	Providing stakeholders with precise analysis to overcome the conflicts and enabling sensitive communities to participate
	Resource management	Managing urban financial, energy, and natural resources by monitoring and optimizing consumption and carbon footprint
Urban planning	Emergency response	Mitigating the damage of bad climatic conditions by detecting hazards, early warning, and facilitating rescue and recovery
	Environmental protection	Safeguarding natural habitats and protecting urban greenery by enabling the monitoring and maintaining of biodiversity
	Traffic and land use planning	Reducing urban carbon footprint and promoting walkability by optimizing mobility integration and land use compatibility
Urban design	Site [vulnerability] analysis	Mitigating urban vulnerability and citizens' exposure rate by accurately analysing the crisis patterns and access to amenities
	Design simulation	Revealing the deficiencies and environmental/climatic impact of urban projects by simulating under different scenarios
	Building optimization	Promoting urban energy efficiency and thermal comfort by informing parametric design through form analysis and BIM

3.4 Requirements for and challenges of implementing AI applications

Although integrating AI and its subsets into urban governance, planning, and design can greatly assist decision-makers in building climatically resilient cities and communities, there are significant considerations in the forms of ethical and technical requirements and challenges that should be addressed [66, 67]. These requirements and challenges are great points of discussion within the literature, as AI-based tools are evolving in various fields, especially the policymaking domain, which can widely affect the citizens' interests and quality of life [68]. This section elaborates on these requirements and challenges to clarify the general considerations limiting the applications of AI in climatically resilient urban decision-making.

The effectiveness of AI systems hinges greatly on the quantity and quality of training data they receive. Training data is the foundation for neural networks to recognize patterns and make informed decisions. AI models may perpetuate biases and inaccuracies without robust and representative training data, leading to suboptimal outcomes [69]. Therefore, accurate datasets are a primary requirement in developing reliable AI systems for urban environments. Expanding the volume, velocity, and variety of training data can enhance the model's predictive accuracy and responsiveness to real-time changes. However, a balance must be struck to prevent diluting the model with irrelevant or noisy data that could distort the learning process or cause bias [70]. Also, the convergence of data collection technologies and monitoring systems, including sensors, cameras, and tracking devices powered by AI, highlight the ethical implications of data collection, privacy concerns, and energy consumption that must be carefully managed to ensure responsible AI deployment [71].

Another critical aspect of AI integration in urban environments is the need for ongoing training and upskilling of the workforce. As AI technologies become more prevalent, local governments need to invest in equipping employees with the necessary skills and knowledge to leverage AI effectively. By fostering a culture of innovation, collaboration, and learning, cities can maximize the potential of AI to build more sustainable and climatically resilient communities [29, 72]. Moreover, citizen participation and engagement are key components of responsible urban AI deployment. Involving residents in the planning, developing, and evaluating AI-driven initiatives can help build trust, ensure transparency, and address community needs more effectively. By creating avenues for feedback, collaboration, and co-creation, cities can empower residents to shape their urban environment's future in an inclusive manner that is responsive to diverse needs [73]. These collaborations and participations highly require transparency in the process of training and implementing AI-based tools [74].

However, one main challenge to AI systems is the lack of transparency and accountability. The decision-making process of AI is often opaque, making it difficult to explain or understand the basis of their decisions. This can lead to issues such as algorithmic bias, privacy violations, and the removal of human responsibility for the outcomes [70]. Additionally, the complexity and unpredictability of AI behaviour further contribute to the challenges faced in urban settings. Moreover, data privacy and security concerns are significant hurdles in adopting AI-based technologies. Issues such as misuse of personal information, data breaches, and lack of transparency in data collection practices can hinder public acceptance of AI-driven urban services [41]. The availability and quality of data, as well as storage and processing capabilities, also present obstacles to effectively implementing AI solutions in urban environments [75].

Furthermore, the societal impact of AI on jobs and livelihoods is a critical concern. The fear of unemployment, job displacement, and economic inequality due to automation and AI-driven technologies create tension among citizens and policymakers [76]. The equitable distribution of benefits and risks associated with AI adoption remains a key challenge for ensuring inclusive urban development. Besides, fragmented governance structures and a lack of regulatory frameworks pose challenges to the seamless integration of AI in urban services. The absence of comprehensive strategies, political will, and coordination among local governments hinders the advancement of AI initiatives in smart cities. Moreover, cultural resistance to change, budget constraints, and limited technical expertise constrain the progress of AI projects in urban environments [72].

3.5 CONCLUSION AND RECOMMENDATIONS

In conclusion, integrating AI into urban governance, planning, and design holds immense potential for building more resilient cities to climate change and its consequences. The systematic literature review conducted in this study identified various applications of AI in enhancing urban climate resilience. These applications range from service/energy monitoring and infrastructure management to citizen engagement, resource management, emergency response, environmental protection, traffic and land use planning, site vulnerability analysis, design simulation, and building optimization. These applications contribute significantly to urban climate mitigation and adaptation capacities, promoting sustainable urban development.

However, the effective implementation of AI in climatically resilient urban planning requires several prerequisites and faces various challenges. Key requirements include having robust and representative training data for AI models, providing ongoing workforce training and upskilling, fostering citizens participation and engagement, ensuring transparency in decision-making processes, and developing regulatory frameworks. Challenges include algorithmic bias, lack of transparency and accountability, data privacy and security concerns, societal impact on jobs and livelihoods, fragmented governance structures, regulatory frameworks, cultural resistance to change, budget constraints, and limited technical expertise. To address these challenges, we recommend the following actions:

- *Invest in data quality* to ensure the availability of accurate, diverse, and representative training data to enhance AI models' predictive accuracy and responsiveness.
- *Prioritize workforce training* by investing in training programmes to equip urban employees with the necessary skills to leverage AI effectively in urban decision-making processes.
- *Promote citizen participation* by fostering a culture of collaboration and inclusivity by educating and involving residents in the planning and evaluating AI-driven initiatives.
- *Enhance transparency* by implementing transparent decision-making processes to address algorithmic bias and privacy violations and foster public trust in AI systems.
- *Address data privacy concerns* by developing robust data privacy and security protocols to mitigate the risks associated with personal information and data breaches.
- *Mitigate societal impacts* by addressing job displacement and economic inequality concerns and ensuring the equitable distribution of benefits and risks associated with AI.
- *Improve governance structures* by establishing comprehensive strategies, regulatory frameworks, and coordination mechanisms among local governments to facilitate the integration of AI in urban services.
- *Respond to technical challenges* by addressing technical requirements related to data storage, processing capabilities, and limited technical expertise to implement AI solutions in urban environments effectively.

While we aimed to comprehensively review how AI can be used to build more resilient cities in the face of climate change and its consequences, this study encountered limitations due to gaps in the existing body of knowledge. The main limitation is that many studies focus solely on the applications of AI in building climate-resilient cities, with discussions on challenges often being generalized to the public policy domain. In other words, current literature concentrates broadly on AI applications without delving into specific challenges and solutions for individual cases. Moreover, the need for case-specific assessments is emphasized to better understand how AI can be effectively implemented in different urban settings. There is a noticeable lack of empirical evidence or case studies illustrating specific instances of AI application in urban planning, detailing both successes and hurdles encountered during and after implementation. The chapter points to gaps in understanding how AI technologies can be integrated into existing urban governance frameworks and the potential need for new coordination mechanisms. Ultimately, although societal impacts such as job displacement are discussed, there is still a need for more studies evaluating both the positive and negative effects of AI integration in urban planning on various population segments.

NOTES

1. AI is a broad field comprising various subsets, such as machine learning (ML), deep learning (DL), natural language processing (NLP), and robotics. ML is a data analysis technique that involves building analytical models automatically. It employs iterative algorithms that learn from data, enabling computers to identify hidden insights. DL, a subset of ML, employs multi-layered neural networks (hence "deep") to model high-level abstractions in data. NLP focuses on the interactions between computers and human language, particularly on programming computers to process and analyse vast amounts of natural language data. On the other hand, robotics involves the creation, construction, operation, and utilization of robots to perform tasks traditionally carried out by humans [12].
2. Preferred reporting items for systematic reviews and meta-analyses

REFERENCES

1. Serdeczny, O., Adams, S., Baarsch, F., Coumou, D., Robinson, A., et al. (2016). Climate change impacts in Sub-Saharan Africa: from physical changes to their social repercussions. *Regional Environmental Change*, 17(6), 1585–1600. https://doi.org/10.1007/s10113-015-0910-2
2. Li, J., Yang, L., & Long, H. (2018). Climatic impacts on energy consumption: Intensive and extensive margins. *Energy Economics*, 71, 332–343. https://doi.org/10.1016/j.eneco.2018.03.010
3. Murage, P., Kovats, S., Sarran, C., Taylor, J., McInnes, R., et al. (2020). What individual and neighbourhood-level factors increase the risk of heat-related mortality? A case-crossover study of over 185,000 deaths in London using high-resolution climate datasets. *Environment International*, 134, 105292. https://doi.org/10.1016/j.envint.2019.105292
4. Mitchell, B., & Chakraborty, J. (2018). Exploring the relationship between residential segregation and thermal inequity in 20 U.S. cities. *Local Environment*, 23(8), 796–813. https://doi.org/10.1080/13549839.2018.1474861
5. WMO. (2021). Climate and weather related disasters surge five-fold over 50 years, but early warnings save lives. United Nations News (World Meteorological Organization [WMO] and UN Office for Disaster Risk Reduction [UNDRR]). https://news.un.org/en/story/2021/2009/1098662
6. Cheshmehzangi, A., He, B., Sharifi, A., & Matzarakis, A. (2023). Climate change, cities, and the importance of cooling strategies, practices, and policies. *Climate Change and Cooling Cities*. 1, 2–19. https://doi.org/10.1007/978-981-99-3675-5_1

7. Jabareen, Y. (2013). Planning the resilient city: Concepts and strategies for coping with climate change and environmental risk. *Cities*, 31, 220–229. https://doi.org/10.1016/j.cities.2012.05.004

8. Brown, A., Jr., Dayal, A., Rumbaitis Del Rio, C., Da Silva, & Da Silva, E. Al. (2014). Urban climate change resilience. Asian Development Bank. https://www.adb.org/sites/default/files/publication/149164/urban-climate-change-resilience-synopsis.pdf

9. Intergovernmental Panel on Climate Change (IPCC). (2023). Cities, Settlements and Key Infrastructure. In Climate Change 2022 – Impacts, Adaptation and Vulnerability: Working Group II Contribution to the Sixth Assessment Report of the Intergovernmental Panel on Climate Change (pp. 907–1040). chapter, Cambridge: Cambridge University Press. https://doi.org/10.1017/9781009325844.008

10. Tyler, S., & Moench, M. (2012). A framework for urban climate resilience. *Climate and Development*, 4(4), 311–326. https://doi.org/10.1080/17565529.2012.745389

11. Rezvani, S. M. H. S., Almeida, N., & Falcão, M. J. (2023). Climate adaptation measures for enhancing urban resilience. *Buildings*, 13(9), 2163. https://doi.org/10.3390/buildings13092163

12. Soori, M., Arezoo, B., & Dastres, R. (2023). Artificial intelligence, machine learning and deep learning in advanced robotics, a review. *Cognitive Robotics*, 3, 54–70. https://doi.org/10.1016/j.cogr.2023.04.001

13. Vinuesa, R., Azizpour, H., Leite, I., Balaam, M., Dignum, V., Domisch, S., Felländer, A., Langhans, S. D., Tegmark, M., & Nerini, F. F. (2020). The role of artificial intelligence in achieving the Sustainable Development Goals. *Nature Communications*, 11(1), 233. https://doi.org/10.1038/s41467-019-14108-y

14. Olawade, D. B., Wada, O. J., David-Olawade, A. C., Kunonga, E., Abaire, O. J., & Ling, J. (2023). Using artificial intelligence to improve public health: a narrative review. *Frontiers in Public Health,* 11. https://doi.org/10.3389/fpubh.2023.1196397

15. Valle-Cruz, D., Criado, J. I., Sandoval-Almazán, R., & Ruvalcaba-Gómez, E. A. (2020). Assessing the public policy-cycle framework in the age of artificial intelligence: From agenda-setting to policy evaluation. *Government Information Quarterly*, 37(4), 101509. https://doi.org/10.1016/j.giq.2020.101509

16. Zuiderwijk, A., Chen, Y., & Salem, F. (2021). Implications of the use of artificial intelligence in public governance: A systematic literature review and a research agenda. *Government Information Quarterly*, 38(3), 101577. https://doi.org/10.1016/j.giq.2021.101577

17. Zhao, J., & Fariñas, B. G. (2022). Artificial intelligence and sustainable decisions. European Business Organization Law Review, 24(1), 1–39. https://doi.org/10.1007/s40804-022-00262-2

18. Chen, P., Wu, L., & Wang, L. (2023). AI Fairness in Data Management and Analytics: A review on challenges, Methodologies and applications. *Applied Sciences*, 13(18), 10258. https://doi.org/10.3390/app131810258

19. Chin, S., & Lloyd, V. (2024). Predicting climate change using an autoregressive long short-term memory model. *Frontiers in Environmental Science*, 12. https://doi.org/10.3389/fenvs.2024.1301343

20. Xiang, X., Li, Q., Khan, S., & Khalaf, O. I. (2021). Urban water resource management for sustainable environment planning using artificial intelligence techniques. *Environmental Impact Assessment Review*, 86, 106515. https://doi.org/10.1016/j.eiar.2020.106515

21. Srivastava, A., & Maity, R. (2023). Assessing the potential of AI–ML in urban climate change adaptation and sustainable development. *Sustainability*, 15(23), 16461. https://doi.org/10.3390/su152316461

22. Grant, M. J., & Booth, A. (2009). A typology of reviews: an analysis of 14 review types and associated methodologies. *Health Information & Libraries Journal*, 26(2), 91–108. https://doi.org/10.1111/j.1471-1842.2009.00848.x

23. Tomor, Z., Meijer, A., Michels, A., & Geertman, S. (2019). Smart governance for sustainable cities: findings from a systematic literature review. *Journal of Urban Technology*, 26(4), 3–27. https://doi.org/10.1080/10630732.2019 .1651178

24. Page, M. J., McKenzie, J. E., Bossuyt, P. M., Boutron, I., Hoffmann, T., Mulrow, C. D., Shamseer, L., Tetzlaff, J., Akl, E. A., Brennan, S., Chou, R., Glanville, J., Grimshaw, J., Hróbjartsson, A., Lalu, M. M., Li, T., Loder, E., Mayo-Wilson, E., McDonald, S., . . . Moher, D. (2021). The PRISMA 2020 statement: an updated guideline for reporting systematic reviews. *The BMJ*, n71. https://doi.org/10.1136/bmj.n71

25. Linneberg, M. S., & Korsgaard, S. (2019). Coding qualitative data: a synthesis guiding the novice. *Qualitative Research Journal*, 19(3), 259–270. https:// doi.org/10.1108/qrj-12-2018-0012

26. Holtz, Y. (2019). Sankey Diagram. data-to-viz. https://www.data-to-viz.com /graph/sankey.html. Accessed June 25, 2022.

27. Kamrowska-Załuska, D. D. (2021). Impact of AI-Based tools and urban big data analytics on the design and planning of cities. *Land*, 10(11), 1209. https://doi.org/10.3390/land10111209

28. Cowls, J., Tsamados, A., Taddeo, M., & Floridi, L. (2021). The AI gambit: leveraging artificial intelligence to combat climate change—Opportunities, challenges, and recommendations. *AI & Society*, 38(1), 283–307. https://doi .org/10.1007/s00146-021-01294-x

29. Maxwell, K., Julius S., Grambsch A., Kosmal, A., Larson, L., & Sonti, N. (2018). Built environment, urban systems, and cities. In *Impacts, Risks, and Adaptation in the United States: Fourth National Climate Assessment, Volume II* [Reidmiller, D.R., C.W. Avery, D.R. Easterling, K.E. Kunkel, K.L.M. Lewis, T.K. Maycock, and B.C. Stewart (eds.)]. U.S. Global Change Research Program, Washington, DC. pp. 438–478.

30. Sanchez, T. W. (2023). Planning on the verge of AI, or AI on the verge of planning. *Urban Science, 7*(3), 70. https://doi.org/10.3390/urbansci7030070

31. Samsurijan, M. S., Ebekozien, A., Azazi, N. a. N., Shaed, M. M., & Firdaus, R. B. R. (2022). Artificial intelligence in urban services in Malaysia: a review. *PSU Research Review*. https://doi.org/10.1108/prr-07-2021-0034

32. Harnal, S., Sharma, G., Malik, S., Kaur, G., Khurana, S., Kaur, P., Simaiya, S., & Bagga, D. (2022). Bibliometric mapping of trends, applications and challenges of artificial intelligence in smart cities. *ICST Transactions on Scalable Information Systems*, 9(4), e76. https://doi.org/10.4108/eetsis.vi.489

33. Suleimany, M. (2023). Urban climate justice in hot-arid regions: Vulnerability assessment and spatial analysis of socio-economic and housing inequality in

Isfahan, Iran. Urban Climate, 51, 101612. https://doi.org/10.1016/j.uclim .2023.101612

34. OECD. (2018). Climate-resilient infrastructure. Oecd Environment Policy Paper [Report]. OECD Publication. https://www.oecd.org/environment/cc/ policy-perspectives-climate-resilient-infrastructure.pdf

35. Salimi, M., & Al-Ghamdi, S. G. (2020). Climate change impacts on critical urban infrastructure and urban resiliency strategies for the Middle East. *Sustainable Cities and Society*, 54, 101948. https://doi.org/10.1016/j.scs .2019.101948

36. Serey, J., Quezada, L. E., Alfaro, M., Fuertes, G., Ternero, R., Gatica, G., Gutiérrez, S., & Vargas, M. (2020). Methodological proposals for the development of services in a smart city: a literature review. *Sustainability*, 12(24), 10249. https://doi.org/10.3390/su122410249

37. Yiğitcanlar, T., Kankanamge, N., Regona, M., Maldonado, A. R., Rowan, B., Ryu, A., Desouza, K. C., Corchado, J. M., Mehmood, R., & Li, R. Y. M. (2020). Artificial intelligence technologies and related urban planning and development concepts: how are they perceived and utilized in Australia? *Journal of Open Innovation: Technology, Market, and Complexity*, 6(4), 187. https://doi.org/10.3390/joitmc6040187

38. Palla, A., Pezzagno, M., Spadaro, I., & Ermini, R. G. A. (2024). Participatory approach to planning urban resilience to climate change: brescia, genoa, and matera—Three case studies from Italy compared. *Sustainability*, 16(5), 2170. https://doi.org/10.3390/su16052170

39. Gupta, S., & Degbelo, A. (2023). An empirical analysis of AI contributions to sustainable cities (SDG 11). *Philosophical Studies Series,* 152, 461–484. https://doi.org/10.1007/978-3-031-21147-8_25

40. Caputo, F., Magliocca, P., Canestrino, R., & Rescigno, E. (2023). Rethinking the role of technology for citizens' engagement and sustainable development in smart cities. *Sustainability*, 15(13), 10400. https://doi.org/10.3390/ su151310400

41. Sherman, S. L. (2022). The Polyopticon: a diagram for urban artificial intelligences. *Ai & Society*, 38(3), 1209–1222. https://doi.org/10.1007/s00146-022 -01501-3

42. - Chen, L., Chen, Z., Zhang, Y., Liu, Y., Osman, A. I., Farghali, M., Hua, J., Al-Fatesh, A. S., Ihara, I., Rooney, D., & Yap, P. (2023). Artificial intelligence-based solutions for climate change: a review. *Environmental Chemistry Letters*, 21(5), 2525–2557. https://doi.org/10.1007/s10311-023-01617-y

43. Golubchikov, O., & Thornbush, M. J. (2020). Artificial intelligence and robotics in smart city strategies and planned smart development. *Smart Cities*, 3(4), 1133–1144. https://doi.org/10.3390/smartcities3040056

44. Herath, H., & Mittal, M. (2022). Adoption of artificial intelligence in smart cities: A comprehensive review. *International Journal of Information Management Data Insights*, 2(1), 100076. https://doi.org/10.1016/j.jjimei .2022.100076

45. Gupta, S., Modgil, S., Kumar, A., Sivarajah, U., & Irani, Z. (2022). Artificial intelligence and cloud-based collaborative platforms for managing disaster, extreme weather and emergency operations. *International Journal of Production Economics*, 254, 108642. https://doi.org/10.1016/j.ijpe.2022 .108642

46. Yiğitcanlar, T., Mehmood, R., & Corchado, J. M. (2021). Green Artificial Intelligence: towards an efficient, sustainable and equitable technology for smart cities and futures. *Sustainability*, 13(16), 8952. https://doi.org/10.3390/su13168952

47. Nishant, R., Kennedy, M., & Corbett, J. (2020). Artificial intelligence for sustainability: Challenges, opportunities, and a research agenda. *International Journal of Information Management*, 53, 102104. https://doi.org/10.1016/j.ijinfomgt.2020.102104

48. European Parliament, Directorate-General for Internal Policies of the Union, Herold, A., Gailhofer, P., Urrutia, C. (2021). The role of artificial intelligence in the European Green Deal, European Parliament. https://data.europa.eu/doi/10.2861/882830

49. Bibri, E. S., Krogstie, J., Kaboli, A., & Alahi, A. (2024). Smarter eco-cities and their leading-edge artificial intelligence of things solutions for environmental sustainability: A comprehensive systematic review. *Environmental Science and Ecotechnology*, 19, 100330. https://doi.org/10.1016/j.ese.2023.100330

50. Karanth, S., Benefo, E. O., Patra, D., & Pradhan, A. K. (2023). Importance of artificial intelligence in evaluating climate change and food safety risk. *Journal of Agriculture and Food Research*, 11, 100485. https://doi.org/10.1016/j.jafr.2022.100485

51. Silvestro, D., Goria, S., Sterner, T., & Antonelli, A. (2022). Improving biodiversity protection through artificial intelligence. *Nature Sustainability*, 5(5), 415–424. https://doi.org/10.1038/s41893-022-00851-6

52. Chong, Y., Villanueva-Libunao, K., Chee, S. Y., Alvarez, M. J., Yau, K. A., & Keoh, S. L. (2022). Artificial intelligence policies to enhance urban mobility in Southeast Asia. *Frontiers in Sustainable Cities*, 4. https://doi.org/10.3389/frsc.2022.824391

53. Deweerdt, T., & Fabre, A. (2022). The role of land use planning in urban Transport to Mitigate climate Change: A literature review. *Advances in Environmental and Engineering Research*, 3(3), 1. https://doi.org/10.21926/aeer.2203033

54. Ye, Z., Yang, J., Zhong, N., Tu, X., Jia, J., & Wang, J. (2020). Tackling environmental challenges in pollution controls using artificial intelligence: a review. *Science of the Total Environment*, 699, 134279. https://doi.org/10.1016/j.scitotenv.2019.134279

55. Ghahramani, M., Galle, N. J., Duarte, F., Ratti, C., & Pilla, F. (2021). Leveraging artificial intelligence to analyze citizens' opinions on urban green space. *City and Environment Interactions*, 10, 100058. https://doi.org/10.1016/j.cacint.2021.100058

56. Sarabi, S., Han, Q., De Vries, B., Romme, A., & Almássy, D. (2022). The nature-based solutions case-based system: A hybrid expert system. *Journal of Environmental Management*, 324, 116413. https://doi.org/10.1016/j.jenvman.2022.116413

57. Jain, H., Dhupper, R., Shrivastava, A., Kumar, D., & Kumari, M. (2023). AI-enabled strategies for climate change adaptation: protecting communities, infrastructure, and businesses from the impacts of climate change. *Computational Urban Science*, 3(1), 25. https://doi.org/10.1007/s43762-023-00100-2

58. AI for Urban Climate: An Eo-Based Approach for High-Resolution Mapping of Human Exposure to Heatwaves. (2022). *IEEE Conference Publication | IEEE Xplore.* https://ieeexplore.ieee.org/document/9883071

59. Zennaro, F., Furlan, E., Simeoni, C., Torresan, S., Aslan, S., Critto, A., & Marcomini, A. (2021). Exploring machine learning potential for climate change risk assessment. *Earth-Science Reviews*, 220, 103752. https://doi.org/10.1016/j.earscirev.2021.103752

60. Long, L. D. (2023). An AI-driven model for predicting and optimizing energy-efficient building envelopes. *Alexandria Engineering Journal*, 79, 480–501. https://doi.org/10.1016/j.aej.2023.08.041

61. Mazzeo, D., Matera, N., Peri, G., & Scaccianoce, G. (2023). Forecasting green roofs' potential in improving building thermal performance and mitigating urban heat island in the Mediterranean area: An artificial intelligence-based approach. *Applied Thermal Engineering*, 222, 119879. https://doi.org/10.1016/j.applthermaleng.2022.119879

62. Shivaprakash, K. N., Swami, N., Mysorekar, S., Arora, R., Gangadharan, A., Vohra, K., Jadeyegowda, M., & Kiesecker, J. M. (2022). Potential for Artificial Intelligence (AI) and Machine Learning (ML) applications in biodiversity conservation, managing forests, and related services in India. *Sustainability*, 14(12), 7154. https://doi.org/10.3390/su14127154

63. Hafez, F. S., Sa'di, B., Safa-Gamal, M., Taufiq-Yap, Y., Alrifaey, M., Seyedmahmoudian, M., Stojcevski, A., Horan, B., & Mekhilef, S. (2023). Energy efficiency in sustainable buildings: a systematic review with taxonomy, challenges, motivations, methodological aspects, recommendations, and pathways for future research. *Energy Strategy Reviews*, 45, 101013. https://doi.org/10.1016/j.esr.2022.101013

64. Ibrahim, Y., Kershaw, T., Shepherd, P., & Coley, D. (2021). On the optimisation of urban form design, energy consumption and outdoor thermal comfort using a parametric workflow in a hot arid zone. *Energies*, 14(13), 4026. https://doi.org/10.3390/en14134026

65. Bagheri, A., Genikomsakis, K. N., Koutra, S., Sakellariou, V. I., & Ioakimidis, C. S. (2021). Use of AI algorithms in different building typologies for energy efficiency towards smart buildings. *Buildings*, 11(12), 613. https://doi.org/10.3390/buildings11120613

66. Ghallab, M. (2019). Responsible AI: requirements and challenges. *AI Perspectives & Advanes*, 1(1), 3. https://doi.org/10.1186/s42467-019-0003-z

67. Trotta, A., Ziosi, M., & Lomonaco, V. (2023). The future of ethics in AI: challenges and opportunities. *AI & SOCIETY*, 38(2), 439–441. https://doi.org/10.1007/s00146-023-01644-x

68. Alhosani, K., & Alhashmi, S. M. (2024). Opportunities, challenges, and benefits of AI innovation in government services: a review. *Discover Artificial Intelligence*, 4(1), 18. https://doi.org/10.1007/s44163-024-00111-w

69. Bratton, B. H. (2021). AI urbanism: a design framework for governance, program, and platform cognition. *AI & Society*, 36(4), 1307–1312. https://doi.org/10.1007/s00146-020-01121-9

70. Zhang, D., Pee, L. G., Pan, S. L., & Liu, W. (2022). Orchestrating artificial intelligence for urban sustainability. *Government Information Quarterly*, 39(4), 101720. https://doi.org/10.1016/j.giq.2022.101720

71. Marvin, S., While, A., Chen, B., & Kovacic, M. (2022). Urban AI in China: Social control or hyper-capitalist development in the post-smart city? *Frontiers in Sustainable Cities*, 4. https://doi.org/10.3389/frsc.2022.1030318

72. Yiğitcanlar, T., Corchado, J. M., Mehmood, R., Li, R. Y. M., Mossberger, K., & Desouza, K. C. (2021). Responsible urban innovation with local government artificial intelligence (AI): a conceptual framework and research Agenda. *Journal of Open Innovation: Technology, Market, and Complexity*, 7(1), 71. https://doi.org/10.3390/joitmc7010071

73. Allam, Z., & Dhunny, Z. A. (2019). On big data, artificial intelligence and smart cities. *Cities*, 89, 80–91. https://doi.org/10.1016/j.cities.2019.01.032

74. Yiğitcanlar, T., & Cugurullo, F. (2020). The sustainability of artificial intelligence: an urbanistic viewpoint from the lens of smart and sustainable cities. *Sustainability*, 12(20), 8548. https://doi.org/10.3390/su12208548

75. Aldoseri, A., Al-Khalifa, K., & Hamouda, A. (2023). Re-Thinking Data Strategy and Integration for Artificial intelligence: Concepts, opportunities, and challenges. *Applied Sciences,* 13(12), 7082. https://doi.org/10.3390/app13127082

76. Yiğitcanlar, T., Agdas, D., & Degirmenci, K. (2022). Artificial intelligence in local governments: perceptions of city managers on prospects, constraints and choices. *AI & Society,* 38(3), 1135–1150. https://doi.org/10.1007/s00146-022-01450-x

Achieving sustainable development goals through knowledge management in Industry 4.0

A critical review from the perspective of Bangladesh

Imran Hossain, A. K. M. Mahmudul Haque, Abdul Kadir, and Md. Sohel Rana

4.1 INTRODUCTION

The sustainable development goals (SDGs) are an ambitious global agenda set forward by the United Nations to solve the most important social, economic, and environmental issues of our era. These 17 interconnected goals, which were adopted in 2015 by all 193 UN members, provide a roadmap for realizing a more sustainable and equitable world by 2030 (United Nations, 2015). The SDGs are important because they are all-inclusive, covering a broad range of topics from combating poverty to addressing climate change, with the goal of leaving no one behind. With an emphasis on the interdependence and connectivity of social, economic, and environmental factors, each objective focuses on particular areas that are essential for sustainable development (Le Blanc, 2015).

Bangladesh, a densely populated country in South Asia, faces many challenges in the areas of socio-economic issues, the environment, and development. These issues have a strong connection to the global agenda represented in the 17 SDGs, which makes the framework extremely relevant to the nation (Hossain et al., 2023c). The government's development plans and goals for a sustainable future are strongly aligned with the SDGs, which are directly tied to Bangladesh's national priorities (Kaiser, 2023; Hossain et al., 2024; GoB, 2019).

Concurrently, Industry 4.0's development, marked by revolutionary technologies, including artificial intelligence (AI), the Internet of Things (IoT), big data analytics, and automation, has revolutionized industrial landscapes globally (Schwab, 2016). This revolutionary shift towards intelligent, interconnected, and data-based production systems offers unprecedented opportunities for innovation, efficiency, and economic growth across diverse sectors. Additionally, the impact of Industry 4.0 on global economies is multidimensional, influencing employment patterns, skill requirements, and

DOI: 10.1201/9781003581246-4

economic growth movements. AI and automation improve workflows, but they also raise questions about job displacement and shifting skill requirements (Manyika et al., 2017). Routine tasks are increasingly automated, necessitating upskilling or reskilling the workforce to meet the demands of a technology-driven economy. However, Industry 4.0 also creates new job opportunities, particularly in technology development, data analytics, cybersecurity, and digital innovation (Brynjolfsson & McAfee, 2014). Moreover, Industry 4.0 contributes to fostering innovation ecosystems and promoting cross-sector collaborations. The interconnectedness of systems encourages collaboration between industries, academia, and governments, fostering innovation hubs and technology clusters (Wang & Hajli, 2017). This collaboration fuels technological advancements, accelerates innovation cycles, and encourages economic growth by fostering an environment conducive to entrepreneurship and technological breakthroughs.

In the context of Bangladesh, a developing nation facing multifaceted challenges, the significance of aligning Industry 4.0 advancements with the pursuit of the SDGs becomes particularly pronounced. Bangladesh, with its rapidly growing population, socio-economic complexities, and environmental vulnerabilities, grapples with challenges like poverty, inadequate healthcare, limited access to education, and environmental degradation (World Bank, 2021). However, the country also demonstrates resilience, innovative capacity, and a strong commitment to sustainable development. In this aspect, knowledge management (KM) emerges as a pivotal element within the confluence of Industry 4.0 and the SDGs. KM, defined as the systematic process of acquiring, organizing, sharing, and utilizing knowledge within organizations, plays a crucial role in facilitating learning, innovation, and decision-making (Alavi & Leidner, 2001). Manufacturing, agriculture, healthcare, energy, transportation, infrastructure, and urban development are important verticals in Bangladesh where Industry 4.0 can have a revolutionary effect (Gabriel & Pessl, 2016; Müller et al., 2018; Abu-Rumman et al., 2023). For example, the application of IoT and AI-enabled smart factories in the manufacturing sector can result in improved product quality, less waste, and higher efficiency (Bonilla et al., 2018; Abu-Rumman et al., 2023; Namchoochai et al., 2020). Similar to this, precision farming methods in agriculture can maximize crop yields and resource usage with the help of drones, sensors, and data analytics (Tripicchio et al., 2015). In the context of Industry 4.0, KM assumes an even more significant role by enabling the effective utilization of data-driven insights and technological innovations to drive sustainable development agendas.

The integration of KM practices within the framework of Industry 4.0 holds immense promise for Bangladesh's journey towards achieving the SDGs. By effectively managing and utilizing knowledge, organizations and industries can enhance their capabilities to address socio-economic challenges while fostering innovation and sustainability. However, the complicated relationship between KM, Industry 4.0, and SDGs within the specific socio-economic context of Bangladesh warrants a comprehensive review and critical analysis.

This study aims to provide a detailed examination of how KM practices intersect with Industry 4.0 initiatives and contribute to the attainment of the SDGs in Bangladesh. This study also looks at existing research, real-life instances, and best practices to find out what the pros, cons, gaps, and suggestions are for using knowledge management in Industry 4.0 to help Bangladesh reach the SDGs. By critically evaluating the current context and integrating key findings, this research intends to offer actionable insights and strategic directions for stakeholders, policymakers, industries, and academia to drive sustainable development agendas in Bangladesh.

4.2 CONCEPTUALIZING INDUSTRY 4.0 AND ITS KEY FEATURES

"Industry 4.0" refers to a significant shift in production and manufacturing processes that integrates AI, advanced automation, digital technology, and data exchange (Kagermann et al., 2013). Industry 4.0, which builds on its predecessors, signifies a paradigm shift in industrial operations by placing an emphasis on data-driven decision-making, interconnection, and the merging of the digital and physical domains (Mourtzis et al., 2022). Cyber-physical systems (CPS), which combine digital and physical processes to form intelligent, adaptable systems, are the core idea behind Industry 4.0 (Jazdi, 2014; Hostos Orjuela, 2023). These systems are made up of linked machines, sensors, and devices that work together in real time to communicate and make decisions, allowing for more autonomous decision-making and improved operational efficiency (Jazdi, 2014; Salkin et al., 2018).

The IoT is a key component of Industry 4.0 architecture. It allows objects and devices integrated with sensors and controls to be connected to one another and interact and share data (Georgios et al., 2019; Zanella et al., 2014). According to Jiang et al. (2017), this interconnection serves as the basis for the development of intelligent networks, which in turn enable a wide range of applications across industries, including smart manufacturing, healthcare, logistics, and more. Industry 4.0 is also distinguished by its use of big data and advanced analytics. Large volumes of data are produced by the widespread usage of digital technologies, which come from a variety of sources such as sensors, machinery, and user interactions (Chen et al., 2014; Bughin et al., 2018). When combined with machine learning and predictive algorithms, big data analytics enables businesses to gain insightful knowledge, streamline operations, and make data-driven decisions quickly (Chen et al., 2014). Furthermore, a fundamental component of Industry 4.0 is the idea of "horizontal" and "vertical" integration (Aoun et al., 2021). In order to promote smooth information flow and resource optimization, horizontal integration entails the connection and cooperation of many departments, stakeholders, and processes inside an organization (Aoun et al., 2021; Qin et al., 2016). On the other hand, vertical integration is the process of integrating different production layers, such as enterprise-level systems and shop floor operations, to provide end-to-end visibility and control (Li et al., 2017; Aoun et al., 2021).

Smart automation and robotics are further concepts that Industry 4.0 supports (Salkin et al., 2018). Intelligent systems that can execute intricate tasks with accuracy and flexibility can be created through the integration of advanced robotics, AI algorithms, and machine learning (Brynjolfsson & McAfee, 2014). In addition to increasing efficiency, these intelligent systems help make workplaces safer and more adaptable. Additionally, the emergence of digital twins – virtual duplicates of actual assets, goods, or procedures – accompanies Industry 4.0 (Leng et al., 2021; Attaran et al., 2023). Digital twins reduce development times and improve product quality by simulating real-world events, thus enabling predictive analysis, performance optimization, and quick prototyping (Leng et al., 2021). Industry 4.0 is essentially a revolutionary vision that goes beyond conventional manufacturing paradigms and places a strong emphasis on connection, data-driven intelligence, self-governing decision-making, and the merging of the digital and physical domains. Cyber-physical systems, the internet of things, big data analytics, layer-by-layer integration, intelligent automation, and the rise of digital twins are some of its prominent characteristics.

4.3 THE RELATIONSHIP BETWEEN INDUSTRY 4.0, KNOWLEDGE MANAGEMENT, AND SUSTAINABLE DEVELOPMENT

Knowledge management, sustainable development, and Industry 4.0 have come together to form a revolutionary nexus that has the ability to address current global challenges and promote equitable and sustainable growth. Industry 4.0 offers previously unheard-of potential to completely transform economies and sectors through the integration of digital technology, automation, and data-driven processes (Schwab, 2016). Organizational landscapes are altered by this technological growth, which makes it easier to generate, gather, and analyse enormous amounts of data and information.

Knowledge management is at the centre of this change and is essential to utilizing Industry 4.0 to achieve sustainable development objectives. By facilitating the efficient use and distribution of information across Industry 4.0 ecosystems, knowledge management methods help businesses innovate, gain actionable insights, and make well-informed decisions (Rao, 2012). Effective knowledge resource management strengthens organizational capacities, encourages teamwork, and quickens learning cycles – all vital components in navigating the challenges of the Fourth Industrial

Revolution. Furthermore, Industry 4.0 frameworks that incorporate knowledge management techniques provide a substantial contribution to sustainable development programmes. A balanced strategy that balances social advancement, economic prosperity, and environmental stewardship is required for sustainable development (United Nations, 2015). By encouraging the sharing of best practices, encouraging innovation in sustainable technologies, and improving resource usage, knowledge management acts as a catalyst for the alignment of these goals (Cash et al., 2003). Effective knowledge leverage enables firms to create and execute solutions that tackle social issues covered by the SDGs.

There are several important variables that define the interaction between Industry 4.0, knowledge management, and sustainable development. First off, Industry 4.0 technologies produce enormous amounts of data, opening up new avenues for the production, extraction, and application of knowledge (Chen et al., 2014; Bughin et al., 2018). By transforming raw data into insightful knowledge, knowledge management strategies help businesses recognize patterns, forecast results, and make well-informed decisions that advance sustainable development goals. Second, Industry 4.0's use of knowledge management encourages an innovative and never-ending learning culture. Knowledge-sharing organizations promote the exchange of ideas, skills, and experiences among members, which speeds up the creation and uptake of sustainable practices and technology (Cash et al., 2003). This cooperative setting improves problem-solving skills, allowing businesses to more successfully handle challenging sustainability issues. Using knowledge management principles strategically also makes organizations more adaptable and stronger, which are important in the fast-paced world of Industry 4.0 and for achieving sustainable development goals (Ghobakhloo et al., 2021; Willcocks et al., 2015). Effective knowledge management enables organizations to anticipate and adapt to shifting consumer needs, legislative requirements, and social expectations, all of which support long-term sustainability initiatives.

Realizing the full potential of the connection between Industry 4.0, knowledge management, and sustainable growth is still difficult. Data privacy, the digital gap, ethical questions about AI and automation, and the need for inclusive technological breakthroughs that benefit all societal segments are some of these concerns (Whittlestone et al., 2019). In order to overcome these obstacles, a comprehensive strategy that incorporates social, ethical, and environmental factors into knowledge management plans within Industry 4.0 frameworks is needed. All things considered, the mutually beneficial connection of Industry 4.0, knowledge management, and sustainable growth presents never-before-seen chances to promote progress. One of the keystones to unlocking the transformative potential of technological breakthroughs towards reaching the SDGs is the strategic integration of knowledge management methods within Industry 4.0 frameworks.

4.4 MATERIALS AND METHOD

In order to examine the function of knowledge management in accomplishing the SDGs in the context of Industry 4.0 in Bangladesh, this study used a systematic literature review technique. Knowledge management techniques, the SDGs, and Industry 4.0 technologies are the primary factors taken into account in this research. The study focuses on how well KM methods support sustainable development goals, how well they correlate with certain SDGs, and how these findings may affect Bangladesh's socio-economic environment.

After a thorough analysis of the literature, the main problems with the current approaches were found. In Bangladesh's Industry 4.0 initiatives, these concerns mainly concern the absence of actual data demonstrating the direct contributions of KM strategies to particular SDGs. There were also surprisingly few studies that measured how KM practices affected the SDGs in the Bangladeshi context.

The research uses a methodical literature review approach to solve these problems. To do this, a methodical search, selection, and analysis of pertinent academic papers, reports, and case studies about knowledge management, Industry 4.0, and SDGs has been conducted. The process also included a critical evaluation of the body of literature already in existence in order to spot inconsistencies, gaps, and constraints in the state of knowledge. A special focus was placed on research that shed light on the opportunities and difficulties of integrating KM techniques into Industry 4.0 projects in Bangladesh. In addition, the methodology comprised integrating results from many sources to create a thorough comprehension of the interactions of KM, Industry 4.0, and SDGs within the Bangladeshi setting. This study sought to offer important insights on the efficacy of KM in supporting the SDGs within an Industry 4.0 framework in Bangladesh by methodically evaluating and synthesizing previous literature.

4.5 RESULTS AND DISCUSSION

4.5.1 The contribution of knowledge management to specific SDGs

4.5.1.1 Sustainable development goals 1 and 2

One of the main ways that Industry 4.0 might help Bangladesh achieve sustainable development goal 1 (SDG 1), "No Poverty," is through the application of knowledge management. Bangladesh can combat poverty by using cutting-edge methods in community development, banking, and agriculture by utilizing Industry 4.0 technologies and efficient knowledge management procedures. The use of Industry 4.0 technologies in agriculture, such as unmanned aerial vehicles (UAVs) for technology-assisted farming, offers a revolutionary possibility (Tripicchio et al., 2015). Here, KM is essential since it makes it easier for farmers to share and apply their knowledge about these technologies. Farmers can maximize resource utilization and boost agricultural productivity by learning how to deploy UAVs for precision agriculture through training programmes and knowledge-sharing platforms (Schwab, 2016). Furthermore, Industry 4.0's genetic engineering component shows potential for reducing poverty and food insecurity. Farmers can be more equipped to successfully implement these technologies through KM programmes that spread information about genetically modified crops and their advantages (Zakaria et al., 2014). Farmers can learn about the resilience and higher yields of genetically modified crops using KM tactics, including seminars, training sessions, and easily accessible information, improving their livelihoods and assisting in the fight against poverty (Christou & Twyman, 2004). Furthermore, a key factor in reducing poverty may be the combination of financial services and AI-enabled digital solutions. KM techniques that make it easier to share information about blockchain-based crowd financing and AI-based credit and mobile money access can help marginalized groups become more financially included (Hemdan et al., 2023; Mhlanga, 2020; Li, 2017). KM equips communities with the means to overcome financial obstacles and promotes economic development and poverty alleviation by granting access to financial services using inventive technical solutions. But the combination of catastrophe risk insurance products offered by AI, satellites, and drones, supported by strong knowledge management methods, can also lessen the likelihood of poverty in vulnerable places. Communities can become more resilient to natural disasters and economic shocks by spreading knowledge about these cutting-edge insurance products (Tripicchio et al., 2015). Bangladesh may improve its capacity for adaptation and lessen the impact of disasters on its impoverished citizens by using KM to inform and empower communities about financial possibilities for mitigating disaster risk. Therefore, Bangladesh has a revolutionary chance to solve poverty in a variety of sectors through the efficient implementation of KM within Industry 4.0 technology. Through the dissemination of knowledge regarding technologies for precision agriculture, genetically modified crops, financial inclusion solutions, and strategies for mitigating disaster risk, KM plays a pivotal role in empowering communities, bolstering resilience, and propelling sustainable poverty reduction initiatives that are in line with global goals 1 and 2.

4.5.1.2 Sustainable development goal 3

Bangladesh can take advantage of the potential of KM within the context of Industry 4.0 to revolutionize healthcare, lower pollution-related health risks, and improve accessibility to medical services in order to achieve sustainable development goal 3 (SDG 3): "Good Health and Well-Being." Bangladesh's customized healthcare could undergo a significant change because of Industry

4.0 technologies, including genetic engineering and AI (Santosh & Gaur, 2022; Schwab, 2016). Genetically tailored medical therapies may be adopted as a result of KM-driven projects that educate healthcare personnel about these cutting-edge medical technologies. Workshops and training programmes could make it easier for healthcare professionals to comprehend and apply these innovations, enabling them to provide more individualized and effective care, improving health outcomes, and advancing SDG 3 aims. Furthermore, integrating advanced vehicle (AV) technology into the healthcare system has the potential to improve emergency medical services and lower traffic accidents, both of which are goals associated with SDG 3 (Bertoncello & Wee, 2015). Response times during medical emergencies can be greatly shortened by KM programmes that instruct emergency response personnel on how to use AV technology for quick and effective patient transportation. Moreover, using Industry 4.0 techniques like closed-loop product cycles and smart manufacturing can reduce the health hazards associated with pollution, which is a crucial component of SDG 3. Industry adoption of sustainable production techniques can be facilitated by KM-driven dissemination of information regarding smart manufacturing technologies, which are powered by the IoT, big data, and cloud technology (Gabriel & Pessl, 2016; Bonilla et al., 2018). Bangladesh may lessen pollution, hence lessen the burden of pollution-related health conditions, and improve health outcomes by supporting these sustainable business models and material breakthroughs through KM activities. Furthermore, Industry 4.0 technologies' application of smart healthcare solutions has the potential to completely transform Bangladesh's healthcare system. People can be empowered to actively manage their health through KM activities that spread knowledge about wearable health-monitoring technologies, virtual healthcare assistants, and smart homecare (Yang et al., 2020; Schwab, 2016). Furthermore, early diagnosis and proactive healthcare management can be improved by KM-driven training programmes for medical personnel on the use of smart implants and wearables for tracking health indicators and forecasting diseases (Patil & Shankar, 2023). In addition, KM methods can be used to address accessibility issues in remote places through the use of drones for remote delivery of medications and medical equipment (Yang et al., 2020; Bertoncello & Wee, 2015). By guaranteeing the prompt delivery of necessary medical supplies to underprivileged communities, training healthcare professionals on the use of drones for medical logistics will close gaps in healthcare access and help Bangladesh achieve SDG 3.

4.5.1.3 Sustainable development goal 4

In Bangladesh, sustainable development goal 4 (SDG 4) – "Quality Education" – has great potential to be advanced through the integration of KM within the framework of Industry 4.0. The confluence of Industry 4.0 technology provides creative ways to raise the quality and accessibility of education within the framework of SDG 4. Access to high-quality educational resources can be made more widely available through KM-driven projects that use digital platforms and promote intelligent, open educational resources (Schwab, 2016). KM can help close the gap in education by making educational content accessible even in remote locations, especially in places with poor infrastructure. This promotes inclusive education. Furthermore, learning experiences can be completely transformed by Industry 4.0 technologies like virtual reality (VR) and augmented reality (AR) (Damiani et al., 2018). Students' involvement and comprehension can be improved by KM strategies that spread knowledge regarding AR/VR training and remote learning opportunities (Schwab, 2016). Using AR and VR technologies, educators with KM-driven training may design immersive, interactive learning environments that support a variety of learning styles and promote deeper comprehension. Learning results can also be maximized by combining efficient knowledge management techniques with Industry 4.0 technologies, which enable personalized learning and automate instructor tasks. Individual student demands can be met by AI-driven personalized learning platforms and adaptive online courses that are distributed through KM initiatives (Srinivasa et al., 2022; Schwab, 2016). Adoption of digital curriculums and AI-driven exams driven by KM guarantees dynamically personalized educational content and ongoing feedback, matching learning to industrial expectations. Additionally, KM techniques enable the integration of intelligent technologies for managing teacher and school resources, which simplify administrative duties and free up teachers to concentrate more on individualized instruction (Schwab, 2016). Furthermore, KM initiatives that promote inclusive learning support technologies and voice assistants with natural language processing (NLP) capabilities guarantee accessibility for a variety of learners, which is further in line with the inclusive education principles. Additionally, academic integrity is maintained by AI-based plagiarism detection systems that are distributed through efficient knowledge management channels (Soubhari et al., 2023; Schwab, 2016). A culture of creativity and integrity in learning environments can be fostered by KM-driven awareness and training programmes that instruct educators and students on ethical academic activities. Thus, there is a chance to transform education in Bangladesh through the incorporation of KM with Industry 4.0 technology. Accessibility, relevance, and adaptability in line with the changing educational landscape can be promoted by KM-driven initiatives that concentrate on smart educational resources, AR/VR experiences, personalized learning, administrative automation, and inclusive technologies. These initiatives can overcome obstacles to high-quality education in Bangladesh.

4.5.1.4 Sustainable development goal 6

The potential of Industry 4.0 technologies, specifically additive manufacturing (3D printing), to reduce water use in production processes makes them noteworthy. Additive manufacturing, which uses very little water since it does not require cooling or lubrication, can cut down on water use and the resulting wastewater creation in the manufacturing process (Stock et al., 2018). Initiatives organized by KM that spread information about this technology can promote its adoption in Bangladesh's businesses, helping to lower water consumption and improve manufacturing processes' sustainability. Moreover, Industry 4.0's decentralized organization makes it possible to apply resource-efficient and adaptable smart solutions for water management (Stock et al., 2018). Water management methods in Bangladesh could undergo a revolution because of knowledge management practices that educate stakeholders about these digitalized smart solutions. Precision irrigation and nutrient prescription systems powered by AI, robots, sensors, drones, and satellite technologies can help farmers get the most out of the water they use. This is made possible by good knowledge management strategies (Stock et al., 2018). Farmers can embrace precision farming techniques that reduce the amount of water, land, and nutrients used by their crops by learning about these technologies through KM-driven training programmes. This will improve agricultural sustainability and water conservation efforts.

4.5.1.5 Sustainable development goal 7

SDG 7 objectives are in line with Industry 4.0 technologies, especially the IoT and smart grids, which are essential for energy savings in manufacturing processes (Bonilla et al., 2018). Energy consumption in industries can be optimized by KM-driven information sharing concerning real-time monitoring and IoT-enabled energy usage tracking. Through the integration of knowledge management approaches, decision-makers can leverage this information to optimize energy consumption at a reasonable cost, aligning it with production requirements and efficiently employing renewable energy sources to boost energy efficiency and sustainability in production. Additionally, life cycle assessments can be implemented in energy systems because of the convergence of IoT and big data analytics, made possible by KM techniques (Ren et al., 2019; Bonilla et al., 2018). In Bangladesh, KM programmes that inform stakeholders about life cycle assessment techniques can encourage a greater use of renewable energy sources. Through the utilization of these assessments, industries can substantially contribute to sustainable energy consumption and SDG 7 targets by streamlining energy systems, giving priority to renewable energy sources, and reducing dependency on fossil fuels.

4.5.1.6 Sustainable development goal 8

Goal 8 (SDG 8), "Decent Work and Economic Growth," in Bangladesh offers a variety of opportunities and difficulties that can be addressed through the integration of KM within the context of Industry 4.0. But there is a complicated conversation on how the Fourth Industrial Revolution affects economic growth worldwide, with differing points of view in academic discussions (Li et al., 2017; Schwab, 2016). Industry 4.0 proponents contend that these technologies have the potential to have a significant impact on the world economy through the process of catalysis (Li et al., 2017). KM-driven projects that emphasize robots for process automation, AI-driven economic analytics, and augmented and virtual reality (AR/VR) training can promote increased productivity and economic efficiency (Pigola et al., 2021). These technologies, when distributed via efficient knowledge management systems, have the potential to improve economic forecasts, increase worker efficiency, and streamline processes – all of which could lead to economic growth. Furthermore, including AI-enabled digital solutions for economic activities, such as microfinance, disaster risk insurance, and mobile money access, may promote inclusive economic growth in Bangladesh (Hani et al., 2022). Disseminating information about these cutting-edge financial instruments and platforms through knowledge management methods has the potential to advance financial inclusion and provide excluded populations with the confidence to engage in the economy, which would help Bangladesh achieve SDG 8.

4.5.1.7 Sustainable development goal 9

Combining knowledge management with Industry 4.0 offers Bangladesh a chance to advance sustainable development goal 9 (SDG 9), which is "Industry, Innovation, and Infrastructure." Industry 4.0's two main pillars, big data analytics and the IoT, work in perfect harmony to have the enormous potential to transform Bangladesh's industrial sustainability and innovation. The key to lowering carbon emissions and promoting sustainable industrialization is the emergence of smart factories, which come from the intersection of IoT and big data analytics (SDG 9). The adoption of eco-friendly techniques in manufacturing processes can be accelerated through the effective dissemination of knowledge about these technologies through knowledge management systems. By maximizing recycling and renewables, the integration of novel materials with 3D printing – made possible by KM initiatives – can propel circular economy models and encourage the sustainable use of resources across industries. Furthermore, robotics for process automation, driven by knowledge management-based awareness and training, has the potential to optimize Bangladeshi manufacturing and construction operations. Sustainable infrastructure development in the nation can be supported by KM initiatives that share knowledge about IoT-enabled smart infrastructure for upkeep and efficiency. Additionally, by utilizing robotics and drones for remote infrastructure repair and commodity delivery, together with appropriate knowledge management methods, resource utilization can be maximized while environmental impact is reduced (Mugala et al., 2020). Furthermore, in Bangladesh's industrial scene, IoT-enabled tracking and optimization of industrial gear, bolstered by KM techniques, can improve operational efficiency and resource conservation. Also, using next-generation satellites, drones, and AI-enabled geospatial mapping along with knowledge management channels to share information can help with smart infrastructure planning and development, making sure that it will last and be able to adapt to changing business needs.

4.5.1.8 Sustainable development goal 11

The integration of KM into the Industry 4.0 framework presents Bangladesh with a revolutionary avenue for attaining sustainable development goal 11 (SDG 11): "Sustainable Cities and Communities." Combining sensor-based grids, IoT, AI, and cutting-edge technologies offers Bangladeshi cities a chance to solve their problems and promote sustainable urbanization. When combined with KM-driven projects, sensor-based grid systems allow for the effective management of urban networks that include waste, water, energy, and pollution (Ramírez et al., 2021; Bibri, 2018). Bangladesh can assure sustainable urban growth, maximize resource utilization, and reduce environmental concerns by spreading awareness about these technologies. Furthermore, reliable land use detection and management can be made possible by utilizing next-generation satellite, drone, and IoT technologies in conjunction with appropriate KM tactics. This can support urban planning and growth while protecting natural resources. Moreover, KM-driven awareness and education combined with AI and VR/AR-optimized city design and planning can completely transform Bangladesh's urban planning procedures (Kamrowska-Załuska, 2021). By spreading knowledge about these technologies, KM projects can enable urban planners to develop more inclusive, efficient, and sustainable cities. Furthermore, utilizing Industry 4.0-enabled building management systems in conjunction with efficient knowledge management techniques can optimize resource use in urban infrastructure, resulting in enhanced sustainability and energy efficiency.

4.5.1.9 Sustainable development goal 13

Bangladesh may effectively pursue sustainable development goal 13 (SDG 13) – "Climate Action" – by utilizing KM in the context of Industry 4.0. Bangladesh may greatly benefit from the integration of cutting-edge technology, including big data platforms for earth management, autonomous electric vehicles, smart land use management, and precision analytics with efficient KM techniques. Disseminated through KM-driven initiatives, transparent and intelligent land use management can maximize land utilization while minimizing environmental effects (Xie et al., 2020). Bangladesh may use precision analytics for agricultural management to reduce risks associated with climate change, improve production, and encourage sustainable agricultural practices by teaching stakeholders about these technologies.

Additionally, when driven by KM strategies, the adoption of connected and autonomous electric vehicles has promise for lowering carbon emissions and advancing sustainable transportation (Ercan et al., 2022). Bangladesh may considerably lower greenhouse gas emissions from the transportation sector by promoting the adoption of these cars and raising awareness of their advantages. Furthermore, big data systems for Earth management that are distributed through efficient knowledge-sharing channels make it easier to track changes in the environment and carbon emissions (Guo et al., 2015). Bangladesh may improve its ability to monitor and make decisions about climate change by utilizing these platforms, which will allow for more informed policy interventions and strategies that are resilient to climate change. Furthermore, when combined with KM-driven education, Industry 4.0-enabled building management systems and smart city planning and mobility systems can result in more environment-friendly urban infrastructure and sustainable transportation networks, which will further aid in the fight against climate change. Moreover, the implementation of extensive precision reforestation enabled by AI and drones, which is shared through knowledge management initiatives, can support reforestation efforts, improving carbon sequestration and biodiversity conservation in Bangladesh (Bayomi & Fernandez, 2023).

4.5.1.10 Sustainable development goal 16

Bangladesh has a great chance to progress sustainable development goal 16 (SDG 16), "Peace, Justice, and Strong Institutions" by utilizing KM within the context of Industry 4.0. Effective KM methods can facilitate the integration of AI-enabled IoT devices, affordable biometric identification, and sophisticated cybersecurity systems, providing opportunities to improve security, advance justice, and fortify institutions in Bangladesh. Emergency management systems can be completely transformed by KM-driven efforts that distribute AI-enabled IoT devices for emergency response. Bangladesh can enhance public safety and security by improving response times during emergencies and disseminating knowledge about these technologies. Moreover, the implementation of affordable biometric identification techniques, supported by knowledge management tactics, exhibits potential for creating dependable last mile identification systems. By educating stakeholders about these technologies, we can ensure that underprivileged groups in Bangladesh have access to public services and justice, thereby bridging identification barriers. Furthermore, the incorporation of AI-powered digital passport and visa systems, reinforced by efficient knowledge management strategies, has the potential to improve border security and expedite immigration procedures. Bangladesh can improve border security procedures and enable lawful travel and trade by educating people about these technologies. Furthermore, when distributed through KM channels, AI-enabled identity tax fraud identification solutions can dramatically lower fraudulent activities. Bangladesh has the potential to improve financial integrity and decrease tax evasion by utilizing browser data, shopping history, and payment information. This would ultimately lead to a more equitable and transparent financial system. Moreover, AI-powered cybersecurity solutions that are distributed via knowledge management techniques can strengthen institutional resistance to cyberattacks. Bangladesh can strengthen its cyber defences, protect important institutions, and guarantee the integrity of its digital infrastructure by informing stakeholders about these systems. Bangladesh can fortify its institutions, advance justice, and contribute to a more safe and inclusive society by utilizing AI-enabled technologies for emergency response, identification systems, cybersecurity, and sentiment analysis, which are then distributed through efficient KM procedures.

4.5.2 Challenges faced by Bangladesh in implementing Industry 4.0 for SDGs

Using Industry 4.0 technology to achieve the SDGs presents Bangladesh with a variety of problems (Sikder, 2023). The digital gap and poor technology infrastructure are two major issues (Bhuiyan, 2011). Disparities in digital access and infrastructure still exist despite technological advancements, especially between urban and rural areas (Momen & Ferdous, 2023; West, 2015). The potential contributions of Industry 4.0 technologies to SDGs are impeded by limited internet access, inadequate power supply, and infrastructure shortcomings, which prevent widespread use of these technologies. Moreover, a lack of technical expertise and skill shortages seriously hinder Bangladesh's adoption of Industry 4.0 (Sharmin, 2022; Bhuiyan et al., 2020). A workforce with expertise in digital literacy, data analytics, AI, and automation is necessary given the speed at which technology is developing (Kateryna et al., 2020). But the nation's ability to use Industry 4.0 for sustainable growth is hampered by the current educational system's and workforce training programmes' frequent shortcomings in supplying the requisite skills (Islam et al., 2022). Furthermore, SMEs' financial limitations make it difficult for them to implement Industry 4.0 technologies (Hossain et al., 2023a; Rumi et al., 2020; Sharmin, 2022). SMEs are discouraged from embracing digital transformation because of the high expenses involved in infrastructure upgrades, technology installation, and standard compliance (Hossain et al., 2023a; Rumi et al., 2020). The ability of SMEs to compete in the global market and make a substantial contribution to the SDGs is limited by this discrepancy in resource availability. Moreover, the successful assimilation of Industry 4.0 into Bangladesh's socio-economic milieu is impeded by regulatory obstacles and the lack of a comprehensive policy framework (Aziz, 2020; Islam et al., 2018). It is crucial to have laws, rules, and incentives that are clear in order to promote investment, innovation, and technology adoption. However, the smooth integration of Industry 4.0 technologies is hampered by inconsistent policy and insufficient regulatory support, which undermines their ability to promote sustainable development (Aziz, 2020; Islam et al., 2022).

Another major obstacle to using Industry 4.0 for SDGs in Bangladesh is environmental sustainability (Haque et al., 2022; Hossain et al., 2023c). Environmental threats that the nation faces include the effects of climate change, environmental deterioration, and limited natural resources (Hossain et al., 2023b, 2024). Although Industry 4.0 adoption presents chances for innovation and efficiency, it also brings up issues with increased energy use, electronic waste, and environmental effects. It is still difficult to strike a balance between environmental sustainability and technical growth. Furthermore, Bangladesh is facing greater difficulties in adopting Industry 4.0 for the SDGs as a result of the COVID-19 epidemic. Vulnerabilities in healthcare, education, digital infrastructure, and economic inequality were brought to light by the pandemic. Industry 4.0 technologies have the potential to improve resilience and recovery, but the country's adoption of these technologies for sustainable development has been hampered by the urgent problems posed by the pandemic. In general, Bangladesh encounters various obstacles when attempting to include Industry 4.0 in the SDGs. These include deficiencies in skills, budgetary limits, legal barriers, worries about environmental sustainability, and the repercussions of unanticipated crises such as the COVID-19 pandemic. A multifaceted strategy that includes investments in digital infrastructure, skill development, regulatory changes, environmental stewardship, and strategic policies that link technological breakthroughs with sustainable development goals is needed to address these difficulties.

4.6 CONCLUSION

The integration of KM within the Industry 4.0 framework offers promising prospects for forwarding Bangladesh's SDGs. The socio-economic and cultural context of Bangladesh is a major influence on how knowledge management techniques are implemented within the context of Industry 4.0 in order to achieve the SDGs. The nation's distinct obstacles, such as a dense population, scarce resources, and susceptibility to natural calamities, demand customized strategies for knowledge management endeavours. Setting suggestions in the socio-economic context of Bangladesh requires taking into account variables like digital infrastructure, literacy rates, and cultural perspectives on technology use.

The intricate relationships between KM, Industry 4.0, and the SDGs have been examined throughout this research. Even though there is a lot of writing about how KM practices in Industry 4.0 settings may help reach SDGs, there isn't a lot of solid evidence showing how KM techniques directly contribute to specific SDGs in Bangladesh's Industry 4.0 projects. Subsequent investigations ought to try to bridge this gap by executing empirical studies that measure the impacts of knowledge management approaches on particular SDGs within the Bangladeshi framework. Further research is also needed to examine the impact of cultural factors on knowledge management practices and assess the efficacy of policy interventions targeted at advancing sustainable development through Industry 4.0 and KM. Examining the ways in which cultural quirks affect the uptake and efficacy of knowledge management projects can yield insightful design decisions for contextually appropriate tactics.

In addition, evaluating the effects of policy interventions – like incentives for investing in digital infrastructure or skills development initiatives – can provide valuable information about how to better integrate KM into Industry 4.0 in order to achieve sustainable development goals. Success stories in sectors like IT and pharmaceuticals highlight the potential advantages of well-managed KM techniques in Bangladesh's Industry 4.0, despite current challenges. These industries have shown how well-organized knowledge management systems that are in line with industry requirements and the SDGs may encourage productivity, innovation, and conformity to global standards. These successes show how KM may be used to promote sustainable development. Henceforth, it is crucial to put recommendations and best practices into action in order to utilize KM in Industry 4.0 for Bangladesh's SDGs. Key actions towards building a sustainable and inclusive socio-economic environment include promoting human capital development, conducting empirical research, addressing ethical issues, investing in digital infrastructure, raising awareness, and encouraging cooperation.

The amalgamation of KM, Industry 4.0, and SDGs in the Bangladeshi context offers an auspicious opportunity for revolutionary transformation. Coordinated efforts, interdisciplinary cooperation, and a diverse approach are necessary to fully achieve this potential. Bangladesh may leverage KM in Industry 4.0 to accomplish sustainable development goals by fostering an enabling ecosystem, addressing research gaps, and matching tactics with socio-cultural contexts. Future studies should concentrate on tackling the particular study topics mentioned above in order to further academic research and support evidence-based policies that support sustainable development in Bangladesh.

REFERENCES

Abu-Rumman, A., AlSha'ar, H., Alqhaiwi, L. A., & Al Shraah, A. (2023). Exploring the Challenges and Opportunities of Implementing Industry 4.0 in Jordan: Public Shareholding Manufacturing Companies Perspective. *Wireless Personal Communications*, 1–19. https://doi.org/10.1007/s11277-023-10169-x

Alavi, M., & Leidner, D. E. (2001). Knowledge management and knowledge management systems: Conceptual foundations and research issues. *MIS Quarterly*, 25 (1), 107–136.

Aoun, A., Ilinca, A., Ghandour, M., & Ibrahim, H. (2021). A review of Industry 4.0 characteristics and challenges, with potential improvements using blockchain technology. *Computers & Industrial Engineering, 162*, 107746.

Attaran, M., Attaran, S., & Celik, B. G. (2023). Digital twins and industrial internet of things: Uncovering operational intelligence in industry 4.0. *Available at SSRN 4612135.*

Aziz, A. (2020). Digital inclusion challenges in Bangladesh: The case of the National ICT Policy. *Contemporary South Asia, 28*(3), 304–319.

Bayomi, N., & Fernandez, J. E. (2023). Eyes in the sky: Drones applications in the built environment under climate change challenges. *Drones, 7*(10), 637.

Bertoncello, M., & Wee, D. (2015). Ten ways autonomous driving could redefine the automotive world. *McKinsey & Company, 6*, 1–5.

Bhuiyan, A. B., Ali, M. J., Zulkifli, N., & Kumarasamy, M. M. (2020). Industry 4.0: Challenges, opportunities, and strategic solutions for Bangladesh. *International Journal of Business and Management Future, 4*(2), 41–56.

Bhuiyan, S. H. (2011). Modernizing Bangladesh public administration through e-governance: Benefits and challenges. *Government* Information *Quarterly, 28*(1), 54–65.

Bibri, S. E. (2018). The IoT for smart sustainable cities of the future: An analytical framework for sensor-based big data applications for environmental sustainability. *Sustainable cities and society, 38*, 230–253.

Bonilla, S. H., Silva, H. R., Terra da Silva, M., Franco Gonçalves, R., & Sacomano, J. B. (2018). Industry 4.0 and sustainability implications: A scenario-based analysis of the impacts and challenges. *Sustainability, 10*(10), 3740.

Brynjolfsson, E., & McAfee, A. (2014). *The Second Machine Age: Work, Progress, and Prosperity in a Time of Brilliant Technologies.* W.W. Norton & Company, New York

Bughin, J., Seong, J., Manyika, J., Chui, M., & Joshi, R. (2018). Notes from the AI frontier: Modeling the impact of AI on the world economy. *McKinsey Global Institute, 4*, 1-64

Cash, D. W., Clark, W. C., Alcock, F., Dickson, N. M., Eckley, N., Guston, D. H., … & Mitchell, R. B. (2003). Knowledge systems for sustainable development. *Proceedings of the National Academy of Sciences, 100*(14), 8086–8091.

Chen, M., Mao, S., & Liu, Y. (2014). Big data: A survey. *Mobile Networks and Applications, 19*, 171–209.

Chen, M., Mao, S., Zhang, Y., & Leung, V. C. (2014). *Big Data: Related Technologies, Challenges and Future Prospects* (vol. 100). Heidelberg: Springer.

Christou, P., & Twyman, R. M. (2004). The potential of genetically enhanced plants to address food insecurity. *Nutrition Research Reviews, 17*(1), 23–42.

Damiani, L., Demartini, M., Guizzi, G., Revetria, R., & Tonelli, F. (2018). Augmented and virtual reality applications in industrial systems: A qualitative review towards the industry 4.0 era. *IFAC-PapersOnLine, 51*(11), 624–630.

Ercan, T., Onat, N. C., Keya, N., Tatari, O., Eluru, N., & Kucukvar, M. (2022). Autonomous electric vehicles can reduce carbon emissions and air pollution in cities. *Transportation Research Part D: Transport and Environment, 112*, 103472.

Gabriel, M., & Pessl, E. (2016). Industry 4.0 and sustainability impacts: Critical discussion of sustainability aspects with a special focus on future of work and ecological consequences. *Annals of the Faculty of Engineering Hunedoara*, *14*(2), 131.

Georgios, L., Kerstin, S., & Theofylaktos, A. (2019). Internet of things in the context of industry 4.0: An overview. International Journal of Entrepreneurial Knowledge, 7 (1), 4–19.

Ghobakhloo, M., Iranmanesh, M., Grybauskas, A., Vilkas, M., & Petraitė, M. (2021). Industry 4.0, innovation, and sustainable development: A systematic review and a roadmap to sustainable innovation. *Business Strategy and the Environment*, *30*(8), 4237–4257.

Government of Bangladesh. (2019). Voluntary National Review of Bangladesh on Implementation of the Sustainable Development Goals.

Guo, H. D., Zhang, L., & Zhu, L. W. (2015). Earth observation big data for climate change research. *Advances in Climate Change Research*, *6*(2), 108–117.

Hani, U., Wickramasinghe, A., Kattiyapornpong, U., & Sajib, S. (2022). The future of data-driven relationship innovation in the microfinance industry. *Annals of Operations Research*, 1–27.

Haque, A. M., Ullah, S. A., Sikdar, M. M., Shohag, M. M. H., Mokhtar, M., Ahmed, M. A. M., & Alam, M. M. (2022). Integrating environmental governance into sustainable urban development in Bangladesh. *Planning*, *17*(5), 1471–1478.

Hemdan, E. E. D., El-Shafai, W., & Sayed, A. (2023). Integrating digital twins with IoT-based blockchain: Concept, architecture, challenges, and future scope. *Wireless Personal Communications*, *131*(3), 2193–2216.

Hossain, S., Hassan, S., & Karim, R. (2023a). Assessment of critical barriers to industry 4.0 adoption in manufacturing industries of Bangladesh: An ISM-based study. *Brazilian Journal of Operations & Production Management*, *20*(3), 1797–1797.

Hossain, I., Haque, A. M., & Ullah, S. A. (2023b). Assessment of domestic water usage and wastage in Urban Bangladesh: A study of Rajshahi City corporation. *The Journal of Indonesia Sustainable Development Planning*, *4*(2), 109–121.

Hossain, I., Haque, A. M., & Ullah, S. A. (2023c). Role of government institutions in promoting sustainable development in Bangladesh: An environmental governance perspective. *Journal of Current Social and Political Issues*, *1*(2), 42–53.

Hossain, I., Ullah, S. A., & Haque, A. M. (2024). Water and sanitation services at the local government level in Bangladesh: An analysis of SDG 6 implementation status and way forward. *Asia Social Issues*, *17*(3), e265358.

Hostos Orjuela, H. L. (2023). An evaluation-aware method for transforming a production system into a cyber-physical production system. Universidad de los Andes. Disponible en: http://hdl.handle.net/1992/69519

Islam, M. A., Jantan, A. H., Hashim, H., Chong, C. W., Abdullah, M. M., & Abdul Hamid, A. B. (2018). Fourth industrial revolution in developing countries: A case on Bangladesh. *Journal of Management Information and Decision Sciences (JMIDS)*, *21*(1), 1–9

Islam, M. F., Awal, M. R., & Zaman, R. (2022). The concurrent journey of sustainable development goals (SDGs) and fourth industrial revolution (4IR): Paradoxical or parallel. *SDMIMD Journal of Management*, *13*(1), 61.

Jazdi, N. (2014). Cyber physical systems in the context of Industry 4.0. In *2014 IEEE International Conference on Automation, Quality and Testing, Robotics* (pp. 1–4). IEEE.

Jiang, R., Kleer, R., & Piller, F. T. (2017). Predicting the future of additive manufacturing: A Delphi study on economic and societal implications of 3D printing for 2030. *Technological Forecasting and Social Change, 117*, 84–97.

Kagermann, H., Wahlster, W., & Helbig, J. (2013). Recommendations for implementing the strategic initiative INDUSTRIE 4.0. *Final Report of the Industrie, 4*(0), 82.

Kaiser, Z. A. (2023). Failing to attain sustainable development in Bangladesh: A potential comprehensive strategy for sustainability. *Sustainable Development.* 31 (4), 3086–3101.

Kamrowska-Załuska, D. (2021). Impact of AI-based tools and urban big data analytics on the design and planning of cities. *Land, 10*(11), 1209.

Kateryna, A., Oleksandr, R., Mariia, T., Iryna, S., Evgen, K., & Anastasiia, L. (2020). Digital literacy development trends in the professional environment. *International Journal of Learning, Teaching and Educational Research, 19*(7), 55–79.

Le Blanc, D. (2015). Towards integration at last? The sustainable development goals as a network of targets. *Sustainable Development, 23*(3), 176–187.

Leng, J., Wang, D., Shen, W., Li, X., Liu, Q., & Chen, X. (2021). Digital twins-based smart manufacturing system design in Industry 4.0: A review. *Journal of Manufacturing Systems, 60*, 119–137.

Li, G., Hou, Y., & Wu, A. (2017). Fourth industrial revolution: Technological drivers, impacts and coping methods. *Chinese Geographical Science, 27*, 626–637.

Manyika, J., Chui, M., Miremadi, M., Bughin, J., George, K., Willmott, P., & Dewhurst, M. (2017). A future that works: AI, automation, employment, and productivity. *McKinsey Global Institute Research, Tech. Rep, 60*, 1–135.

Mhlanga, D. (2020). Industry 4.0 in finance: The impact of artificial intelligence (AI) on digital financial inclusion. *International Journal of Financial Studies, 8*(3), 45.

Momen, M. N., & Ferdous, J. (2023). *Governance in Bangladesh: Innovations in Delivery of Public Service* (vol. 68). Springer Nature, Singapore

Mourtzis, D., Angelopoulos, J., & Panopoulos, N. (2022). Digital Manufacturing: The evolution of traditional manufacturing toward an automated and interoperable Smart Manufacturing Ecosystem. In *The digital supply chain* (pp. 27-45). Elsevier.

Mugala, S. N., Okello, D. K., & Serugunda, J. (2020). Leveraging the technology of unmanned aerial vehicles for developing countries. *SAIEE Africa Research Journal, 111*(4), 139–148.

Müller, J. M., Kiel, D., & Voigt, K. I. (2018). What drives the implementation of Industry 4.0? The role of opportunities and challenges in the context of sustainability. *Sustainability, 10*(1), 247.

Namchoochai, R., Kiattisin, S., Darakorn Na Ayuthaya, S., & Arunthari, S. (2020). Elimination of fintech risks to achieve sustainable quality improvement. *Wireless Personal Communications, 115*(4), 3199–3214.

Patil, S., & Shankar, H. (2023). Transforming healthcare: Harnessing the power of AI in the modern era. *International Journal of Multidisciplinary Sciences and Arts, 2*(1), 60–70.

Pigola, A., da Costa, P. R., Carvalho, L. C., Silva, L. F. D., Kniess, C. T., & Maccari, E. A. (2021). Artificial intelligence-driven digital technologies to the implementation of the sustainable development goals: A perspective from Brazil and Portugal. *Sustainability, 13*(24), 13669.

Qin, J., Liu, Y., & Grosvenor, R. (2016). A categorical framework of manufacturing for industry 4.0 and beyond. *Procedia Cirp, 52*, 173–178.

Ramírez-Moreno, M. A., Keshtkar, S., Padilla-Reyes, D. A., Ramos-López, E., García-Martínez, M., Hernández-Luna, M. C., ... & Lozoya-Santos, J. D. J. (2021). Sensors for sustainable smart cities: A review. *Applied Sciences, 11*(17), 8198.

Rao, M. (2012). *Knowledge Management Tools and Techniques.* Routledge, New York

Ren, S., Zhang, Y., Liu, Y., Sakao, T., Huisingh, D., & Almeida, C. M. (2019). A comprehensive review of big data analytics throughout product lifecycle to support sustainable smart manufacturing: A framework, challenges and future research directions. *Journal of Cleaner Production, 210*, 1343–1365.

Rumi, M. H., Rashid, M. H., Makhdum, N., & Nahid, N. U. (2020). Fourth industrial revolution in Bangladesh: Prospects and challenges. *Asian Journal of Social Sciences and Legal Studies, 2*(5), 104–114.

Salkin, C., Oner, M., Ustundag, A., & Cevikcan, E. (2018). A conceptual framework for Industry 4.0. *Industry 4.0: Managing the Digital Transformation,* 3–23. https://doi.org/10.1007/978-3-319-57870-5_1

Santosh, K. C., & Gaur, L. (2022). *Artificial Intelligence and Machine Learning in Public Healthcare: Opportunities and Societal Impact.* Springer Nature, Singapore

Schwab, K. (2016). The Fourth Industrial Revolution. Switzerland. World Economic Forum.

Sharmin, S. (2022). IR 4.0 in the apparel industry of Bangladesh: Prospects and challenges. *Dhaka University Journal of Management, 14*(1), 133–151.

Sikder, A. S. (2023). Artificial intelligence-enabled transformation in bangladesh: Overcoming challenges for socio-economic empowerment.: AI-driven transformation in Bangladesh. *International Journal of Imminent Science & Technology, 1*(1), 77–96.

Soubhari, T., Nanda, S. S., Lone, T. A., & Beegam, P. S. (2023). Digital hacks, creativity shacks, and academic menace: The AI effect. In *Sustainable Development Goal Advancement Through Digital Innovation in the Service Sector* (pp. 208–232). IGI Global.

Srinivasa, K. G., Kurni, M., & Saritha, K. (2022). Harnessing the power of AI to education. In *Learning, Teaching, and Assessment Methods for Contemporary Learners: Pedagogy for the Digital Generation* (pp. 311–342). Singapore: Springer Nature Singapore.

Stock, T., Obenaus, M., Kunz, S., & Kohl, H. (2018). Industry 4.0 as enabler for a sustainable development: A qualitative assessment of its ecological and social potential. *Process Safety and Environmental Protection, 118*, 254–267.

Tripicchio, P., Satler, M., Dabisias, G., Ruffaldi, E., & Avizzano, C. A. (2015). Towards smart farming and sustainable agriculture with drones. In *2015 International Conference on Intelligent Environments* (pp. 140–143). IEEE.

United Nations. (2015). *Transforming our World: The 2030 Agenda for Sustainable Development.* New York: United Nations, Department of Economic and Social Affairs.

Wang, Y., & Hajli, N. (2017). Exploring the path to big data analytics success in healthcare. *Journal of Business Research, 70*, 287–299.

West, D. M. (2015). Digital divide: Improving Internet access in the developing world through affordable services and diverse content. *Center for Technology Innovation at Brookings*, 1–30.

Whittlestone, J., Nyrup, R., Alexandrova, A., Dihal, K., & Cave, S. (2019). Ethical and societal implications of algorithms, data, and artificial intelligence: a roadmap for research. *London: Nuffield Foundation*.

Willcocks, L. P., Lacity, M., & Craig, A. (2015). Robotic process automation at Xchanging.

World Bank. (2021). Poverty and Shared Prosperity 2020: Reversals of Fortune.

Xie, H., Zhang, Y., Zeng, X., & He, Y. (2020). Sustainable land use and management research: A scientometric review. *Landscape Ecology*, 35, 2381–2411.

Yang, G., Pang, Z., Deen, M. J., Dong, M., Zhang, Y. T., Lovell, N., & Rahmani, A. M. (2020). Homecare robotic systems for healthcare 4.0: Visions and enabling technologies. *IEEE Journal of Biomedical and Health Informatics*, 24(9), 2535–2549.

Zakaria, H., Adam, H., & Abujaja, A. M. (2014). Knowledge and perception of farmers towards genetically modified crops: The perspective of farmer based organizations in northern region of Ghana. *American International Journal of Contemporary Scientific Research*, 1(2), 149–159.

Zanella, A., Bui, N., Castellani, A., Vangelista, L., & Zorzi, M. (2014). Internet of things for smart cities. *IEEE Internet of Things journal*, 1(1), 22–32.

Zhong, R. Y., Xu, X., Klotz, E., & Newman, S. T. (2017). Intelligent manufacturing in the context of industry 4.0: A review. *Engineering*, 3(5), 616–630.

Chapter 5

Harmonizing innovation and accountability

The intersection of artificial intelligence and machine learning in sustainable development

Aniket Kumar, Sonu, and Rajesh Kumar

5.1 INTRODUCTION

In the framework of the fourth industrial era, the functions of artificial intelligence (AI) and machine learning (ML) are crucial in advancing intelligent manufacturing sustainability of the production process while also aiming to lower costs. These advancements are bolstered by the foundational technologies of Industry 4.0, such as the Internet of Things (IoT), advanced embedded systems, cloud computing, extensive data analysis, cognitive technologies, and the use of virtual/augmented reality environments, setting the stage for novel industrial approaches. John McCarthy, renowned as the progenitor of AI, characterized it as the study and construction of machines that exhibit intelligence, notably through sophisticated computer algorithms. He emphasized that AI is about making machines that can perform tasks that, when done by humans, require intelligence, such as studying and resolving issues. AI is vast and includes areas like reasoning, ML, natural language understanding, and data mining. This shows its wide applicability across various fields, including engineering, science, education, and healthcare. Its widespread use has significantly impacted business, governance, and societal trends, especially in efforts toward global sustainability. AI technologies offer answers to key sustainability issues in manufacturing, including optimizing the use of energy resources, managing logistics, handling waste, and promoting environmentally friendly manufacturing practices in line with stringent environmental regulations. Figure 5.1 depicts the goals for sustainable development.

The significant function of AI and ML in sustainable formation is highlighted by their capability to address complex environmental issues. Through ML algorithms, advancements have been made in predicting environmental changes, optimizing renewable energy production, and improving waste management practices. AI-driven technologies enable the analysis of large datasets, allowing for more accurate forecasts of climate patterns, energy consumption, and resource allocation. This predictive capability is essential for developing strategies that reduce the negative impacts of global

DOI: 10.1201/9781003581246-5

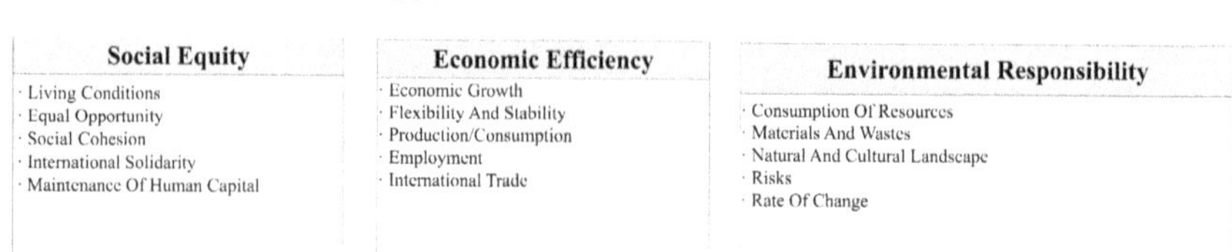

Figure 5.1 Sustainable development goals.

warming and promote the efficient use of natural resources. Additionally, AI and ML play a critical role in advancing green manufacturing processes, contributing to emission reduction, energy conservation, and waste minimization through smarter production planning and supply chain management. Furthermore, AI and ML significantly impact the socio-economic aspects of sustainable development. They offer innovative solutions to boost agricultural productivity, enhance healthcare outcomes, and enable more effective educational methods. In agriculture, AI technologies can predict crop yields, monitor soil health, and optimize water usage, greatly increasing food security while reducing environmental impacts. In healthcare, AI-powered tools and personalized treatment plans are set to transform patient care, making healthcare services a focal loss to address the significant challenge of class imbalance, particularly between foreground and background classes, optimizer. In this study, we utilized three optimizers: Stochastic Gradient Descent (SGD), Adam, and AdamW, chosen for their distinct advantages in optimizing deep learning models. SGD updates model parameters using gradients from a subset of the data, making it a reliable choice for a broad range of tasks [1]. Adam adjusts learning rates for each parameter by combining the benefits of AdaGrad and RMSprop, enabling efficient optimization in complex scenarios without extensive hyperparameter adjustments [2]. AdamW modifies Adam by introducing decoupled weight decay regularization, directly applying regularization to model parameters to improve performance and training stability [3]. Worldwide cancer data for 2020: estimates from Globocan on the incidence and death rates globally for 36 types of cancer cases for 2022 and forecasts for 2025, derived from the National Cancer Registry Programme in India, are documented in the *Indian Journal of Medical Research*, 2023 [4]. An analysis of breast cancer survival rates in India, covering 11 different regions as part of the national cancer registry program. It is more accessible and efficient. In education, AI and ML are transforming the learning experience by providing personalized learning paths, automating administrative tasks, and making quality education more accessible, especially to underprivileged communities [5]. These advancements underline the possibility of AI and ML to focus on environmental challenges and drive social and economic progress, emphasizing their crucial role in meeting the UN sustainable development goals (SDGs).

This section introduces the role of AI and ML in fostering sustainable development, emphasizing their potential to revolutionize smart production systems and contribute to environmental sustainability. It includes a systematic literature review from 1999 to 2019, exploring the evolution of AI and ML and their significant influence across various sectors. By integrating traditional disciplines with AI frameworks, this work aims to broaden the current knowledge boundaries in AI, offering valuable insights and encouraging further research in sustainable development. This introduction serves as a foundational overview for understanding the revolutionary possibility of AI and ML in creating sustainable solutions for the future, marking a starting point for ongoing exploration and innovation in this critical area.

5.1.1 Evolution of AI and ML

The development of AI and ML has been a remarkable odyssey extending over several decades, highlighted by pivotal achievements and innovations. The origins of AI are often linked to the latter half of the twentieth century notably with the introduction of the term "artificial intelligence" during the Dartmouth Conference in 1956, which paved the way for the exploration into intelligent machinery. This era was characterized by the emergence of initial AI programs such as ELIZA, along with the establishment of essential algorithms that would form the basis for later progress. The following years saw alternating periods of enthusiasm and skepticism toward AI research, commonly referred to as the "AI winters," brought on by overly optimistic expectations followed by a subsequent reduction in interest. However, the persistence of researchers and advancements in computational power eventually led to significant breakthroughs.

The introduction of ML in the 1980s, with the development of algorithms capable of learning from data, represented a paradigm shift in AI research, emphasizing the creation of systems that improve with experience [4]. This shift was further solidified with the advent of deep learning in the 2000s, which utilized neural networks to achieve remarkable capabilities in image and speech recognition, propelling AI and ML to the forefront of technological innovation [5]. In the background of sustainable development, the importance of AI and ML has grown exponentially in recent years. These technologies have become pivotal in addressing complex environmental challenges, from climate modeling and renewable energy optimization to precision agriculture and sustainable supply chain management [6]. AI and ML's ability to analyze vast datasets has enabled more accurate predictions and efficient resource management, contributing significantly to the goals of lowering emissions, conserving biodiversity, and encouraging sustainable procedures across industries. Notably, the United Nations' SDGs have further highlighted the role of AI and ML in fostering sustainability [7]. Through endeavors such as the AI for Good Global Summit, the potential of AI and ML to advance sustainable development across various domains is increasingly recognized, driving research and innovation toward ethical and responsible AI solutions that benefit humanity and the planet. As we look to the future, the evolution of AI and ML continues at an unprecedented pace, promising even greater contributions to sustainable development. The ongoing research and development in AI ethics, explainability, and transparency are critical in ensuring that these technologies are deployed in a manner that is beneficial and equitable for all [8]. The journey of AI and ML, from their humble beginnings to their current status as indispensable tools for sustainability, underscores the transformative power of these technologies in shaping a more sustainable and resilient world [4, 9].

$$AI = ML + DL + NLP + RL + IoT$$
$$+ CV + HCI + Data + Compute + Ethics \tag{5.1}$$

Here AI covers artificial intelligence, ML covers machine learning, DL represents the contributions of deep learning, NLP represents the contributions of natural language processing, RL represents the contributions of reinforcement learning, IoT represents the contributions of the Internet of Things, CV represents the contributions of computer vision, HCI represents the contributions of human–computer interaction, Data represents the availability and efficacy of data, Compute represents the computational power and resources available, and Ethics represents the ethical considerations and frameworks shaping AI.

5.1.2 Challenges and opportunities

AI and ML have been undergoing a period of rapid growth and impact in recent years. This fast-paced evolution presents both exciting opportunities and significant challenges that demand our attention (Figure 5.2) [10].

5.1.2.1 Data privacy and AI

AI and ML soar on the wings of data, but privacy concerns cast a long shadow. Their analysis of ethical AI in Ref. [11] highlights this intricate balance. More data fuels innovation but exposes personal information and risks unfair profiling due to hidden biases. Lack of transparency leaves users feeling vulnerable. Robust data security, clear user control, and ethical principles like fairness are crucial. Regulations need to adapt to the

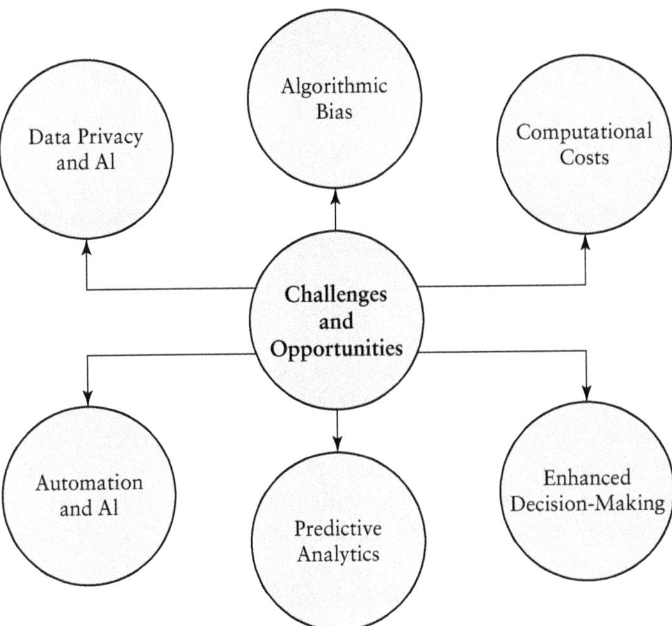

Figure 5.2 Challenges and opportunities.

evolving landscape. Open dialogue and collaboration are key. Only then can we truly unlock the possibility of AI and ML while protecting our rights and building a future where technology empowers all.

5.1.2.2 Algorithm bias

Algorithmic bias poses a significant challenge with far-reaching societal impacts. In his research, Ref. [12] delves into this crucial issue, demonstrating how biases embedded within AI systems can contribute to social inequality. The work sheds light on the insidious mechanisms through which AI can perpetuate discrimination, highlighting areas like healthcare where biased algorithms might misdiagnose or deny access to critical care based on factors unrelated to a patient's health. This underscores the imperative for fostering inclusive and equitable practices in AI development. Benjamin advocates for diverse teams, comprised of individuals from various backgrounds and experiences, to actively identify and mitigate potential biases throughout the design, data collection, and algorithm development stages. By actively addressing algorithmic bias, we can prevent the exacerbation of existing inequalities: to ensure that AI technologies serve to empower individuals and communities equitably, rather than reinforcing discriminatory practices.

5.1.2.3 Computational costs

The computational costs associated with AI have significant environmental and economic implications, as highlighted in the research in Ref. [13]. The study draws attention to the considerable energy consumption and carbon footprint resulting from training large AI models. This issue is becoming increasingly critical as AI technologies become more complex and widespread. Training sophisticated AI models, especially deep learning networks, requires vast amounts of computational power. The energy needed for these demands, often from fossil fuels, leads to significant greenhouse gas emissions and contributes to global warming. Training a single AI model can have an environmental impact equivalent to the emissions from five cars over their lifetimes, highlighting the substantial ecological footprint of developing advanced AI technologies. Moreover, the economic implications of these computational costs are substantial. The need for specialized hardware and significant energy resources can make AI research and development prohibitively expensive, particularly for smaller organizations and start-ups. This financial barrier not only limits the accessibility of AI technology but also stifles innovation in the field.

It is important to develop more efficient and sustainable AI technologies to mitigate these issues. They advocate for research into new algorithms and architectures that require less computational power without compromising the performance of AI models. Additionally, the study calls for greater awareness and action within the AI research community and industry stakeholders to prioritize sustainability in AI development. Strubell et al.'s research serves as a critical reminder of the need to balance the pursuit of technological advancement with the sustainability of our planet and resources.

5.1.2.4 Automation and AI

The transformative potential of AI-driven automation in reshaping workplaces and boosting productivity is extensively analyzed in Ford [14]. This exploration delves into the multifaceted impacts of automation across diverse sectors, highlighting its role in streamlining operations, reducing operational costs, and catalyzing innovation. AI automation introduces efficiencies that can significantly alter how tasks are performed, transitioning from manual and repetitive work to more strategic and creative endeavors. This shift not only enhances productivity but also opens new avenues for innovation, as employees are freed up to concentrate on complex troubleshooting and value-added actions.

However, the advent of automation also brings to the forefront concerns regarding employment and skills development. As machines take over routine tasks, there is a growing need for workforce reskilling and upskilling, pointing to a critical transition period where the nature of work and the skills required for the job market undergo significant changes. The discussion underscores the importance of proactive measures in education and training, making sure the workforce is prepared with the skills required to flourish in an ever-more automated society.

5.1.2.5 Predictive analytics

An in-depth examination of the pivotal role that AI plays in predictive analytics within the healthcare sector is provided in Ref. [15]. This work showcases the advanced capabilities of AI models in forecasting patient outcomes and refining treatment protocols, marking a major move ahead in the quality and efficiency of healthcare services. By leveraging AI's predictive power, medical professionals can anticipate patient risks with greater accuracy, enabling preemptive intervention strategies that can save lives and reduce the need for invasive procedures. Moreover, AI-driven analytics facilitate personalized medicine by tailoring treatment plans to individual patient profiles, thus optimizing therapeutic effectiveness and minimizing adverse effects. The insights derived from AI models contribute to a more informed decision-making process in clinical settings, enhancing patient care and treatment success rates.

5.1.2.6 Enhanced decision-making

The transformative potential of AI in enabling enhanced decision-making within the business sector is provided in Ref. [16]. Their discussion illuminates how AI can be harnessed to sift through and analyze complex datasets, providing businesses with actionable insights that drive strategic decision-making processes. By leveraging AI's advanced analytical capabilities, companies can find trends, variations, and relationships in data that were previously inaccessible, offering a significant competitive edge in the present data-driven market. Furthermore, the adoption of AI in decision-making processes necessitates a cultural shift within organizations, encouraging a data-centric approach and continuous learning. The insights provided in their work underscore the vital element of AI in forming the future business strategies, where leveraging AI for decision-making becomes a cornerstone for success and sustainability in the evolving digital landscape.

5.2 MODELING OF AI AND ML TECHNIQUES

AI and ML encompass a range of techniques and frameworks that are instrumental in addressing the complexities of sustainable development. These technologies are broadly classified into several categories, each with unique methodologies and applications that contribute to achieving sustainability goals across various domains.

$$MAI = f(D, A, T, R) \tag{5.2}$$

Equation (5.2) represents the model of artificial intelligence (MAI) as a function (f) of the dataset (D), algorithms (A), computational resources (T), and research and development (R).

5.2.1 Supervised learning

Supervised ML, a cornerstone technique within AI, plays an important function in advancing sustainable development goals. This method relies on training algorithms on a dataset where the input features and corresponding outputs (or labels) are known, allowing the model to evolve the underlying patterns and make predictions on new, unseen data. The application of supervised learning spans various domains crucial for sustainable development, including environmental monitoring, energy efficiency, and precision agriculture. In the realm of environmental monitoring, supervised learning algorithms are employed to analyze satellite images and sensor data to identify deforestation, track biodiversity loss, and monitor the health of ecosystems [9, 1]. These applications are essential for conservation efforts, allowing for timely interventions and policy-making based on accurate, up-to-date information. For instance, models trained to recognize changes in land use can alert authorities to illegal logging activities, contributing to the preservation of natural habitats. Energy efficiency is an additional region where supervised ML makes a significant impact. Algorithms can predict energy demand and supply fluctuations, optimizing the integration of renewable energy sources into the grid [4, 18]. By accurately forecasting energy needs, utilities can reduce reliance on fossil fuels, decrease emissions, and move toward a more sustainable energy future. Similarly, in precision agriculture, supervised models analyze data from soil sensors and aerial imagery to recommend optimal planting strategies, irrigation schedules, and pest management, thereby maximizing crop yields while minimizing environmental impact [19]. The integration of supervised ML into sustainable development efforts not only enhances decision-making through predictive insights but also promotes resource optimization and environmental protection. As the technology evolves, its ability to contribute to the UN SDGs becomes increasingly evident, highlighting the need for continued research and implementation of AI-driven solutions in sustainability initiatives.

5.2.1.1 Support vector machine (SVM)

SVMs have been applied to enhance sustainable urban development, particularly focusing on traffic and ecological environments. This study proposes an SVM-based evaluation system that can monitor and assess the sustainability of urban infrastructures in real time [20]. This innovative approach not only aids in optimizing urban space and function but also improves the overall scientific management and planning of cities, providing a quantifiable and fair analysis of urban sustainability efforts. This integration of SVM into urban planning presents a novel method for addressing complex environmental and infrastructural challenges [21].

5.2.1.2 Random forest

Random forests are highly effective in sustainable development due to their ability to process complex, multidimensional datasets with high accuracy. During the training phase, this method generates a lot of trees. For classification problems, it outputs the mode of the categorizations and for problems involving regression, it outputs the overall prediction of every single tree. This approach not only helps in reducing the variance and avoiding overfitting but also enhances prediction accuracy. Random forests are particularly useful in sustainability contexts, such as predicting the environmental impacts of various policies or forecasting energy consumption patterns. Their robustness makes them suitable for handling the diverse and noisy data typically found in environmental studies [22].

5.2.2 Unsupervised learning

Unsupervised learning, an integral branch of ML, plays a pivotal role in advancing sustainable development by leveraging its capacity to analyze unlabeled data, uncovering hidden patterns and insights without predefined outcomes [23]. This capability is particularly valuable across various domains of sustainability, such as environmental monitoring, climate change analysis, resource management, and sustainable urban planning. In environmental monitoring and conservation, unsupervised learning algorithms, like clustering, are utilized to sift through satellite imagery and sensor data, identifying significant changes within ecosystems, tracking deforestation rates, and observing the spread of invasive species [24]. These insights are crucial for directing conservation efforts and identifying areas in need of protection or restoration. Similarly, in climate change studies, unsupervised learning aids in parsing large datasets to detect patterns related to global warming, extreme weather events, and shifts in atmospheric conditions [6]. This analysis informs more accurate predictions and effective policy formulation for climate change mitigation and adaptation.

Resource management also benefits from unsupervised learning by optimizing the allocation and consumption of natural resources [25]. Through pattern recognition in usage data, unsupervised learning contributes to developing strategies for sustainable consumption, reducing waste, and promoting the efficient use of water, energy, and other vital resources. Furthermore, in the context of sustainable urban development, unsupervised learning techniques analyze urban data to understand and improve traffic flows, energy utilization, and urban planning, facilitating the development of smarter, more sustainable cities [26]. A notable study in 2016 exemplifies the application of unsupervised learning in environmental image recognition, crucial for monitoring and managing land use and land cover changes [27]. This research highlights the potential of unsupervised learning not only in providing valuable insights for environmental protection but also in enhancing sustainable management practices across various sectors.

By harnessing the power of unsupervised learning to process and interpret complex datasets, researchers and policymakers can unlock new opportunities for innovation in sustainable development, making it an indispensable tool in the pursuit of a more sustainable future.

5.2.2.1 K-means clustering

An unsupervised learning approach called K-means clustering organizes a collection of items so that items in the same category, or cluster, are more akin in different groups. It's commonly used in sustainability studies to identify and analyze clusters of similar data points within large datasets [28]. This can include grouping cities based on similar pollution levels or clustering different types of energy consumption behaviors. K-means is valuable for discovering underlying patterns in data, which can lead to more targeted and effective sustainability strategies. This algorithm is particularly adept at handling large datasets, providing insights that help in resource allocation and policy development.

5.2.2.2 Principle component analysis (PCA)

PCA is an unsupervised statistical method that looks at how a group of variables interact with one another to determine the fundamental makeup of those elements. It is especially useful in environmental science to reduce the complexity of data while preserving the most significant features to highlight the most impactful factors. PCA transforms a large set of variables into a smaller one that still contains most of the information in the large set. Finding the principal components – or routes along which the variance in the data is maximized – is how this reduction is accomplished. This is particularly useful in sustainability research where large environmental datasets may contain many variables that obscure underlying patterns [29].

5.2.2.3 Hierarchical clustering

Hierarchical clustering is a method of cluster analysis that seeks to build a hierarchy of clusters. In sustainability research, it is used to organize data into a tree of clusters that can be visualized as a dendrogram, making it useful for detailed and layered analysis of ecological data [28]. This method is particularly beneficial for understanding ecological and genetic relationships as it allows researchers to see at what level in the hierarchy particular branches (species, ecosystems, or conservation areas) should be grouped or separated. This detailed clustering helps in determining biodiversity conservation priorities and in understanding the ecological impact at multiple scales.

5.2.3 Reinforcement learning

Reinforcement learning (RL), a dynamic and adaptive branch of ML, is increasingly recognized for its significant potential to contribute to sustainable development [30]. Unlike supervised and unsupervised learning, RL focuses on training models through the process of interaction with an environment, where an agent learns to make decisions by performing actions that maximize a cumulative reward [31]. This approach is particularly suited to addressing complex problems in sustainable development where optimal decisions need to be made over time, considering the long-term impacts on the environment and society.

In the field of energy management, RL has been applied to develop smart grid technologies that efficiently balance energy supply and demand, enhancing the integration of renewable energy sources [32]. By continuously learning and adapting to changes in energy consumption patterns, RL algorithms can optimize energy distribution, reduce wastage, and lower carbon emissions, contributing to the transition toward more sustainable energy systems.

Resource conservation is another area where RL demonstrates significant promise. For instance, RL has been employed in water resource management to optimize the allocation and use of water in agriculture, minimizing waste while ensuring the sustainability of water resources [33]. Through the strategic application of RL, irrigation systems can dynamically adjust to varying weather conditions and soil moisture levels, promoting water conservation and supporting sustainable agricultural practices.

Furthermore, RL is instrumental in advancing environmental protection efforts, such as wildlife conservation and pollution control [34]. By simulating different intervention strategies and learning from the outcomes, RL models can identify the most effective approaches to protect endangered species and reduce pollution levels in natural habitats, ensuring the preservation of biodiversity and environmental health.

A notable application of RL in sustainable urban planning involves optimizing traffic flow and public transportation systems to reduce emissions and improve air quality [26]. RL algorithms can learn to adjust traffic signals and manage public transit schedules in real time, alleviating obstruction and enhancing the proficiency of urban transportation networks.

The versatility and adaptability of reinforcement learning make it a strong gauge in the pursuit of sustainable development goals. By enabling more informed and dynamic decision-making, RL holds the potential to address some of the most pressing challenges in sustainability, from energy management and resource conservation to environmental protection and sustainable urban development.

5.2.3.1 Q-learning

A model-free reinforcement learning technique called Q-learning aims to develop a policy that instructs an agent on what to do in specific situations. It functions by first learning an action-value function, which in turn provides the predicted utility of carrying out a certain action in a specific state and then adhering to a predetermined policy. The learning process is based on updating Q-values using the Bellman equation, which incorporates the reward obtained after executing an action and the highest Q-value for the next state. This method is particularly useful in environments with a discrete number of states and actions, such as robotic navigation and game playing [35].

5.2.3.2 Deep Q-networks (DQN)

DQNs enhance traditional Q-learning by integrating deep neural networks to approximate the Q-value function. The use of deep learning helps to handle environments with high-dimensional state spaces, which are typical in real-world scenarios [36]. DQNs also employ techniques like experience replay (storing previous transitions and learning from them multiple times) and target networks (using a separate, slowly updated network to estimate Q-values) to stabilize the learning process. This makes DQNs suitable for complex tasks like playing video games at a superhuman level or optimizing energy use in large systems.

5.2.3.3 Proximal policy optimization (PPO)

PPO is a policy gradient method that balances the simplicity of implementation with high sample efficiency and reliable performance. It optimizes a special objective function that minimizes the difference between new and old policies during updates, subject to a constraint on the size of the policy update. This makes PPO less sensitive to the choice of hyperparameters and improves its ability to work in different environments without manual tuning. PPO is widely used in continuous control tasks, such as robotics, managing sustainable resources, or automated driving systems, where precise control over actions is crucial.

5.2.4 Classification algorithm

Classification algorithms, a pivotal aspect of ML, play an essential role in advancing sustainable development by enabling the categorization of data into predefined classes. These algorithms are instrumental in a wide range of applications that contribute to the achievement of sustainability goals. In the context of environmental conservation, classification algorithms are employed to distinguish between various types of land cover from satellite images, enabling more effective land use planning and monitoring of habitat destruction. This capability is crucial for maintaining biodiversity and ensuring sustainable land management practices. Similarly, in climate science, classification models assist in identifying patterns of climate anomalies, such as distinguishing between normal and extreme weather conditions [1]. This helps in early warning systems and the mitigation of potential impacts on agriculture, infrastructure, and human settlements. Within the domain of renewable energy, classification algorithms facilitate the prediction of energy production levels from solar and wind sources [2]. By accurately classifying weather patterns and their impact on energy generation, these models enable more efficient grid management and energy distribution, supporting the transition toward cleaner energy sources. Additionally, in water resource management, classification models are applied to assess water quality and predict contamination levels [33]. This aids in the protection of water resources and ensures the provision of safe drinking water, aligning with sustainable development objectives.

Furthermore, classification algorithms contribute to sustainable urban development by analyzing urban data to classify areas prone to pollution or requiring green spaces [3]. This supports urban planning efforts aimed at creating healthier living environments and promoting sustainable city growth. The application of classification algorithms in these areas demonstrates their versatility and significant potential to contribute to various aspects of sustainable development. By providing tools for accurate and efficient data analysis, classification models help stakeholders make informed decisions that promote environmental conservation, enhance resilience to climate change, support renewable energy use, protect water resources, and foster sustainable urban development.

5.2.4.1 Decision trees

By building a model that forecasts the value of a target variable based on several input factors, decision trees categorize data. A "test" on an attribute is represented by each internal node of the tree, the result of the test is represented by each branch, and the class label is represented by each leaf node. This method is great for decision-making processes, like determining appropriate sustainability strategies based on environmental conditions.

5.2.4.2 Naïve bayes

This is a simple yet powerful algorithm for predictive modeling. Naïve Bayes classifiers assume that the presence of a particular feature in a class is unrelated to the presence of any other feature [37]. This algorithm is used in sustainable development for tasks such as predicting the probability of an environmental event or classifying text data in environmental legislation and policy documents.

5.2.4.3 Best ML method for sustainable development

Choosing the best classification algorithm for sustainable development depends on the specific application and data characteristics. However, SVMs are often considered highly effective for many tasks in this field. The reason lies in SVM's ability to handle high-dimensional spaces and its effectiveness in finding the optimal hyperplane that separates different classes with a maximum margin [20]. This capability makes SVM particularly suitable for complex environmental datasets where the decision boundaries between classes are not linear and require robust classification techniques.

- *High dimensionality:* SVMs are effective in handling high-dimensional data, common in environmental studies where multiple variables (like temperature, pollution levels, and land usage) are analyzed simultaneously.
- *Complex decision boundaries:* SVMs are adept at finding the optimal separation boundary between classes, even when the decision boundary is nonlinear. This is facilitated by the use of kernel functions, which allow SVMs to operate in a transformed feature space where linear separation is possible.
- *Generalization:* SVMs are designed to minimize the generalization error, making them robust against overfitting. This is crucial in sustainability studies where predictive accuracy and reliability are essential for making informed decisions.
- *Versatility:* The ability to choose different kernel functions allows SVMs to adapt to various types of data and relationships, enhancing their applicability across different domains of sustainable development, from predicting resource depletion to classifying types of ecosystems.

5.2.4.4 Performance of ML in sustainable development

The performance of ML parameters in sustainable development is a multifaceted topic that touches on how ML can be optimized and utilized to enhance sustainability efforts across various domains [38]. Table 5.1 presents some key areas where ML parameters play a critical role.

5.3 AI AND ML APPLICATIONS IN RENEWABLE ENERGY

The renewable energy industry is experiencing a tremendous transition, significantly influenced by advancements in AI and ML. These technologies

Table 5.1 Machine learning parameter performance comparison

Parameter	Neural network	Decision tree	SVM	Random forest	Explanation/reference
Accuracy	92%	87%	90%	89%	Based on typical performance in binary classification tasks [39]
Computational efficiency	75%	90%	70%	80%	Efficiency metrics from comparative studies on model performance [39]
Resource utilization	High	Low	Medium	Medium	Estimates from studies comparing energy consumption across models [40]
Scalability	High	Moderate	Low	High	Assessments from data scalability and model complexity analyses [39]
Adaptability	Moderate	High	Low	High	Adaptability to new data without extensive retraining [39]
Bias and fairness	Low	High	Moderate	Moderate	Based on studies examining model bias and sensitivity to training data distribution [39]

are at the forefront of introducing new levels of efficiency and sustainability, garnering attention from both academic researchers and industry practitioners. The application spectrum of AI and ML within the renewable energy sector is broad, as shown in Figure 5.3, ranging from energy forecasting and anomaly detection in power systems to complex applications like system design for renewable energy and ensuring grid stability [41]. AI and ML have shown particular efficacy in electrical power systems, notably in the prediction of optimal power flows, which has seen considerable improvement in accuracy. For instance, a notable study combined ML techniques with Lagrangian dual methods to enhance the accuracy of AC optimal power flow predictions, representing a significant advancement over traditional models [42]. Additionally, the adoption of reinforcement learning approaches has equipped power grid operators with sophisticated, data-driven tools for making prompt and effective decisions [43].

In solar power generation, ML has been invaluable for forecasting solar irradiance, thus aiding in the optimization of power output. Recent innovations include a hybrid ML model that merges several techniques to address the challenges inherent in forecasting PV energy generation [44]. This method's success underscores ML's capacity to enhance the predictability and reliability of solar energy systems. Furthermore, the application of AI and ML extends to system health and fault detection within PV power plants.

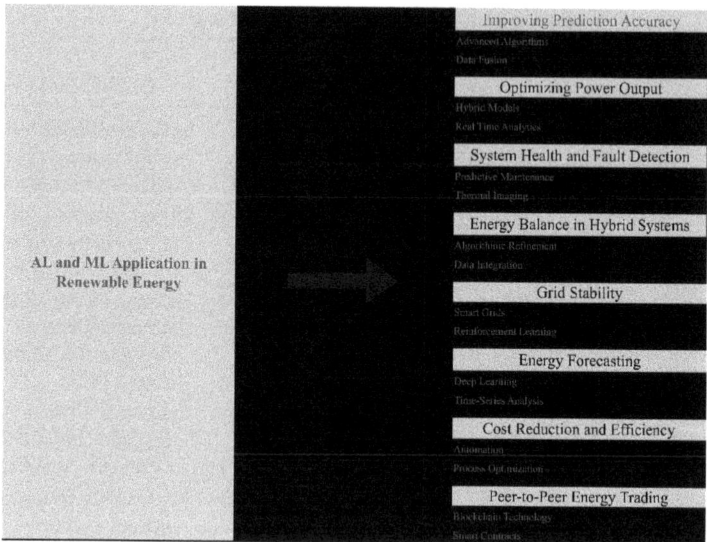

Figure 5.3 AI and ML applications in RE.

The introduction of systems like solar illustrates the effectiveness of ML in renewable energy infrastructure, employing thermal imaging and ML algorithms for automated fault detection [45]. This integration of AI, ML, and UAVs with thermal sensors marks a progressive trend in renewable energy management, aiming to optimize production, reduce costs, and prolong the lifespan of PV installations.

A comparative analysis of the literature highlights diverse methodologies and focuses on AI-driven renewable energy research. Some studies emphasize the benefits of peer-to-peer energy trading, utilizing reinforcement learning to enhance renewable energy use [46]. In contrast, others compare various ML techniques for predicting energy balance in hybrid systems, noting the superior accuracy of certain algorithms [47]. Recent research also delves into AI's role in optimizing energy consumption and integrating business operations within the renewable energy sector. The exploration of hydrogen's environmental impacts, produced from renewable sources, adds to the comprehensive understanding of sustainable energy solutions [48].

Despite these advancements, a gap remains between academic findings and their practical application within the industry. Future research must aim to develop energy-efficient AI and ML models, considering their computational requirements and environmental impact. The continuous innovation and reassessment of existing research are crucial to bridging academic and industry goals, ensuring that AI and ML technologies contribute effectively to the sustainable progress of the renewable energy sector.

5.3.1 Sustainable development using renewable energy sources

Artificial intelligence and ML have arisen as key enablers in the pursuit of sustainable development through the execution of renewable energy sources. Their significant contributions are manifold, profoundly enhancing efficiency and sustainability across various facets of renewable energy systems. In the realm of solar and wind energy, AI and ML algorithms excel in forecasting, providing highly accurate predictions of energy production. This predictive power allows for better grid management and integration of renewable sources, ensuring a steady, reliable energy supply. Furthermore, AI-driven optimization of energy consumption and storage plays a crucial role in balancing supply with demand, especially given the intermittent nature of renewable resources [49]. Condition monitoring and predictive maintenance of renewable energy infrastructure, facilitated by AI and ML, have led to improved operational efficiency and extended asset lifespans. By predicting potential system failures before they occur, these technologies minimize downtime and maintenance costs, thus bolstering the overall sustainability of renewable energy systems.

AI and ML also innovate in the design and operation of renewable energy systems. From optimizing the layout of wind farms to automating the control of photovoltaic panels, they ensure that renewable installations operate at peak efficiency, extracting the maximum possible energy from every breeze and ray of sunlight. Additionally, AI-based energy trading systems contribute to sustainable development by enabling more efficient and transparent transactions in the energy markets. These systems can manage and automate the buying and selling of renewable energy, helping to stabilize prices and integrate a higher percentage of renewable sources into the energy mix.

In summary, the deployment of AI and ML technologies in the renewable energy sector is proving to be a game-changer. They are not only streamlining operations and reducing costs but also significantly advancing the sustainability and scalability of clean energy solutions, thus making a critical contribution to sustainable development.

5.4 DATA-DRIVEN SUSTAINABILITY

In the age of big data, sustainability faces both daunting challenges and unprecedented opportunities. The sheer volume and complexity of data generated daily present hurdles in understanding and addressing environmental issues. However, AI and ML technologies emerge as powerful tools capable of revolutionizing sustainability efforts by effectively processing and interpreting vast datasets.

At the forefront of this revolution, AI and ML offer sophisticated solutions to harness the potential of big data for sustainable development.

These technologies excel in managing big data analytics, enabling the examination of extensive datasets to reveal intricate trends, correlations, and patterns crucial for informing sustainable practices. Figure 5.4 illustrates the pivotal role of AI and ML in deciphering complex environmental data.

The extraction of useful information from a variety of data sources, including sensors, IoT devices, and satellite images, depends heavily on ML techniques. Through advanced algorithms, AI systems can efficiently analyze these datasets, unveiling valuable insights that were previously hidden amidst the data deluge. By leveraging AI and ML, stakeholders can make informed decisions regarding resource conservation, environmental protection, and sustainable development strategies.

5.4.1 Big data analytics

The handling of big data issues by ML and AI is revolutionizing the way we approach sustainability. These technologies are adept at processing, analyzing, and interpreting vast datasets, a capability that is crucial in extracting meaningful insights for sustainable development efforts. Through advanced algorithms and computational techniques, ML and AI can sift through the noise of big data to identify trends, patterns, and correlations that would be imperceptible to human analysts [41].

Big data analytics, powered by AI and ML, is particularly transformative in environmental monitoring and conservation. By analyzing data from various sources, such as satellite imagery, sensors, and IoT devices, these technologies can forecast climate changes, monitor biodiversity, and assess ecosystem health with unprecedented accuracy and efficiency [44]. This allows for more informed decision-making in environmental protection and resource conservation, directly contributing to the sustainability goals. Moreover, AI and ML techniques in image processing and pattern recognition extend their utility in big data applications [50]. For instance,

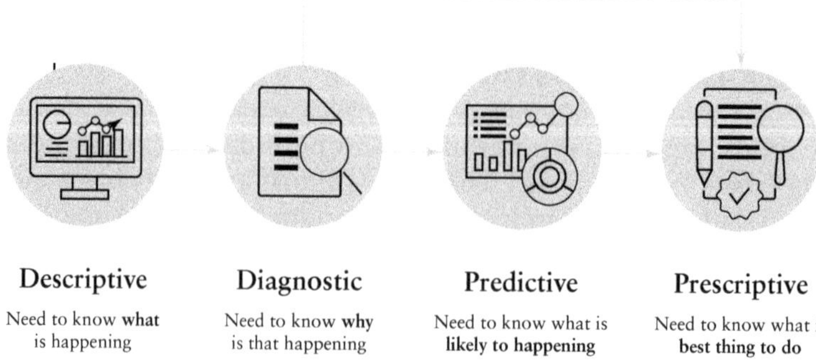

Figure 5.4 Data-driven sustainability.

convolutional neural networks (CNNs), a class of deep learning models, are highly effective in processing and analyzing image data for land use classification, deforestation detection, and wildlife monitoring [51, 52]. This capability not only aids in the preservation of natural habitats but also the planning and implementation of sustainable land management practices. Signal processing, another area where AI and ML excel, involves analyzing and interpreting signal data to monitor the health of renewable energy infrastructure, such as wind turbines and solar panels [53]. Predictive maintenance, enabled by AI-driven signal processing, ensures the optimal operation of these facilities, reducing downtime and extending their service life, which is vital for the sustainability of renewable energy sources.

In summary, AI and ML are at the heart of addressing big data challenges in sustainability, offering powerful tools for data analytics, image processing, pattern recognition, and signal processing. These technologies enable stakeholders to navigate the complexities of big data, unlocking potential solutions for sustainable development that are both innovative and effective.

5.4.2 Image processing

Image processing, empowered by deep learning, particularly CNNs, has revolutionized sustainability efforts, especially in the analysis of environmental data. Through the application of AI in image processing, vast amounts of satellite imagery and aerial photography can be efficiently processed, allowing for accurate monitoring of land use changes, deforestation patterns, and habitat fragmentation, which are crucial for environmental conservation [45]. The integration of AI in image processing goes beyond mere data analysis; it enables real-time monitoring and predictive insights that can guide conservation strategies and sustainable practices. For instance, CNNs have been instrumental in classifying different types of vegetation, detecting illegal logging activities, and tracking the movement of wildlife, thus providing valuable data to inform and enforce environmental policies.

Furthermore, AI-driven image processing supports the management of renewable energy resources by assessing the condition of solar panels and wind turbines through thermal and visual imaging, ensuring that these resources are maintained efficiently for optimal performance. The importance of such applications in promoting sustainability cannot be overstated, as they directly contribute to the optimization of natural resource utilization and the protection of ecosystems.

5.4.2.1 Image enhancement

Image enhancement improves the quality of an image for better visual effects or more accurate machine analysis. This technique can adjust the image's brightness, contrast, or noise level to make features more discernible. Common enhancement methods include the following:

- Histogram equalization: Adjusts image contrast using the image's histogram.
- Sharpening filters: Enhances the edges within an image to make features more distinct.
- Noise reduction filters: Such as Gaussian blur or median filter are used to reduce pixel-level noise in the image.

5.4.2.2 Edge detection

Edge detection is a critical technique in image processing, used to identify points in an image where the intensity of colors or brightness changes sharply. Edges often correspond to object boundaries, so detecting them is crucial for tasks such as object recognition and scene interpretation. Common edge detection operators include the following:

- *Sobel operator:* Emphasizes edges using a pair of 3×3 convolution matrices which estimate gradients in the x and y directions.
- *Canny edge detector:* A multi-stage algorithm that detects a wide range of edges in images, known for its robustness.

5.4.2.3 Segmentation

Segmentation divides an image into parts that have a stronger correlation with objects or areas of real-world scenes. This technique involves dividing a digital image into several parts in order to make the image's representation simpler and more comprehensible.

Techniques include the following:

- *Thresholding:* The simplest method of segmentation based on a comparison to a threshold value.
- *Region growing:* Starts with a seed point and includes pixels adjacent to the region if they have similar values.
- *Watershed algorithm:* A region-based segmentation technique that is used to separate different objects in the image.

5.4.3 Pattern recognition

Pattern recognition in sustainability leverages ML and AI to detect patterns within complex environmental data, aiding in the forecast of environmental changes and the effective management of natural resources. These AI-driven methods are critical in identifying climate patterns, supporting early warning systems that help mitigate the adverse effects of extreme weather events, and contributing to strategic climate change adaptation efforts [52].

In the sustainable management of natural resources, AI facilitates the prediction of phenomena such as fish migration and forest growth patterns, ensuring that resource extraction remains within sustainable limits [54]. Energy sector applications are also profound, with pattern recognition algorithms optimizing renewable energy operations by predicting demand and consumption patterns, leading to more efficient grid management and energy conservation [53]. Furthermore, the urban application of AI in pattern recognition analyzes traffic flow to optimize transportation systems, reducing emissions and promoting sustainable urban development [55]. These applications of pattern recognition demonstrate AI's transformative potential in decoding complex data for the advancement of sustainability objectives.

5.4.3.1 Feature extraction

Feature selection is a critical process in pattern recognition and ML that involves identifying the most relevant variables for use in model construction42. The goal of feature selection is to improve the model's performance by eliminating redundant, overfitting, in which the model operates well on data used for training but badly on unknown data, might result from irrelevant or noisy data. Here is a breakdown of why feature selection is important:

- *Improves model accuracy:* By keeping only the most relevant features, models are less likely to learn from noise and more likely to generalize from training to test data.
- *Reduces overfitting:* Fewer data features reduce the complexity of the model, which can help in minimizing the risk of overfitting.
- *Decreases training time:* Fewer features mean less computation, thereby speeding up the model training process.
- *Enhances interpretability:* A model with fewer variables is often easier to understand and interpret, which is important in many applications such as healthcare or finance where decision clarity is critical.

5.4.3.2 Methods of feature selection

Feature selection techniques can be broadly classified into three categories.

5.4.3.2.1 Filter methods

Statistical measures for filter methods involve assessing the relationship between each feature and the target variable. A commonly used statistical measure is the Pearson correlation coefficient, calculated as follows:

$$\rho_{X,Y} = \frac{\text{cov}(X,Y)}{\sigma_X \sigma_Y} \tag{5.3}$$

where $\text{cov}(X,Y)$ is the covariance between the feature X and the target Y, and σ_X LaGrange σ_Y are the standard deviations of X and Y, respectively.

Features with a correlation coefficient close to +1 or −1 are highly correlated with the target, suggesting their importance for the model. This method is fast and effective for preliminary feature selection but doesn't account for interactions between features.

5.4.3.2.2 Wrapper methods

A common wrapper method is Recursive Feature Elimination (RFE), which involves recursively building models and removing features with the least significance in prediction. RFE can be mathematically represented by its dependency on the model's coefficients, often those derived from linear models:

$$\min_{\beta}\left\{\sum_{i=1}^{n}\left(y_i - x_i^T\beta\right)^2 + \lambda\sum_{j=1}^{p}|\beta_j|\right\} \tag{5.4}$$

where $x_i^T\beta$ is the linear predictor, β represents the coefficients, y_i are the target values, and λ is a tuning parameter that controls the strength of the penalty imposed on the size of the coefficients. Features whose coefficients shrink to zero are eliminated, thus simplifying the model while attempting to maintain its predictive power.

5.4.3.2.3 Embedded methods

An example of an embedded method is the Lasso regression. It modifies the cost function to include a penalty term that enforces sparsity:

$$\min_{\beta}\left\{\frac{1}{2_n}\sum_{i=1}^{n}\left(y_i - x_i^T\beta\right)^2 + \alpha\sum_{j=1}^{p}|\beta_j|\right\} \tag{5.5}$$

Here, α is a turning parameter that controls the level of regularization. A higher α leads to more coefficients being set to zero, effectively selecting fewer features. This method is particularly useful in scenarios where the number of features far exceeds the number of observations.

- *Application in pattern recognition*

In pattern recognition, the mathematical formulations help in identifying the most discriminative features that are crucial for accurate classification or prediction. The efficiency of these methods can dramatically impact the computational efficiency and performance of pattern recognition systems.

5.4.4 Signal processing

Signal processing, empowered by AI and ML, plays a crucial role in enhancing sustainability initiatives. Through the advanced analysis of signal data,

AI and ML facilitate the monitoring and maintenance of various systems critical to sustainable practices. In the renewable energy sector, signal processing techniques are applied to ensure the optimal performance of wind turbines and solar panels, through predictive maintenance that anticipates failures before they occur, significantly reducing downtime and extending equipment lifespan [53].

AI-driven signal processing is also pivotal in water quality monitoring, where it analyzes data from sensors to detect contaminants instantly. This immediate detection allows for swift action to mitigate pollution and ensure the safety of water resources [54]. Additionally, in the context of sustainable urban development, signal processing aids in the real-time monitoring of infrastructure health, identifying potential issues in buildings, bridges, and roads, thereby preventing catastrophic failures and promoting safety [55]. Moreover, signal processing contributes to environmental conservation by detecting and classifying sounds within ecosystems, such as the calls of specific animal species. This application is invaluable for biodiversity monitoring, enabling conservationists to assess the health of ecosystems and the presence of species without intrusive methods [52]. These applications underscore the importance of signal processing in the realm of sustainability, offering precise and efficient tools for environmental protection, resource management, and infrastructure maintenance. By leveraging AI and ML in signal processing, stakeholders can make informed decisions that contribute to the resilience and sustainability of our planet.

5.4.5 Contribution in energy market and forecasting

AI and ML's contribution to the energy market encapsulates a transformative approach toward demand analysis, forecasting, and the optimization of renewable energy generation, which is illustrated in Figure 5.5. These technologies have become indispensable in navigating the complexities of the energy sector, offering both precision and efficiency in managing and predicting energy flows [41]. Demand analysis benefits significantly from AI and ML, which can parse through vast amounts of data to understand consumption patterns and predict future demand with high accuracy. This capability is crucial for energy providers to ensure a balance between supply and demand, thereby enhancing grid stability and reducing the need for excessive energy storage or wasteful excess generation [42].

Forecasting plays a pivotal role in the energy market, where AI and ML algorithms excel in predicting energy production from renewable sources. These technologies analyze weather data, historical energy production levels, and other relevant factors to forecast energy availability, helping simplify the insertion of renewable sources into the electricity system. Such predictive capabilities are essential for planning and can significantly reduce reliance on non-renewable energy sources, contributing to a more sustainable energy future [44]. Furthermore, AI and ML facilitate the optimization

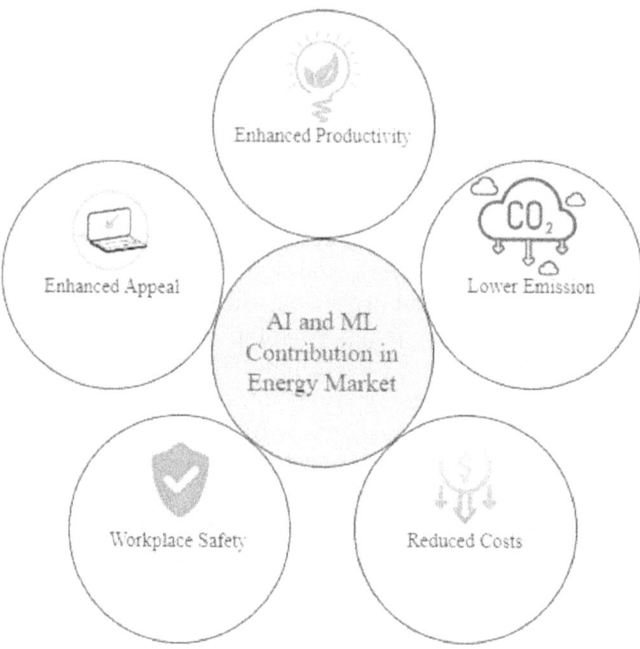

Figure 5.5 AI and ML contribution.

of renewable energy generation by identifying the most efficient ways to capture, convert, and store energy. For instance, deep learning models can optimize the angle of solar panels throughout the day to maximize solar energy capture or determine the optimal configuration for wind farms to increase wind energy generation [52]. This optimization not only improves the efficiency of renewable energy systems but also enhances their cost-effectiveness, making renewable energy a more viable and attractive option for energy production.

In the broader context, AI and ML's impact on the energy market extends to the development of smart grids that can dynamically adjust to changes in energy demand and supply, fostering greater resilience and sustainability in energy infrastructures. By leveraging AI and ML for demand analysis, forecasting, and optimization, the energy sector is better equipped to face the challenges of transitioning to a renewable-based system, ensuring energy security, and promoting environmental sustainability.

5.4.6 Predictive analytics

Predictive analytics is a type of advanced analytics that determines the probability of future events based on previous data using statistical procedures,

ML approaches, and past information. This tool is widely used across various fields to make more informed decisions by forecasting potential trends and behaviors. Here's how predictive analytics is applied in different domains:

5.4.6.1 Business and marketing

In the business sector, predictive analytics is pivotal in understanding and predicting consumer behavior to enhance decision-making regarding marketing strategies and sales initiatives. Companies utilize historical purchase data and customer interaction data to develop models that predict future buying behaviors, likelihood of customer churn, and potential for cross-selling and up-selling. For instance, predictive models might analyze past purchase data and browsing history to identify which customers are likely to buy a new product or upgrade their service package, enabling more targeted and effective marketing campaigns.

5.4.6.2 Finance

Predictive analytics in finance is utilized for credit scoring, risk assessment, and investment forecasting. Banks use predictive models to evaluate the risk of loan default based on a customer's credit history, current financial status, and economic conditions. This assessment helps in categorizing borrowers according to their risk level, thus influencing loan approval decisions and interest rates. Additionally, in the realm of investments, predictive algorithms analyze market trends and historical data to forecast stock performance, aiding investors in making more informed decisions.

5.4.6.3 Healthcare

Predictive analytics is revolutionizing healthcare by enhancing disease management, optimizing resource allocation, and improving patient outcomes. By analyzing historical health data, patient records, and real-time health monitoring, predictive models can forecast an individual's risk of developing chronic diseases such as diabetes or heart disease, allowing for earlier interventions that can mitigate severity and improve quality of life. Furthermore, these models are crucial in predicting patient outcomes post-treatments, helping hospitals plan more effective treatment regimens and manage patient recovery processes efficiently.

Resource optimization is another significant benefit, as predictive analytics helps healthcare facilities anticipate demand spikes – like those during flu seasons or pandemics – enabling better staffing, equipment, and bed management to handle increased patient loads effectively. The field of personalized medicine also benefits from predictive analytics by using genetic data and patient histories to tailor treatments to individual patients, significantly increasing treatment efficacy and reducing adverse effects.

Moreover, predictive analytics supports preventive healthcare measures by identifying at-risk individuals early, thereby enabling targeted preventive interventions that can avert disease development or lessen its impact. Additionally, the technology aids in detecting fraud and abuse within healthcare systems by analyzing billing and claims data to identify unusual patterns that may indicate fraudulent activities, saving substantial resources. Overall, the application of predictive analytics in healthcare not only streamlines operational efficiency but also profoundly impacts patient care delivery and health outcomes, illustrating its indispensable role in modern healthcare systems.

5.4.6.4 Manufacturing

Predictive maintenance in manufacturing uses data from sensors and machine logs to predict equipment failures before they occur. By detecting anomalies and patterns that precede a breakdown, maintenance can be scheduled proactively during nonpeak times, thus reducing operational disruptions and extending the lifespan of machinery [56]. This approach not only reduces maintenance costs but also improves overall manufacturing efficiency.

5.5 SMART INFRASTRUCTURE FOR SUSTAINABLE CITIES

The idea of smart infrastructure within sustainable cities encapsulates the integration of cutting-edge technologies, particularly AI and ML, to foster urban environments that are efficient, resilient, and conducive to the well-being of their citizens. Figure 5.6 illustrates smart infrastructure for smart cities. This approach leverages the power of AI and ML to optimize city operations, enhance the delivery of public services, and ensure sustainable urban development. Figure 5.5 illustrates smart infrastructure for smart cities.

$$Sustainability = \frac{Social\,Impact + Environmental\,Impact}{Economic\,Cost} \tag{5.6}$$

The equation represents the sustainability index as the ratio of the sum of social impact and environmental impact to the economic cost associated with developing and deploying AI/ML solutions.

5.5.1 Smart weather

Smart weather monitoring, utilizing AI and ML, represents a significant advancement in urban planning and environmental management. This technology is pivotal in enhancing the ability of cities to predict, prepare for, and respond to various weather conditions and climate change impacts, thereby

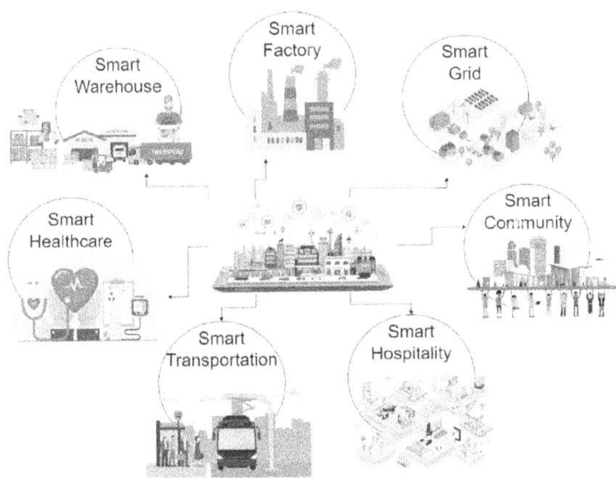

Figure 5.6 Smart infrastructure.

safeguarding infrastructure, ecosystems, and public health. The core of smart weather monitoring rests in the capacity to process and evaluate extensive datasets derived from a multitude of sources, including satellite imagery, weather stations, and ground-based sensors. AI and ML algorithms, particularly deep learning models like CNNs and recurrent neural networks (RNNs), are trained on these datasets. They learn to identify patterns and anomalies in weather systems, enabling accurate forecasts ranging from immediate weather conditions to long-term climate trends [52].

One of the main advantages of smart weather monitoring is its predictive capability, which allows for advanced warning of extreme weather events such as hurricanes, floods, heatwaves, and cold spells. These forecasts provide critical information that can be used by city planners, emergency response teams, and citizens to take preemptive actions, minimizing potential damage and enhancing community resilience. Moreover, smart weather monitoring contributes to more informed decision-making in various sectors affected by weather conditions [57]. For agriculture, it enables precision farming practices by advising on the optimal times for planting, watering, and harvesting based on weather predictions. In the energy sector, it assists in managing renewable energy sources by forecasting solar and wind power generation, thus optimizing the balance between supply and demand. The integration of AI and ML in weather monitoring also supports efforts to understand and mitigate the impacts of climate change. By analyzing historical and current climate data, these technologies can model future climate scenarios, providing invaluable insights for developing effective climate adaptation and mitigation strategies [58]. The effectiveness of smart weather monitoring in enhancing urban resilience and sustainability underscores the importance of continued investment in AI and ML technologies. As these tools become more sophisticated and widely adopted, cities will be better equipped to face the challenges posed by changing weather patterns and a warming planet.

5.5.2 Smart healthcare

Smart healthcare, utilizing the advancements in AI and ML, is converting the medical industry by offering more personalized, efficient, and accessible healthcare solutions. Through the integration of AI algorithms and ML models, smart healthcare systems can analyze vast datasets, enabling the prediction of health outcomes, optimization of treatment plans, and early detection of diseases [15]. This approach not only enhances patient care but also improves the efficiency of healthcare providers by automating routine tasks, managing patient data, and supporting decision-making processes.

One of the pivotal contributions of smart healthcare is in diagnostics, where AI-powered tools can interpret medical images with remarkable accuracy, aiding in the early detection of conditions such as cancer, heart disease, and diabetic retinopathy. Additionally, AI-driven predictive analytics play a crucial role in preventive medicine, identifying at-risk individuals and facilitating early interventions to prevent diseases from developing or progressing. Furthermore, smart healthcare extends its benefits to remote patient monitoring and telemedicine, offering continuous care and medical consultation to patients, regardless of their location [59]. This is particularly valuable in underserved or rural areas, where access to healthcare services is limited [15]. By leveraging AI and ML, smart healthcare promises to revolutionize the medical field, making healthcare more personalized, efficient, and accessible, and significantly contributing to the overall well-being of the global population.

5.5.3 Smart transportation

Smart transportation, powered by AI and ML, is revolutionizing urban mobility by optimizing traffic flow, reducing emissions, and enhancing the efficiency of public transit systems. This innovative approach to transportation leverages AI and ML algorithms to analyze vast amounts of data from traffic cameras, sensors, and GPS devices, enabling cities to manage their transportation networks more effectively. One of the key applications of smart transportation is in traffic management, where AI algorithms predict traffic congestion and dynamically adjust signal timings to improve traffic flow. This not only reduces travel times but also lowers vehicle emissions, contributing to cleaner air quality [44]. Furthermore, ML models are used to analyze patterns of public transit usage, allowing transit authorities to optimize routes and schedules to meet passenger demand more accurately, thereby enhancing the public transit experience and encouraging its use over private vehicles.

Smart transportation also includes the deployment of autonomous vehicles, which rely on AI for navigation and decision-making. These vehicles have the potential to transform urban mobility by reducing the need for parking spaces, decreasing traffic congestion, and providing flexible transportation options for all citizens, including those unable to drive. Additionally, AI and ML contribute to the development of intelligent transportation systems that support vehicle-to-vehicle (V2V) and vehicle-to-infrastructure (V2I) communication [60]. This technology improves road safety by alerting drivers to potential hazards and optimizing routing in real time based on current traffic conditions [44]. The integration of AI and ML in smart transportation systems demonstrates a commitment to advancing sustainable urban mobility. By leveraging these technologies, cities can achieve significant improvements in traffic management, public transit efficiency, and overall urban sustainability, paving the way for smarter, more livable urban environments.

5.5.4 Smart grid

Smart grids, empowered by AI and ML, are revolutionizing the energy sector by enhancing the efficiency and reliability of electricity distribution. These advanced grids utilize AI and ML to predict energy demand, optimize energy flow, and integrate renewable energy sources seamlessly into the power grid [32]. By analyzing data from sensors and meters in real time, smart grids can dynamically adjust to changing energy needs, reducing waste and improving grid stability. One of the key features of smart grids is their ability to manage distributed energy resources, including solar panels and wind turbines, facilitating a more sustainable energy ecosystem. Through the use of reinforcement learning algorithms, smart grids can make informed decisions on how best to distribute and store renewable energy, ensuring a consistent and reliable energy supply [32].

Moreover, smart grids contribute to reducing carbon emissions by optimizing the use of renewable energy sources and encouraging the adoption of electric vehicles by managing charging stations efficiently. They also enhance consumer engagement by providing users with real-time information on their energy consumption, promoting energy-saving behaviors [61]. The implementation of smart grids marks a significant step toward achieving a sustainable and resilient energy future, demonstrating the critical role of AI and ML in advancing smart energy solutions and contributing to the goals of sustainable development [52].

5.5.5 Water management system

The integration of AI and ML into water management systems represents a significant leap toward sustainable and efficient water resource management. These technologies offer innovative solutions for water quality monitoring, distribution efficiency, and wastewater treatment, solving some of the most serious concerns in ensuring access to clean water and maintaining healthy ecosystems.

5.5.5.1 Water quality monitoring

AI and ML algorithms play a crucial role in monitoring water quality in real time. By analyzing data from sensors deployed in rivers, lakes, and reservoirs, these systems can detect contaminants, predict pollution levels, and identify sources of pollution with high accuracy. This immediate insight allows for swift action to prevent waterborne diseases and protect aquatic life, ensuring the safety and cleanliness of water bodies [41].

5.5.5.2 Distribution efficiency

In urban water distribution, AI and ML contribute to identifying leaks and inefficiencies within the network. Through predictive maintenance, AI can forecast potential breakdowns and pinpoint areas that require repairs, significantly reducing water loss and enhancing the overall efficiency of water distribution systems. Moreover, AI-driven optimization models ensure that water supply meets demand across different sectors and residential areas, optimizing pump schedules and valve operations to minimize energy consumption and operational costs [44].

5.5.5.3 Wastewater treatment

The application of AI and ML in wastewater treatment plants enhances the treatment process, enabling the removal of pollutants more effectively and ensuring that the water released back into the environment meets safety standards. ML algorithms optimize the operations of treatment facilities, adjusting processes in response to incoming water quality and flow rates, thus improving the efficiency and cost-effectiveness of wastewater treatment [52].

5.5.5.4 Smart irrigation

In agriculture, AI-driven smart irrigation systems use data from soil moisture sensors, weather forecasts, and crop types to optimize watering schedules. This precision irrigation approach ensures that crops receive the exact amount of water they need, reducing water wastage and supporting sustainable agricultural practices [53].

The execution of AI and ML in water operation systems not only enhances the sustainability and efficiency of water use but also supports global efforts to achieve water security and resilience in the face of climate change. By leveraging these advanced technologies, stakeholders can make informed decisions that promote the conservation and responsible use of water resources.

5.5.6 Smart farming

Smart farming through AI and ML is revolutionizing agricultural practices, making them more efficient, productive, and sustainable. By applying data-driven insights, smart farming enhances decision-making processes, from precision agriculture techniques to crop yield optimization [62]. AI and ML enable precise monitoring and management of crop health, utilizing drones and satellite imagery for detailed field analysis. These technologies assess vegetation indices, detect plant diseases early, and monitor crop growth, allowing for timely interventions to ensure optimal crop health and yield [19].

In precision agriculture, AI-driven systems analyze data from soil sensors, weather information, and historical crop performance to make informed decisions about planting, watering, and harvesting. This tailored approach ensures that resources are used efficiently, reducing waste and environmental impact while maximizing agricultural output [19]. AI and ML also play a crucial role in predictive analytics for agriculture. By forecasting weather conditions, pest invasions, and crop diseases, farmers can proactively take measures to protect their crops, thereby reducing losses and enhancing productivity [18].

Water management in agriculture benefits significantly from AI, with intelligent irrigation systems optimizing water use based on soil moisture content, crop type, and weather predictions. This not only conserves water but also ensures that crops receive adequate hydration, promoting healthy growth and sustainability [54]. Furthermore, AI facilitates the automation of agricultural machinery, from tractors to drones, streamlining operations such as planting, fertilizing, and harvesting. This automation reduces labor costs and enhances precision in agricultural practices [41]. The integration of AI and ML in farming practices represents a significant advancement toward achieving sustainable agriculture. By leveraging these technologies, smart farming aims to address the global challenge of increasing food production while minimizing environmental impacts and ensuring food security for future generations.

5.6 STRATEGIES FOR SUSTAINABLE DEVELOPMENT

Achieving sustainable development goals (SDGs) requires innovative and strategic approaches that leverage the capabilities of AI and ML. These technologies offer unparalleled opportunities for integration across sectors, enhancing stakeholder engagement, and informing policy formulation, thus driving progress toward sustainability objectives.

$$SD = F\left(E,S,P,T,C\right) \tag{5.7}$$

Equation (5.7) represents the level of sustainable development (SD) achieved as a function (f) of environmental considerations (E), social factors (S), economic factors (P), technological factors (T), and cultural factors (C).

5.6.1 Integration across sectors

AI and ML facilitate the cross-sectoral integration necessary for sustainable development by enabling the analysis of large datasets from diverse fields such as agriculture, energy, healthcare, and urban planning. For instance, AI can optimize renewable energy distribution in the power sector while ensuring water conservation in agriculture, demonstrating the interconnectedness of SDGs related to clean energy, sustainable cities, and responsible consumption [7]. By harnessing data-driven insights, stakeholders can identify synergies and trade-offs among goals, ensuring a holistic approach to sustainability.

5.6.2 Ethical considerations and responsible AI

As AI and ML are increasingly deployed to address sustainability challenges, it is crucial to ensure these technologies are developed and used ethically and responsibly. Addressing concerns related to privacy, bias, and transparency is essential for building trust and ensuring that AI contributes positively to sustainable development [11]. Policymakers must establish regulatory frameworks that promote ethical AI practices while encouraging innovation and collaboration across sectors.

5.6.3 Stakeholder engagement

Engaging a broad spectrum of stakeholders is critical for the successful implementation of SDGs. AI and ML can enhance this engagement through platforms that facilitate dialogue, collaboration, and the sharing of knowledge and resources. For example, AI-driven tools can match investors with sustainable projects or enable citizen scientists to contribute to environmental monitoring efforts [6]. These technologies empower communities, businesses, and governments to participate actively in sustainable development processes, fostering a sense of ownership and commitment to achieving SDGs.

5.6.4 Policy formulation

AI and ML provide policymakers with robust analytical tools to assess the impact of various policy options, predict future trends, and monitor the implementation of SDGs. By analyzing data on economic activities, social dynamics, and environmental changes, AI models can offer evidence-based recommendations for effective policy interventions [8]. Furthermore, AI can simulate the outcomes of policy decisions in virtual environments, allowing policymakers to explore different scenarios and their implications for sustainability objectives.

In conclusion, strategic approaches leveraging AI and ML have the potential to significantly accelerate progress toward achieving SDGs. By fostering integration across sectors, enhancing stakeholder engagement, and informing policy formulation with data-driven insights, AI and ML can help navigate the complexities of sustainable development. However, the success of these technologies in contributing to SDGs also depends on addressing ethical considerations and ensuring responsible AI use.

5.7 CHALLENGES AND ETHICAL CONSIDERATIONS IN AI AND ML

The integration of AI and ML into sustainable development initiatives offers immense potential for progress, yet it is accompanied by formidable challenges and ethical dilemmas, as depicted in Figure 5.7. While these technologies hold promise in optimizing resource management, enhancing agricultural practices, and mitigating environmental risks, they also raise concerns regarding data privacy, algorithmic bias, and socio-economic disparities.

As AI and ML increasingly underpin endeavors toward attaining the SDGs, it becomes imperative to navigate these complexities with vigilance and foresight. Addressing these issues demands a multifaceted approach encompassing robust regulatory frameworks, transparent algorithms, and inclusive stakeholder engagement. Furthermore, efforts must be directed toward fostering digital literacy and ensuring equitable access to AI-driven solutions, particularly in marginalized communities.

By taking action to address these difficulties and moral issues, we can harness the transformative potential of AI and ML to drive sustainable development responsibly and inclusively, advancing toward a more equitable and resilient future.

5.7.1 Challenges in deployment

Deploying AI and ML in sustainable development presents challenges, including data privacy, algorithmic bias, and the environmental impact of AI systems. Ensuring these technologies are ethically and sustainably implemented requires careful consideration of their long-term effects on society and the environment, demanding transparency, fairness, and inclusivity in their application.

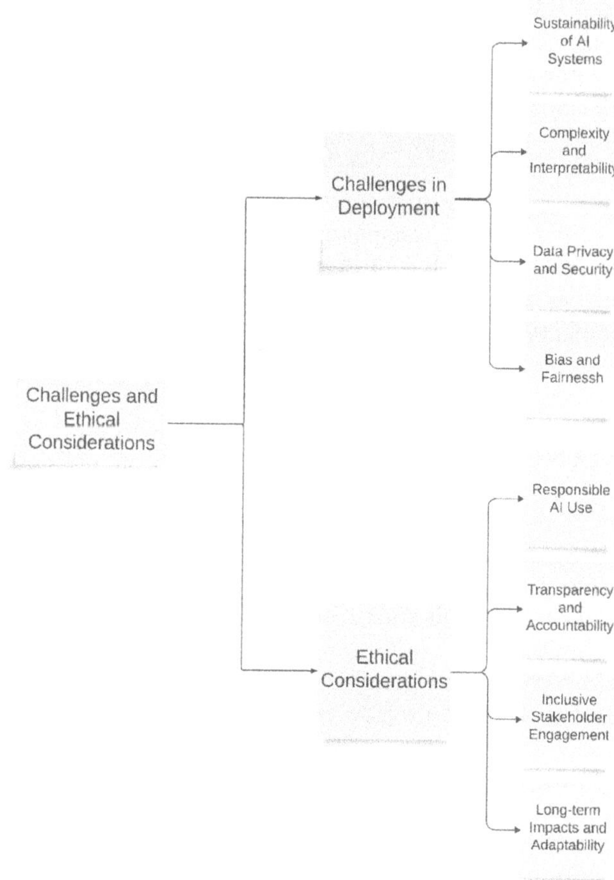

Figure 5.7 AI and ML: ethical maze.

5.7.1.1 Data privacy and security

AI and ML systems' reliance on extensive datasets raises significant concerns regarding data privacy and security. The challenge lies in safeguarding sensitive information while harnessing big data's power for sustainability initiatives. Protecting this data against unauthorized access and ensuring its ethical use is paramount, necessitating robust security measures and ethical guidelines to prevent misuse and violations of privacy [11].

5.7.1.2 Bias and fairness

The risk of AI algorithms perpetuating or exacerbating existing biases is a critical concern. Such biases in training data can lead to unfair outcomes, especially in critical areas like resource allocation, healthcare, and service access. Addressing these biases is essential for fostering equitable sustainable development and ensuring that AI-driven solutions do not reinforce societal inequalities but rather contribute to fair and just outcomes [12].

5.7.1.3 Complexity and interpretability

The "black box" nature of many AI models, particularly those based on deep learning, poses significant challenges in terms of complexity and interpretability. The difficulty in understanding how these models make decisions undermines stakeholder trust and complicates the evaluation of their impact on sustainability. Enhancing model transparency and interpretability is crucial for building trust and ensuring the responsible use of AI in sustainable development [8].

5.7.1.4 Sustainability of AI systems

The sustainability of AI systems themselves is a growing concern, with the environmental impact of training and operating complex models coming under scrutiny. The substantial energy consumption required for these processes can counteract the sustainability benefits they aim to support. Developing more energy-efficient AI technologies and considering the environmental costs of AI applications is critical for aligning AI with sustainability goals [13].

5.7.2 Ethical considerations

Ethical considerations in AI and ML for sustainable development encompass ensuring fairness, transparency, and accountability. It's crucial to address biases in algorithms, protect data privacy, and ensure inclusivity in benefits. Ethically deploying AI demands rigorous standards to prevent harm and ensure technology serves the broader goals of societal well-being and environmental stewardship.

5.7.2.1 Responsible AI use

Ethical AI use involves developing and deploying AI solutions that are not only effective but also equitable and just. This includes ensuring that AI applications in sustainable development do not exacerbate inequalities or harm vulnerable populations [11].

5.7.2.2 Transparency and accountability

Establishing transparency in AI processes and accountability for AI-driven decisions is essential. Stakeholders should be able to understand how AI solutions contribute to sustainable development goals and who is accountable for the outcomes of these interventions [8].

5.7.2.3 Inclusive stakeholder engagement

Engaging a broad range of stakeholders in the development and deployment of AI systems ensures that diverse perspectives are considered, and solutions are inclusive. This is crucial for aligning AI initiatives with the needs and priorities of all sectors of society, particularly those most impacted by sustainability challenges [7].

5.7.2.4 Long-term impacts and adaptability

That is vital to address the long-term consequences of AI and ML on sustainability, including potential unintended consequences. Systems should be designed to be adaptable and responsive to changing environmental conditions and societal needs [6].

Addressing these challenges and ethical considerations requires a multifaceted technique, including robust regulatory frameworks, ongoing stakeholder engagement, and the development of sustainable and responsible AI technologies. By navigating these complexities thoughtfully, AI and ML can play a pivotal role in advancing sustainable development globally.

5.8 CASE STUDY ON FAULT CLASSIFICATION IN INDUCTION MOTORS USING DEEP LEARNING AND THERMAL IMAGING

5.8.1 Challenges in deployment

5.8.1.1 Context and relevance

Induction motors are critical components in various industrial applications due to their robustness, reliability, and efficiency. However, their continuous and optimal operation is often challenged by the occurrence of various faults, which can lead to significant downtime, reduced efficiency, and increased maintenance costs [63]. Traditional fault detection methods, while useful, often lack the precision and timeliness required for modern industrial environments.

The advent of advanced technologies such as thermal imaging and deep learning has opened new avenues for more effective fault detection and classification. Thermal imaging provides a noninvasive and accurate means of monitoring the thermal characteristics of induction motors, which are indicative of their operating conditions [64]. When combined with deep learning techniques, which excel in pattern recognition and anomaly detection, it becomes possible to detect and classify faults with high accuracy and reliability [65].

This case study examines the application of deep learning-based fault classification using thermal imaging in induction motors, emphasizing its contributions to sustainable development. By leveraging these advanced technologies, the proposed method not only enhances fault detection capabilities but also promotes energy efficiency, resource optimization, and environmental sustainability. This aligns with the broader objectives of integrating AI and ML into sustainable industrial practices, ultimately contributing to the achievement of global sustainability goals.

5.8.1.2 Objectives of the case study

The primary objectives of this case study are as follows:

- *Objective 1: Enhance fault detection accuracy*

To develop and implement a deep learning model that leverages thermal imaging data to accurately detect and classify faults in induction motors. This objective aims to improve the precision and reliability of fault detection compared to traditional methods.

- *Objective 2: Promote energy efficiency and resource optimization*

To demonstrate how early and accurate fault detection can lead to more efficient operation of induction motors, thereby reducing energy consumption and optimizing the use of resources. This objective aligns with the principles of sustainable development by minimizing waste and extending the lifespan of industrial equipment.

- *Objective 3: Contribute to environmental sustainability*

To assess the environmental impact of implementing deep learning-based fault classification in induction motors, focusing on the reduction of emissions and waste. This objective seeks to highlight the role of advanced AI and ML technologies in achieving sustainable industrial practices and contributing to global sustainability goals.

5.8.2 Methodology

5.8.2.1 Proposed technique

Figure 5.8 depicts the process of fault classification in induction motors utilizing thermal imaging and deep learning techniques, encompassing several stages from initial image acquisition to final fault classification. The

Figure 5.8 Proposed methodology.

process begins with the capture of thermal images of the induction motor, which reflect the temperature distribution indicative of the motor's operating condition.

The next stage is preprocessing, where the captured thermal images undergo several transformations. Initially, these images are converted to grayscale to simplify the data while retaining essential thermal information. Subsequently, the images are resized to a standard dimension suitable for input into the deep learning model. Each image is then labeled with the corresponding fault type or healthy condition, an essential step for supervised learning. Additionally, data augmentation techniques such as rotation, flipping, and scaling are applied to artificially expand the dataset, thereby improving the model's robustness and generalization capabilities. Following preprocessing, the images are divided into three sets: the training set, validation set, and testing set. The training set is used to train the deep learning model, the validation set is employed for tuning hyperparameters and preventing overfitting, and the testing set is reserved for evaluating the final model's performance.

Hyperparameter tuning involves adjusting the model's parameters based on the validation set to optimize performance, experimenting with different configurations to find the best settings. Once tuned, the deep learning model is trained using the training set, where it learns to recognize patterns associated with different fault types and healthy conditions from the thermal images.

After training, the model is capable of classifying faults in new, unseen thermal images. The faults are categorized into various types, including 1-phase 10\% stator fault, 1-phase 50\% stator fault, 2-phase 10\% stator fault, 2-phase 50\% stator fault, 3-phase 10\% stator fault, rotor fault, bearing fault, and cooling fan fault. Additionally, the model can also identify healthy motor conditions.

In conclusion, the flowchart outlines a comprehensive workflow for fault classification in induction motors using thermal imaging and deep learning. This approach leverages advanced image processing and ML techniques to accurately detect and classify various faults, thereby enhancing maintenance practices and promoting operational efficiency in industrial settings. By ensuring early fault detection and accurate classification, this method supports sustainable development goals by improving energy efficiency, reducing downtime, and minimizing resource consumption.

5.8.2.2 Deep learning model

Figure 5.9 illustrates a deep learning model for classifying faults in induction motors using thermal imaging. The model processes input thermal images through several stages. It starts with a *Convolution* layer for feature extraction, followed by three *Dense Blocks* for enhanced information flow and feature extraction. Interleaved are *Convolutional Block Attention Modules* (CBAM) to focus on important features. Each dense block is followed by additional *Convolution* and *Pooling* layers to refine features and reduce spatial dimensions. A final *Pooling* layer down-samples the data before passing it to a *Linear* layer, which classifies the fault type. The model predicts specific faults, such as the depicted *2-Phase 50\% Stator Fault*. This architecture, combining convolutional layers, dense blocks, attention mechanisms, and pooling, enables accurate fault classification in induction motors, supporting efficient maintenance and sustainability in industrial operations.

5.8.3 Experimental setup and results

5.8.3.1 Dataset description

The dataset used for fault classification in induction motors consists of 250 thermal images, categorized into eight distinct classes. These classes represent various fault types and conditions of the induction motor, including Rotor Fault, 1-Phase 10\% Stator Fault, 2-Phase 10\% Stator Fault, 3-Phase 10\% Stator Fault, Healthy, 1-Phase 50\% Stator Fault, 2-Phase 50\% Stator Fault, and Cooling Fan Failure. Each class contains approximately 31 images, ensuring a balanced dataset that provides a robust basis for training and evaluating the deep learning model. The Rotor Fault class

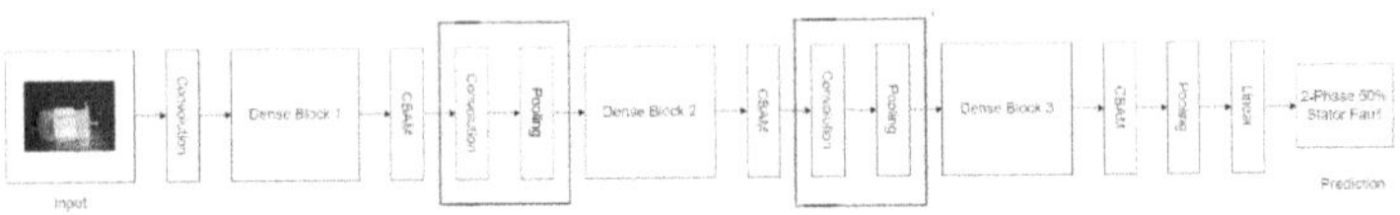

Figure 5.9 The architecture of deep learning model.

includes images depicting faults related to the rotor, while the Healthy class consists of images of properly functioning motors without any faults. The various stator fault classes (1-Phase 10\%, 2-Phase 10\%, 3-Phase 10\%, 1-Phase 50\%, and 2-Phase 50\%) include images showing specific percentages of faults in one, two, or three phases of the stator. Additionally, the Cooling Fan Failure class comprises images depicting failures in the cooling fan of the induction motor. This diverse and well-balanced dataset ensures that the model is exposed to a wide range of examples for each fault type, promoting accurate and reliable fault classification.

5.8.3.2 Performance evaluation of deep learning model for fault classification in induction motors

The performance of the deep learning model for fault classification in induction motors is evaluated using two primary metrics: accuracy and loss, for both training and validation phases.

Training and validation accuracy: Accuracy measures the proportion of correctly classified images out of the total number of images. It provides an indication of how well the model is learning the relationships between the input thermal images and the corresponding fault labels. Training accuracy refers to the accuracy achieved on the training dataset, while validation accuracy is computed on the validation dataset. High training and validation accuracy values indicate that the model is performing well in learning and generalizing from the data, respectively.

Training and validation loss: Loss quantifies the difference between the predicted and actual fault labels. It serves as a measure of the model's prediction error. Training loss is calculated on the training dataset, and validation loss is calculated on the validation dataset. Lower values of training and validation loss indicate better model performance. Monitoring both training and validation loss helps in identifying issues such as overfitting, where the model performs well on training data but poorly on validation data.

Throughout the training process, both accuracy and loss are tracked for each epoch, providing insights into the model's learning dynamics and guiding the tuning of hyperparameters to achieve optimal performance.

5.8.3.3 Results

The performance of the deep learning model for fault classification in induction motors was evaluated using two primary metrics: accuracy and loss, for both training and validation phases. Figure 5.10 illustrates the progression of these metrics over 50 epochs.

The *Training and Validation Accuracy* graph (Figure 5.10, left) shows the accuracy values for both the training and validation datasets over the

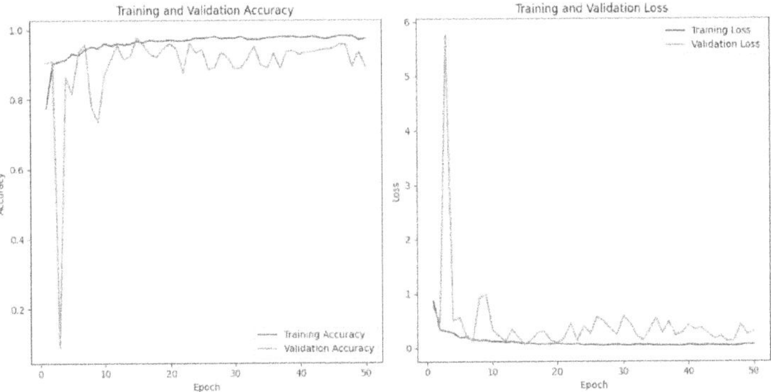

Figure 5.10 Training and validation result.

epochs. The training accuracy, represented by the blue line, starts low but quickly increases, stabilizing above 90\% as the epochs progress. The validation accuracy, shown by the red line, exhibits more variability but generally follows an upward trend, reaching values around 90\% by the end of the training process. This indicates that the model is learning effectively and generalizing well to unseen data.

The *Training and Validation Loss* graph (Figure 5.10, right) presents the loss values for the training and validation datasets. The training loss, depicted by the blue line, decreases steadily, indicating that the model is minimizing the error on the training data. The validation loss, represented by the red line, shows a significant initial spike but then decreases, although with more fluctuations compared to the training loss. The eventual decline and stabilization of validation loss suggest that the model is improving its performance on the validation dataset, albeit with some variability likely due to the limited size and complexity of the data.

In summary, the training and validation accuracy and loss metrics demonstrate that the model is effectively learning to classify faults in induction motors, with high accuracy and reducing loss over time. The fluctuations in validation metrics highlight the importance of continuous monitoring and the potential for further tuning to achieve optimal performance.

5.8.3.4 Sustainability impact

The implementation of deep learning for fault classification in induction motors using thermal imaging has significant sustainability impacts. By accurately detecting and classifying faults early, the approach ensures the optimal operation of induction motors, which are critical components in many industrial applications. This early detection leads to several key sustainability benefits:

Energy efficiency: Early fault detection allows for timely maintenance and repairs, ensuring that motors operate at peak efficiency. This reduces unnecessary energy consumption associated with faulty or inefficiently running motors, thereby lowering overall energy usage and contributing to energy conservation.

Resource optimization: By preventing severe damage through early fault detection, the lifespan of induction motors is extended. This reduces the frequency of replacements and the demand for new motors, conserving raw materials and reducing manufacturing and disposal-related environmental impacts.

Reduction of greenhouse gas emissions: Efficiently operating motors consume less energy, which translates to reduced greenhouse gas emissions, especially in industries relying on fossil fuels for electricity. This contributes to lower carbon footprints and aligns with global efforts to combat climate change.

Minimized downtime and improved productivity: Predictive maintenance enabled by accurate fault classification minimizes unplanned downtime. This enhances productivity and ensures continuous industrial operations, reducing the environmental impact of production halts and restarts.

Waste reduction: Early detection of faults prevents catastrophic failures that often result in significant waste generation. By maintaining motors in good working condition, the approach minimizes the generation of scrap materials and hazardous waste associated with motor breakdowns.

In conclusion, the use of deep learning for fault classification in induction motors promotes sustainable industrial practices by enhancing energy efficiency, optimizing resource use, reducing greenhouse gas emissions, minimizing downtime, and reducing waste. This aligns with the principles of sustainable development, contributing to the long-term goal of achieving a more sustainable and efficient industrial sector.

5.9 CONCLUSION

In conclusion, the exploration of AI and ML within the context of sustainable development reveals landscape rich with potential and fraught with challenges. The integration of these technologies across various sectors – from renewable energy to smart infrastructure and water management – underscores their transformative power in advancing the SDGs. AI and ML not only enhance efficiency and innovation but also offer novel solutions to some of the most pressing environmental and societal challenges of our time. The strategic approaches leveraging AI and ML, including cross-sector integration, stakeholder engagement, and policy formulation, are pivotal for harnessing these technologies toward sustainable ends. They emphasize the importance of a collaborative and interdisciplinary framework, where technology serves as a bridge between diverse fields of knowledge, contributing to a holistic vision of sustainability.

However, the deployment of AI and ML solutions is not without its ethical dilemmas and practical challenges. Issues of data privacy, algorithmic bias, interpretability, and the sustainability of AI systems themselves necessitate a cautious and responsible approach. The ethical considerations surrounding AI and ML compel us to prioritize transparency, fairness, and inclusivity, ensuring that these technologies benefit all segments of society without exacerbating existing disparities.

In this case study, we explored the application of deep learning for fault classification in induction motors using thermal imaging. The comprehensive workflow, from thermal image acquisition to fault classification, demonstrated the effectiveness of leveraging advanced image processing and ML techniques to enhance the reliability and efficiency of industrial motor operations. The sustainability impact of this approach is significant. By enabling early and accurate fault detection, the model promotes energy efficiency, optimizes resource utilization, reduces greenhouse gas emissions, minimizes downtime, and reduces waste. These benefits align with the principles of sustainable development, contributing to more sustainable and efficient industrial practices. The integration of deep learning and thermal imaging for fault classification in induction motors not only enhances maintenance practices but also supports broader sustainability goals. Future research can further refine this approach, exploring additional fault types and improving model robustness to ensure even greater industrial applicability and sustainability impact.

As we stand on the brink of a technological revolution with the potential to reshape our approach to sustainable development, it is imperative to navigate this terrain with an ethical compass and a commitment to long-term well-being. The path forward requires not only technological innovation but also a profound consideration of the values we wish to uphold in our pursuit of a sustainable future. The journey toward sustainable development, powered by AI and ML, is one of immense promise and profound responsibility. By embracing these technologies while remaining vigilant to their challenges, we can unlock unprecedented opportunities for progress toward a more sustainable, equitable, and resilient world.

REFERENCES

1. Kondylatos, S., Prapas, I., Ronco, M., Papoutsis, I., Camps-Valls, G., Piles, M., ... & Carvalhais, N. (2022). Wildfire danger prediction and understanding with deep learning. *Geophysical Research Letters*, 49(17), e2022GL099368.
2. DE, R. (1986). Learning representations by back-propagation errors. *Nature*, *323*, 533–536.
3. Heaton, J. (2018). Ian Goodfellow, Yoshua Bengio, and Aaron Courville: Deep learning: The MIT Press, 2016, 800 pp, ISBN: 0262035618. *Genetic Programming and Evolvable Machines*, 19(1–2), 305–307.

4. Rolnick, D., Donti, P. L., Kaack, L. H., Kochanski, K., Lacoste, A., Sankaran, K., ... & Bengio, Y. (2022). Tackling climate change with machine learning. *ACM Computing Surveys (CSUR), 55*(2), 1–96.

5. Vinuesa, R., Azizpour, H., Leite, I., Balaam, M., Dignum, V., Domisch, S., ... & Fuso Nerini, F. (2020). The role of artificial intelligence in achieving the sustainable development goals. *Nature Communications, 11*(1), 1–10.

6. Arrieta, A. B., Díaz-Rodríguez, N., Del Ser, J., Bennetot, A., Tabik, S., Barbado, A., ... & Herrera, F. (2020). Explainable Artificial Intelligence (XAI): Concepts, taxonomies, opportunities, and challenges toward responsible AI. *Information Fusion, 58*, 82–115.

7. Stahl, B. C., & Wright, D. (2018). Ethics and privacy in AI and big data: Implementing responsible research and innovation. *IEEE Security & Privacy, 16*(3), 26–33.

8. Benjamin, R. (2023). Race after technology. In *Social Theory Re-Wired* (pp. 405–415). Routledge. Taylor and Francis.

9. Strubell, E., Ganesh, A., & McCallum, A. (2019). Energy and policy considerations for deep learning in NLP. *arXiv preprint arXiv:1906.02243.*

10. Ford, M. (2021). *Rule of the ROBOTS: HOW Artificial Intelligence will Transform Everything.* Hachette UK. ISBN: 1529346010, 9781529346015.

11. Hashimoto, D. A., Rosman, G., Rus, D., & Meireles, O. R. (2018). Artificial intelligence in surgery: Promises and perils. *Annals of Surgery, 268*(1), 70.

12. Davenport, T. H., & Ronanki, R. (2018). Artificial intelligence for the real world. *Harvard Business Review, 96*(1), 108–116.

13. Dallaqua, F. B., Faria, F. A., & Fazenda, A. L. (2018, October). Active learning approaches for deforested area classification. In *2018 31st SIBGRAPI Conference on Graphics, Patterns and Images (SIBGRAPI)* (pp. 48–55). IEEE.

14. Khan, P. W., Byun, Y. C., Lee, S. J., & Park, N. (2020). Machine learning based hybrid system for imputation and efficient energy demand forecasting. *Energies, 13*(11), 2681.

15. Sishodia, R. P., Ray, R. L., & Singh, S. K. (2020). Applications of remote sensing in precision agriculture: A review. *Remote Sensing, 12*(19), 3136.

16. Srinivas, T., Aditya Sai, G., & Mahalaxmi, R. (2022). A comprehensive survey of techniques, applications, and challenges in deep learning: A revolution in machine learning. *International Journal of Mechanical Engineering, 7*(5), 286–296.

17. Singh, G. (2023). Environmental monitoring with machine learning. *EPRA International Journal of Multidisciplinary Research (IJMR), 9*(5), 208–212.

18. Drogkoula, M., Kokkinos, K., & Samaras, N. (2023). A comprehensive survey of machine learning methodologies with emphasis in water resources management. *Applied Sciences, 13*(22), 12147.

19. Atitallah, S. B., Driss, M., Boulila, W., & Ghézala, H. B. (2020). Leveraging deep learning and IoT big data analytics to support the smart cities development: Review and future directions. *Computer Science Review, 38*, 100303.

20. Xie, J., Girshick, R., & Farhadi, A. (2016). Unsupervised deep embedding for clustering analysis. In *International Conference on Machine Learning* (pp. 478–487). PMLR.

21. Sutton, R. S., & Barto, A. G. (2018). *Reinforcement Learning: An Introduction.* MIT Press, Cambridge, Massachusetts, London, England.

22. Sarker, I. H. (2021). Machine learning: Algorithms, real-world applications and research directions. *SN Computer Science, 2*(3), 160.

23. Vamvakas, D., Michailidis, P., Korkas, C., & Kosmatopoulos, E. (2023). Review and evaluation of reinforcement learning frameworks on smart grid applications. *Energies, 16*(14), 5326.

24. Ghobadi, F., & Kang, D. (2023). Application of machine learning in water resources management: A systematic literature review. *Water, 15*(4), 620.

25. Chapman, M., Xu, L., Lapeyrolerie, M., & Boettiger, C. (2023). Bridging adaptive management and reinforcement learning for more robust decisions. *Philosophical Transactions of the Royal Society B, 378*(1881), 20220195.

26. Koko, A. F., Yue, W., Abubakar, G. A., Hamed, R., & Alabsi, A. A. N. (2020). Monitoring and predicting spatio-temporal land use/land cover changes in Zaria City, Nigeria, through an integrated cellular automata and Markov chain model (CA-Markov). *Sustainability, 12*(24), 10452.

27. McCarthy, J., Minsky, M. L., Rochester, N., & Shannon, C. E. (2006). A proposal for the dartmouth summer research project on artificial intelligence, August 31, 1955. *AI Magazine, 27*(4), 12–12.

28. Kondylatos, S., Prapas, I., Ronco, M., Papoutsis, I., Camps-Valls, G., Piles, M.,... & Carvalhais, N. (2022). Wildfire danger prediction and understanding with deep learning. *Geophysical Research Letters, 49*(17), e2022GL099368.

29. Ifaei, P., Karbassi, A., Jacome, G., & Yoo, C. (2017). A systematic approach of bottom-up assessment methodology for an optimal design of hybrid solar/wind energy resources–Case study at middle east region. *Energy Conversion and Management, 145*, 138–157.

30. Rylott, E. L., & Bruce, N. C. (2022). Create digital content connect content with people empower marketing efforts. *Science, 1360*(23), 04.

31. Fan, Z., Yan, Z., & Wen, S. (2023). Deep learning and artificial intelligence in sustainability: A review of SDGs, renewable energy, and environmental health. *Sustainability, 15*(18), 13493.

32. Zhou, Y., Zhang, B., Xu, C., Lan, T., Diao, R., Shi, D.,... & Lee, W. J. (2020). A data-driven method for fast AC optimal power flow solutions via deep reinforcement learning. *Journal of Modern Power Systems and Clean Energy, 8*(6), 1128–1139.

33. Zhang, Z., Zhang, D., & Qiu, R. C. (2019). Deep reinforcement learning for power system applications: An overview. *CSEE Journal of Power and Energy Systems, 6*(1), 213–225.

34. Rządkowska, A. (2021). Artificial intelligence assisted smart photovoltaics. In Proceedings-ISES Solar World Congress 2021 (pp. 362–375).

35. Franki, V., Majnarić, D., & Višković, A. (2023). A Comprehensive Review of Artificial Intelligence (AI) Companies in the Power Sector. *Energies, 16*(3), 1077.

36. Morstyn, T., & McCulloch, M. D. (2018). Multiclass energy management for peer-to-peer energy trading driven by prosumer preferences. *IEEE Transactions on Power Systems, 34*(5), 4005-4014.

37. Zhao, N., Zhang, H., Yang, X., Yan, J., & You, F. (2023). Emerging information and communication technologies for smart energy systems and renewable transition. *Advances in Applied Energy, 9*, 100125.

38. Taiebat, M., Brown, A. L., Safford, H. R., Qu, S., & Xu, M. (2018). A review on energy, environmental, and sustainability implications of connected and automated vehicles. *Environmental Science & Technology, 52*(20), 11449–11465.

39. He, T., & Li, C. (2020). Harness the power of genomic selection and the potential of germplasm in crop breeding for global food security in the era with rapid climate change. *The Crop Journal*, *8*(5), 688–700.

40. Fallah, S. N., Deo, R. C., Shojafar, M., Conti, M., & Shamshirband, S. (2018). Computational intelligence approaches for energy load forecasting in smart energy management grids: State of the art, future challenges, and research directions. *Energies*, *11*(3), 596.

41. Saheb, T., Dehghani, M., & Saheb, T. (2022). Artificial intelligence for sustainable energy: A contextual topic modeling and content analysis. *Sustainable Computing: Informatics and Systems*, *35*, 100699.

42. Fraske, T., & Bienzeisler, B. (2020). Toward smart and sustainable traffic solutions: A case study of the geography of transitions in urban logistics. *Sustainability: Science, Practice and Policy*, *16*(1), 353–366.

43. Nixon, M., & Aguado, A. (2019). *Feature Extraction and Image Processing for Computer Vision*. Academic Press, Foundary, London.

44. Fan, J., Han, F., & Liu, H. (2014). Challenges of big data analysis. *National Science Review*, *1*(2), 293–314.

45. Tian, S., Yang, W., Le Grange, J. M., Wang, P., Huang, W., & Ye, Z. (2019). Smart healthcare: Making medical care more intelligent. *Global Health Journal*, *3*(3), 62–65.

46. Rahut, Y., Afreen, R., Kamini, D., & Gnanamalar, S. S. (2018). Smart weather monitoring and real time alert system using IoT. *International Research Journal of Engineering and Technology*, *5*(10), 848–854.

47. Ahmed, N. K., Atiya, A. F., Gayar, N. E., & El-Shishiny, H. (2010). An empirical comparison of machine learning models for time series forecasting. *Econometric Reviews*, *29*(5–6), 594–621.

48. Strubell, E., Ganesh, A., & McCallum, A. (2019). Energy and policy considerations for deep learning in NLP. *arXiv preprint arXiv:1906.02243*.

49. Molina-Gómez, N. I., Díaz-Arévalo, J. L., & López-Jiménez, P. A. (2021). Air quality and urban sustainable development: The application of machine learning tools. *International Journal of Environmental Science and Technology*, *18*(4), 1029–1046.

50. Kodinariya, T. M., & Makwana, P. R. (2013). Review on determining number of cluster in K-Means Clustering. *International Journal*, *1*(6), 90–95.

51. Liu, X. Y., Yang, H., Gao, J., & Wang, C. D. (2021). FinRL: Deep reinforcement learning framework to automate trading in quantitative finance. In *Proceedings of the Second ACM International Conference on AI in Finance* New York, NY, United States (pp. 1–9).

52. Østergaard, P. A., Duic, N., Noorollahi, Y., Mikulcic, H., & Kalogirou, S. (2020). Sustainable development using renewable energy technology. *Renewable Energy*, *146*, 2430–2437.

53. Wang, H., Lei, Z., Zhang, X., Zhou, B., & Peng, J. (2019). A review of deep learning for renewable energy forecasting. *Energy Conversion and Management*, *198*, 111799.

54. Çınar, Z. M., Abdussalam Nuhu, A., Zeeshan, Q., Korhan, O., Asmael, M., & Safaei, B. (2020). Machine learning in predictive maintenance towards sustainable smart manufacturing in industry 4.0. *Sustainability*, *12*(19), 8211.

55. Wolfert, S., Ge, L., Verdouw, C., & Bogaardt, M. J. (2017). Big data in smart farming–A review. *Agricultural Systems*, *153*, 69–80.

56. Kataray, T., Nitesh, B., Yarram, B., Sinha, S., Cuce, E., Shaik, S., ... & Roy, A. (2023). Integration of smart grid with renewable energy sources: Opportunities and challenges–A comprehensive review. *Sustainable Energy Technologies and Assessments, 58*, 103363.
57. Zantalis, F., Koulouras, G., Karabetsos, S., & Kandris, D. (2019). A review of machine learning and IoT in smart transportation. *Future Internet, 11*(4), 94.
58. Tsolaki, K., Vafeiadis, T., Nizamis, A., Ioannidis, D., & Tzovaras, D. (2023). Utilizing machine learning on freight transportation and logistics applications: A review. *ICT Express, 9*(3), 284–295.
59. Breiman, L. (2001). Random forests. *Machine Learning, 45*, 5–32.
60. Maćkiewicz, A., & Ratajczak, W. (1993). Principal components analysis (PCA). *Computers & Geosciences, 19*(3), 303–342.
61. Watkins, C. J., & Dayan, P. (1992). Q-learning. *Machine Learning, 8*, 279–292.
62. Singh, S. K., Taylor, R. W., Pradhan, B., Shirzadi, A., & Pham, B. T. (2022). Predicting sustainable arsenic mitigation using machine learning techniques. *Ecotoxicology and Environmental Safety, 232*, 113271.
63. Parekh, R. (2003). AC induction motor fundamentals. *Microchip Technology Inc,* (DS00887A), 1–24.
64. Jacob, A., Jose, V., & Sebastian, D. (2014). Stator fault detection in induction motor under unbalanced supply voltage. In *2014 Annual International Conference on Emerging Research Areas: Magnetics, Machines and Drives (AICERA/iCMMD)* (pp. 1–6). IEEE.
65. Al-Musawi, A. K., Anayi, F., & Packianather, M. (2020). Three-phase induction motor fault detection based on thermal image segmentation. *Infrared Physics & Technology, 104*, 103140.

Chapter 6

A new innovative methodology for photovoltaic integration on rooftops for cost reduction and reduced grid dependency

Edisson Villa-Ávila, Antonio Cano, Paul Arévalo, Danny Ochoa-Correa, and Francisco Jurado

6.1 INTRODUCTION

6.1.1 Context and motivation

Solar energy has emerged as a key force in the global energy landscape, driven by the deterioration of conventional grids and the rising prices of fossil fuels. In this context, it is considered essential to address the growing energy demands worldwide. Adopting photovoltaic (PV) solar systems, especially on rooftops, is crucial in this scenario. However, it faces challenges in optimization due to the complexity of spatial panel design and urban constraints [1]. Simultaneously, the protection of cultural heritage becomes essential in the context of an energy crisis and climate change [2]. Integrating PV generation into existing structures, such as rooftops, raises questions about balancing sustainability and the preservation of cultural values. Despite advancements, the lack of consideration for the spatial design of PV panels concerning roof structure has been a critical gap in research [3]. Accuracy in estimating solar energy generation potential is crucial, as imprecise estimates impact the efficiency and profitability of PV systems. The application of remote sensing technologies, such as earth observation, offers a promising perspective. This chapter addresses these gaps by comprehensively integrating the spatial design of PV panels with roof structures, using remote sensing technologies and innovative approaches. By reviewing methods in scientific literature, the goal is to boost the efficiency of PV systems and contribute to sustainable planning in the energy and environmental domain.

6.1.2 Literature review

Recent research, like that conducted in Ref. [4], has addressed the existing gap by considering various aspects and angles of rooftops to determine optimal conditions for PV system installation. These studies have focused on evaluating the optimal tilt angle of PV panels, explored two-axis solar

DOI: 10.1201/9781003581246-6

tracking, and analyzed different scenarios of received solar irradiance. Regarding the application of remote sensing technologies, such as earth observation, they offer a promising perspective in assessing the PV energy potential and understanding the development status of PV systems, which are crucial aspects for formulating sustainable planning strategies. A recent study [1] conducted a systematic review of remote sensing technology progress applied at various stages of PV system development, structured into four main parts: PV potential estimation, PV array detection, monitoring and diagnosis of PV failures, and other cross-cutting areas where remote sensing can facilitate PV system development. Additionally, Ref. [5] presented an innovative approach to assess the potential of PV solar energy using light detection and ranging (LiDAR) datasets and geographic information systems (GIS). The results obtained are crucial for various applications, such as financial and urban planning, policy formulation in future energy projects, and the analysis of mechanisms to promote installing PV systems on publicly accessible rooftops. In another perspective, Ref. [6] proposed a novel approach to evaluate the potential of solar energy in extensive areas, focusing on rooftops. It is based on integrating global solar irradiance data from Solargis with detailed building information in the form of polygons. Using LiDAR and AW3D technology, the research was conducted in the west of Aichi, Japan, using GIS to calculate the received solar irradiance on rooftops. It is essential to highlight that, during the LiDAR data estimation process, fundamental aspects such as roof inclination and orientation angle were considered. On the other hand, Ref. [7] developed a mixed-integer programming (MIP) model to address limitations associated with PV systems installed on flat roofs. This model focuses on optimizing the net present value (NPV) and can generate multiple azimuth arrangements. To ensure the effectiveness and profitability of PV systems on flat roofs, the model incorporates practical considerations, such as shadow mitigation and roof accessibility, which are fundamental aspects to maximize the efficiency and profitability of such systems.

An et al. [8] analyzes the implementation potential of PV energy in residential buildings in Shenzhen, particularly focusing on electricity consumption. Using GIS data combined with local meteorological information and electrical demand, it examines how urban configuration influences the solar irradiance received by each building individually. Meanwhile, Ref. [9] has directed its research to Nanning, selecting it as a case study to explore suitable options for installing PV systems on various types of rooftops. In addition to estimating the electrical generation potential of rooftop PV systems, the study also evaluates additional returns derived from these installations. In an innovative approach, Ref. [10] has developed a method based on deep learning to generate three-dimensional models of buildings from high-resolution satellite image to estimate the available PV energy potential. This method employs two convolutional neural networks, significantly enhancing the basic architecture of DeepLabv3+ by integrating specialized layers, adaptive activation functions, and hybrid losses. On the other hand, Ref. [11] uses a model-based approach to examine solar irradiance resources and the potential integration of PV systems in residential buildings in various climatic regions of China. Its results reveal that roofs are the preferred location for integrated PV systems (BIPV) installation, followed by south facades, especially in urban areas at higher latitudes. It also suggests considering east and west facades for PV system installation, even in urban contexts where clouds are frequent.

In designing and evaluating the feasibility of integrated PV solutions on rooftops and facades connected to the electrical grid, Ref. [12] focuses on residential buildings within a university environment. It studies three specific groups of residential typologies, classified according to the built area's size and the residents' historical energy consumption. In electrical engineering, integrating PV systems into historical urban structures is a key challenge and opportunity in the context of smart cities and positive energy districts. Hubinsky et al. [2] thoroughly examines the potential and limits of this integration, emphasizing its relevance for the viability of sustainable urban concepts. It highlights the importance of assessing the visual impact of such systems and selecting appropriate solutions, considering factors such as the azimuth of the normal vector and the slope of the roofs. It uses a datafication process to analyze the solar irradiance received by rooftops, employing a detailed 3D morphological model and the open-source solar irradiance model.

On the other hand, Ref. [13] focuses on estimating the solar energy potential using satellite images and advanced deep learning-based segmentation techniques. It compares convolutional neural network architectures to segment images and calculate the area available for PV panel installation. This technique considers the average tilt of solar panels, their efficiency, the inclined global solar irradiance, and the loss coefficient. It uses these factors to estimate the annual energy production from the available area for solar panel installation. Kafle et al. [14] used a hierarchical geospatial technique based on open-source data to estimate the potential PV energy production in various cities in Nepal, finding significant variations in the theoretical potential PV energy production. Subsequently, Ref. [15] presents the first comprehensive maps of solar roof footprints and potential in Lebanon using deep learning-based instance segmentation from satellite images. It proposes a PV panel placement algorithm that considers each roof's morphology. The results indicate that the average solar potential of rooftops can cover the annual electricity needs of a single-family home using only 5% of the roof area. Additionally, using 50% of the area of a residential apartment roof would ensure energy security for up to eight homes. Nasrallah et al. [15] also calculates the average and total solar potential of roofs per district, considering factors such as size, land coverage ratio, and PV production for each district. Furthermore, Ref. [16] has introduced an innovative method that allows a comprehensive and efficient assessment of rooftop PV potential, with a spatial resolution of up to 1 meter. This approach stands out for its ability to perform a complete census assessment in a short period, using minimal computational resources. It employs a specialized algorithm to calculate shadows projected on rooftops and perform solar irradiance integration over time, thus offering an effective tool for accurate PV potential assessment.

Pinna and Massidda [17] also proposes an optimal planning strategy for distributed PV systems at the municipal level in densely populated cities. Based on learned integer programming techniques, it identifies the most suitable roofs for PV system installation, aiming to maximize electricity generation. By accurately characterizing solar energy potentials and considering budgetary and peak power export constraints, this strategy seeks to optimize the distribution of PV systems at the municipal level, thereby promoting energy efficiency in densely populated urban environments. On the other hand, Ref. [18] presents a low-cost evaluation framework that analyzes the PV potential of rooftops, with a specific focus on a case study on Fernando de Noronha Island. This framework uses economic aero-photogrammetry techniques and geospatial analysis to assess the effectiveness of PV systems on rooftops, providing practical and applicable results. From a broader perspective, Ref. [19] proposes an innovative approach to estimate the spatial distribution of energy generation potential on rural rooftops, using publicly accessible satellite images. This study segments the roofs of rural buildings based on their orientation and tilt angle, using a revised U-Net deep learning network to extract detailed roof images. Additionally, it develops a calculation method to determine the potential installation area of PV panels at a micro level, considering various panel types and their maintenance requirements. This methodology provides an effective tool for assessing PV potential in rural areas and optimizing the design of PV systems on rooftops. To address the common overestimation in solar energy estimates, Ref. [3] presents a new spatially explicit optimization framework. This approach considers not only the roof's structural configuration but also the shape and size of solar panels, based on a maximum coverage spatial optimization model. Table 6.1 summarizes the representative research of the last three years, detailing the methodology used for calculating PV panels on rooftops.

In a related domain, the challenge of overestimation in solar energy estimates, often resulting from the general consideration of total roof area, is addressed in Ref. [3]. This study introduces an innovative solution through a new spatially explicit optimization framework designed to enhance accuracy in rooftop solar energy assessments. This pioneering approach takes into account not only the structural configuration of the roof but also the shape and size of the solar panels, all based on a spatial optimization model for maximum coverage.

Table 6.1 Information from previous research for the estimation of PV generation potential on rooftops

Methodology	*Ref.*
Remote Sensing (RS) for: (i) PV potential estimation, (ii) PV array detection, (iii) monitoring and diagnosis of PV faults, and (iv) other cross-cutting areas where RS can facilitate PV development	[1]
Evaluation of PV solar energy potential based on the combination of LiDAR and GIS datasets	[5]
Mixed-Integer Programming (MIP) model to address limitations in flat rooftop PV systems, focusing on optimizing the Net Present Value (NPV) with considerations for multiple azimuths	[7]
Mathematical model to assess solar irradiance resources and the potential for PV integration in buildings across different climatic zones in China	[11]
Analysis of the technical potential and economic benefits of PV systems in seven different scenarios on three university campuses located in different solar zones in China	[25]
Proposal of a rooftop PV solar panel diagram using a NEM meter installed in the ring distribution system at PSAS	[26]
Comprehensive maps of solar rooftop footprints and potential in Lebanon using deep learning-based instance segmentation to extract building footprints from satellite images, with a PV panel placement algorithm considering each roof's morphology	[15]
Design and evaluation of the feasibility of a PV solution integrated into roofs and facades, connected to the electrical grid, to meet the energy demand of residential buildings on an academic campus	[12]
Learned integer programming, based on high-precision solar energy potential characterization to select suitable rooftops for solar energy with a spatial resolution of 1 meter	[17]
Datafication process, analysis of irradiance of tilted and flat roof polygons in the selected area based on the azimuth of the normal vector and roof slope. Utilization of a detailed 3D morphological model in LOD3 and the open-source solar irradiance model r.sun implemented in GRASS GIS/QGIS	[2]
Use of an algorithm for efficient calculation of shadows on rooftops and integration of solar irradiance over time with a spatial resolution of 1 meter	[16]
Use of GIS systems and 3D models, considering roof angles, overall estimation of the optimal angle, solar tracking, and solar irradiance scenarios	[4]
Modeling of remote sensing data based on GIS, solar irradiance data with SolarGIS, and LiDAR and AW3D light detection and ranging data, to estimate the solar power potential in Aichi, Japan.	[6]
Improvement of the accuracy of rooftop solar energy generation estimates by incorporating the spatial design of solar panels, considering their size, orientation, and roof structure, through a new optimization framework	[3]
Based on GIS data, typical local weather data, and electricity demand, the solar potential of Shenzhen is compared with different characteristics of urban morphology, which influences the solar irradiance received by individual buildings. Use of 3D architectural modeling software Rhinoceros with its plugins Grasshopper and Ladybug	[8]
Analysis of optimal installation options for PV on different types of roofs and estimation of the electricity generation potential of rooftop PV and additional returns, as well as the optimal annual azimuth angle and inclination	[9]

(Continued)

Table 6.1 (Continued) Information from previous research for the estimation of PV
generation potential on rooftops

Methodology	Ref.
Detail-oriented deep learning, constructing three-dimensional building models from high-resolution satellite images for the first time and estimating PV potential, developing convolutional neural networks, roof segmentation model, and height prediction model using the basic architecture of DeepLabv3+	[10]
Use of deep learning segmentation techniques, architectures such as UNet, UNet++, Feature Pyramid Network (FPN), Pyramid Scene Parsing Network (PSPNet), and DeepLabV3, with different encoders. Converts pixels of segmented image into square meters based on the average inclination of the panels	[13]
Use of a hierarchical geospatial technique based on open-source data to estimate the potential PV energy production in various cities in Nepal.	[14]
Application of low-cost aerophotogrammetry, geospatial analysis, scenario creation, mapping of buildings, open spaces, and solar obstacles using an unmanned aerial photogrammetric survey providing a 3D digital surface	[18]
Novel approach to estimate the spatial distribution of the overall power generation potential on rural rooftops from publicly accessible satellite images using a U-Net deep learning network to extract rooftop images from macro-level satellite images, using a calculation method for the potential installation area of PV panels at the micro-level	[19]
Use of GF-2 satellite images, Points of Interest (POI) data, and meteorological data to accurately assess the potential long-term reduction of carbon emissions through rooftop PV installation. DeepLabv3+ model of building rooftops in GF-2 images through multisensory fields	[27]
Measurement of the attractiveness of PV solar modules using a Categorical-Based Evaluation Technique (MACBETH)	[28]
It uses advanced electrical modeling techniques with Monte Carlo Ray Tracing simulation. A detailed analysis of a representative rooftop in Canberra, Australia, is conducted, incorporating real-world conditions and variations to accurately assess potential energy yield gains when implementing bifacial solar modules. Typical mechanical mounting components, installation orientations, and module characteristics are considered	[20]
The use of geospatial techniques and the high-resolution Building Integrated Solar Energy (BISE) supply model. These tools were employed to estimate the main spatial and temporal characteristics of rooftop PV solar energy production potential	[21]
Construction of an assessment framework to analyze the generation potential of energy through rural rooftop photovoltaic panels (RTSPV) in rural areas, focusing on Jiangsu Province, China. The RTSPV generation potential was evaluated and compared with the existing electricity demand to describe the spatial disparity between supply and demand	[22]
Deep learning and geographic information systems (GIS) are used to assess the solar energy potential on urban rooftops. GIS is employed to extract information on land use types and classify buildings based on these types	[23]
Develops a technical framework to optimize the scale of development and spatial design of rooftop solar installations. It relies on high-resolution generation simulations and load-oriented electricity dispatch. It examines how the gradual expansion of rooftop solar development affects its penetration into the electric grid and associated curtailment	[24]

The underexplored potential of bifacial solar modules in rooftop applications, despite their success in utility-scale photovoltaic systems, is tackled in Ref. [20]. The feasibility of integrating these modules on rooftops is examined through advanced modeling and simulation techniques, with a focus on potential energy yield enhancements. The findings reveal a significant boost in solar energy production, emphasizing the pivotal role of module and system design, and underscores the promising potential of bifacial technology in maximizing energy generation on rooftops, with substantial implications for efficiency and sustainability.

An analysis of the rooftop PV energy potential as part of the construction sector's decarbonization strategy to meet the European Union's 2050 climate objectives is presented in Ref. [21]. Despite acknowledging the significance of this energy source in the transition, there exists a dearth of detailed analysis on bridging the gap between technical potential and actual PV energy generation on rooftops, influenced by policies. To address this gap, the authors have employed geospatial techniques and a high-resolution building integrated solar energy (BISE) supply model to estimate rooftop solar energy production potential at spatial and temporal scales. Furthermore, the ramifications of the European Commission's Solar Rooftop Initiative on future PV electricity supply are examined.

The potential for electricity generation through rooftop solar photovoltaic (RTSPV) panels in rural areas to mitigate regional energy consumption conflicts is assessed in Ref. [22]. That survey has developed a sophisticated assessment framework for Jiangsu Province, China, where the RTSPV energy generation potential is scrutinized and juxtaposed with electricity demand to identify spatial disparities. A methodology to assess the solar energy potential on urban rooftops using deep learning and geographic information systems (GIS) is introduced in Ref. [23]. Buildings are categorized based on land use, and rooftop availability is computed. This study validates the approach in a Wuhan area, demonstrating that classification enhances accuracy by 12.68%. A technical framework to optimize the development of rooftop solar installations through generation and electricity dispatch simulations is proposed in Ref. [24]. The results show that rooftop solar expansion decelerates as curtailment rises. Regional integration and grid flexibility marginally augment development scale, whereas energy storage facilitates larger scales. Optimization varies depending on flexibility and storage capacity. In China, substantial flexibility and storage levels are imperative to attain high solar penetrations and mitigate curtailment.

6.1.3 Identification of gaps and contributions

As the adoption of small-scale PV systems increases among households and businesses, there is a need for more accurate methods to assess the available solar potential, as indicated in Ref. [5]. Despite advances in research on PV systems installed on rooftops, significant gaps that require more detailed attention persist. An identified gap lies in the lack of comprehensive consideration of the spatial design of PV panels concerning the specific structure and characteristics of the roof. Although studies have advanced in evaluating the potential of solar energy generation, the lack of integration can lead to inaccurate estimates [3]. It is essential to address this gap to obtain more precise estimates and optimize the arrangement of PV panels on rooftops [11, 25, 26]. Another challenge arises in optimizing PV systems in densely populated urban environments. Implementation in urban neighborhoods faces space and shading constraints and diversity of rooftop availability [12, 17]. These obstacles require specific strategies and detailed studies to maximize efficiency in such environments, a need that has not been fully addressed in the literature. Additionally, despite efforts to consider solar irradiance and PV panel efficiency, research in PV solar energy has addressed the estimation of generation potential limitedly until now [7]. There is a critical gap in the lack of comprehensive studies that address the spatial design of PV panels with roof structure. This gap can lead to a lack of precision in estimates of solar energy generation potential, emphasizing the need for research that addresses this detailed relationship [3].

To fill the gaps mentioned above, this study contributes to the field of PV solar energy by addressing and closing significant gaps in current research on PV systems installed on rooftops. The key contributions are as follows:

- A novel model is introduced to categorize rooftops based on their morphology, considering shape, size, and orientation factors. This model enhances the understanding of roof structures, facilitating the precise placement of PV systems.
- The chapter provides information on identifying optimal rooftops for PV panel installation. By considering the complex interaction of building morphology, location, and the surrounding environment, the methodology facilitates the selection of suitable rooftops for efficient energy generation.
- The roof-solar methodology offers a robust framework for determining the optimal spatial distribution of PV panels. This contribution is crucial for maximizing energy production and minimizing shadows and overlaps, ensuring improved performance of urban PV systems.
- Leveraging geographic information system technologies and three-dimensional models, the chapter presents innovative assessment tools. These technologies provide accurate estimates of PV energy generation potential, contributing to more precise planning and decision-making.
- The practical application of the roof-solar methodology in real urban environments validates its utility. The results demonstrate substantial potential for PV energy generation, with capacities of up to 343 kW, reinforcing the practicality and effectiveness of the methodology.

- The implementation of PV systems on residential rooftops is shown to be an effective strategy to reduce CO_2 emissions and address climate change. This finding highlights the chapter's contribution to a cleaner, more sustainable urban energy matrix.

6.2 MATERIALS AND METHODS

This study focuses on enhancing the arrangement of PV solar panels on urban rooftops to increase electricity production. The proposed methodology is divided into four interrelated stages, as illustrated in Figure 6.1. In the first phase, contiguous roof segments are identified, taking into account their different orientations. For this purpose, a numerical calculation and image processing program, which includes a grayscale filter and edge-smoothing techniques, is employed. This approach facilitates the identification of rooftops and the delineation of the contours of each roof segment.

Once rooftops are identified, the next step is to determine suitable areas within them for installing PV solar panels. Since solar irradiance is

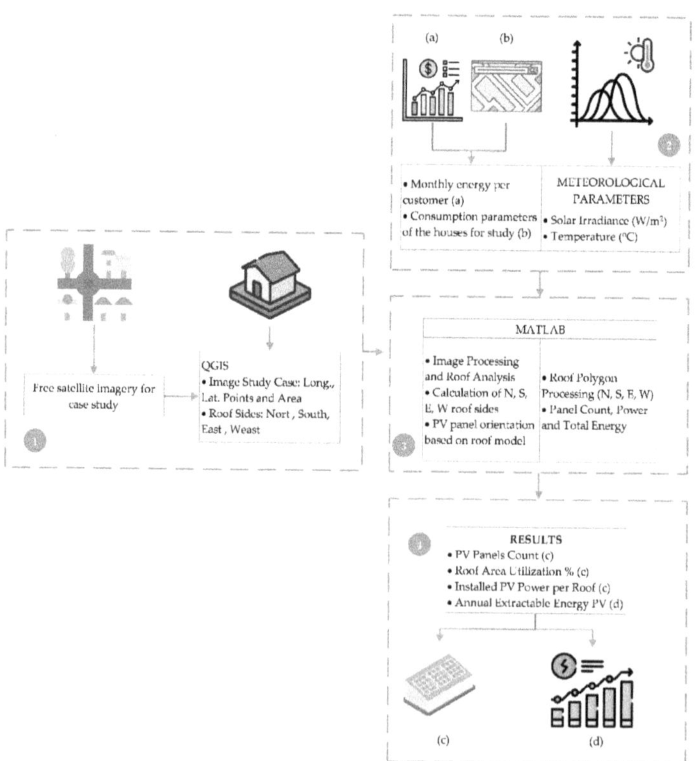

Figure 6.1 Proposed methodology overview.

distributed continuously throughout the rooftop area, a comprehensive calculation is performed to identify a finite number of candidate locations for the PV panels. This process is based on georeferenced areas and locations, aiming to find the optimal arrangement for each panel and determine the maximum number that can constitute the PV arrays on each roof. To demonstrate the applicability of this methodology, a case study based on an urban district located in Cuenca, Ecuador, an equatorial city at high altitude, is employed.

After defining suitable areas, annual solar irradiance is calculated using real data from a meteorological station near the study area. These data, obtained from the Micro-Grid Laboratory of the University of Cuenca, Ecuador [29], comprise solar irradiance records for 2022, with a resolution of 1 second, averaged to an hour. Finally, an algorithm has been developed to address the complexity of irregular polygons that may be found in the structure of urban rooftops. This algorithm adapts to any geometric configuration and aims to determine optimal locations for the placement of PV solar panels. Unlike previous approaches, the algorithm is guided by three fundamental criteria:

- Ensures that each panel is entirely within the boundaries of the rooftop polygon, avoiding any uncovered areas.
- Guarantees that each panel fits perfectly into the corresponding rooftop segment.
- Ensures that adjacent panels do not overlap, ensuring a realistic and feasible design of the PV installations. Additionally, conflict zone parameters have been incorporated to prevent any interference in placing panels within an array.

6.3 ALGORITHM IMPLEMENTATION

6.3.1 Roof-Solar algorithm flowchart

The devised algorithm, denoted as "Roof-Solar," is engineered for the purpose of strategically situating PV panels on rooftops to optimize solar energy generation. The algorithm determines the most effective geospatial configuration of PV solar panels on rooftops. Figure 6.2 delineates the procedural flow of the algorithm employed to ascertain the optimal placement of PV solar panels on rooftops. The initiation involves segmenting the roof into distinct geographical areas based on cardinal orientations: north, south, east, and west. Subsequently, individual roof polygons' data, including identification numbers, areas, latitudes, and longitudes, is meticulously retrieved for each geographical sector, accounting for their unique geospatial attributes. Concurrently, the algorithm defines the physical dimensions (height and width) and electrical parameters (voltage, current, and power)

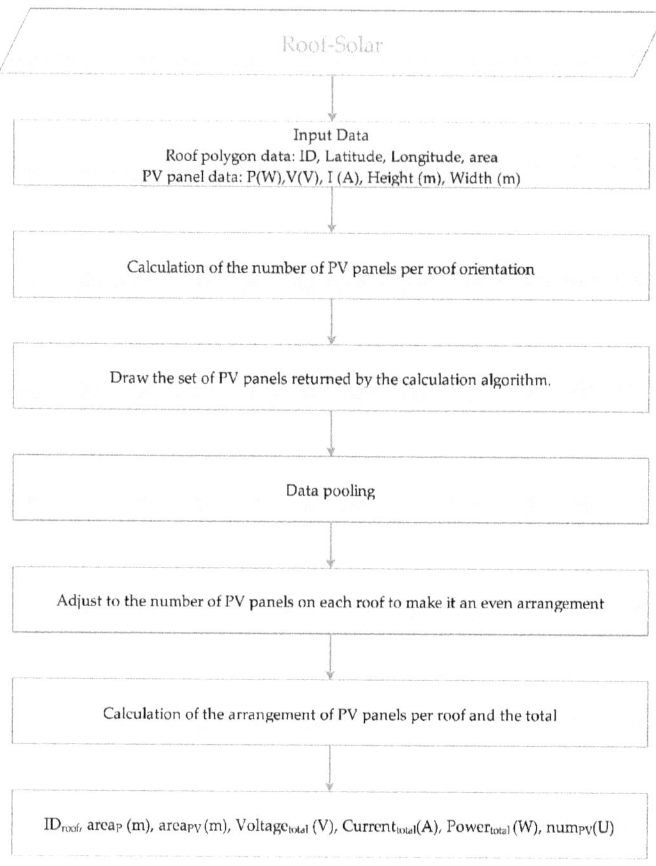

Figure 6.2 Proposed flowchart for optimal placement of PV panels considering the morphology of each roof.

of the PV panel. Using the "Roof-Solar" program, geographical areas are visually represented, showcasing the roof polygons via geospatial visualization functions. A systematic calculation process ensues to ascertain the optimal number of PV solar panels accommodatable within each geographical area, and the outcomes are presented graphically.

Upon the processing of all geographical areas, the program amalgamates and computes data to derive comprehensive insights into the arrangement of PV panels for each roof. Notably, the algorithm incorporates an adjustment mechanism to ensure an even number of panels, streamlining the configuration of PV panel arrays in series and parallel configurations, thereby enhancing the maximization of installable power. In the final stages, voltage, current, and power values are computed for the collective PV solar panel ensemble using specialized functions and divisible options tailored to the calculated number of PV panels per roof. The algorithm, characterized by its modular framework and sophisticated functionalities, is an instrumental tool, addressing pivotal facets of the optimal placement of solar panels on rooftops. The algorithm underscores its adaptability and utility across diverse scenarios, from the acquisition of geospatial data to the computation of panel arrangements and amalgamated options.

6.3.2 Optimal placement of PV panels

This section outlines the implementation of the Roof-Solar algorithm, designed to enhance the positioning of solar panels on geographical polygons. Through calculations and geometric transformations, the algorithm aims to determine the optimal location for the panels to maximize their coverage area and avoid overlaps. Additionally, it incorporates functions to integrate data and explore shared configurations for solar panels. The primary objective of this approach is to increase efficiency in the distribution of solar panels on roofs, translating into higher solar energy production. The pseudocode for this algorithm is presented below:

- A central function is created to process the georeferenced coordinates of a polygon representing the roof where PV solar panels will be installed. The main purpose of this function is to improve the geometric distribution of the defined areas to ensure the correct placement of solar panels on the roof.
- Geographic coordinates are converted to local east, north, and up coordinates. Based on these coordinates, a polygonal object representing the roof is generated to facilitate subsequent analysis.
- The limits of the polygon representing the roof are determined based on minimum and maximum UTM coordinates. These limitations are crucial to ensure optimal placement of solar panels on the roof.
- An analysis is performed on the vertices of various sections of multiple roofs to identify suitable points for placing PV panels that meet the established criteria. Subsequently, solar panels are evaluated, and those without overlaps are enumerated.
- The maximum number of nonoverlapping PV panels is recorded.
- The number of solar panels placed on each roof section is calculated and stored for energy analysis.
- A function is implemented to determine the most suitable arrangement of the number of solar panels based on the voltage and current requirements needed to adjust the size of the corresponding inverter.

This integrated approach has the potential to significantly enhance the efficiency of solar deployments on urban roofs, leading to increased solar energy production. A pseudocode of the algorithm is provided to facilitate a detailed understanding and potential practical implementation.

PSEUDOCODE

Calculation function to maximize the number of PV panels on each roof

1. **Input Data**

   1.1. Constant data: $width_{PV}$, $height_{PV}$, lon_p, lat_p, Id_p, $area_p$

2. **Conversion of georeferenced units for the roof polygons in the four orientations (north, south, east, and west)**

   2.1. Given $\left(lon_p, lat_p\right)$ coordinates of a roof polygon, convert them

   to $\left(x_{meter}, y_{meter}\right)$

3. **Calculate the limits of the roof polygon**

   3.1. $\left(x_{min}, x_{max}, y_{min}, y_{max}\right) = \begin{pmatrix} min\left(x_{meter}\right), max\left(x_{meter}\right), \\ min\left(y_{meter}\right), max\left(y_{meter}\right) \end{pmatrix}$

   3.2. Iterate over points within the polygon limits

   3.3. $x = \left(x_{min} : width_{PV} : x_{max} - width_{PV}\right)$

   3.4. $y = \left(y_{min} : height_{PV} : y_{max} - height_{PV}\right)$

   3.5. Define the coordinates of the vertices of the PV panel within the vector x_{PV}

   3.6. $x_{PV} = \begin{bmatrix} x & x + width_{PV} & x + width_{PV} & x \end{bmatrix}$

   3.7. $y_{PV} = \begin{bmatrix} y & y & y + height_{PV} & y + height_{PV} \end{bmatrix}$

   3.8. Check if the number of PV panels found is greater than the current maximum

      a.   3.8.1. If $num_{PV} > max_{PV}$
      b.   3.8.2. $max_{PV} = num_{PV}$
      c.   3.8.3. End

   a.   3.9. If all $\left(x_{PV}, y_{PV}\right)$ inside the roof polygon $\left(x_{meter}, y_{meter}\right)$

   b.   3.10. Draw the PV panels in UTM coordinates

      a.   3.10.1. $draw\left(x_{PV_{new}}, y_{PV\,new}, 'blue'\right)$

      b.   3.10.2. $num_{PV} = num_{PV} + 1$

   c.   3.11. End

1. **4. Output information**

   a.   4.1. Id_{PV}, $area_p$, num_{PV}, $lat_{PV_{new}}$, $lon_{PV\,new}$

6.4 CASE STUDY

6.4.1 Description and solar irradiance levels

Cuenca is a highland city situated in the Andes mountains of Ecuador, at a latitude of 2°53′00″ south and a longitude of 79°00′00″ west, maintaining an average elevation of 2,500 meters above sea level. The chosen study area encompasses an urban district with consumers from a medium to high socioeconomic stratum, indicating a considerable energy demand. This specific district is near the Laboratory of Micro-Grid building at the University of Cuenca, which houses a meteorological station. The records from this station will be utilized as input information for applying the proposed methodology in the study. Due to its close proximity to the Earth's equator, Cuenca experiences significantly high levels of solar irradiance. The average annual global horizontal solar irradiance is substantial (see Figure 6.3), presenting an exceptional potential for PV energy generation in the region [30].

6.4.2 Electrical grid in the research area

The identification of Transformer No. 33443, with a capacity of 100 kVA, connected to the primary feeder of the city's electrical company, stands out. Additionally, the monthly consumption information of each subscriber was reviewed to obtain accurate data supporting this analysis. This initial procedure is crucial to reflect the real conditions of the environment accurately.

6.4.3 Energy consumption in the study district

Based on the information provided in Ref. [31], it was concluded that the distribution transformer supplied energy to 23 consumers. Figure 6.4 displays actual consumption records from residential users in the analyzed district. In the context of our research, we have access to the annual records

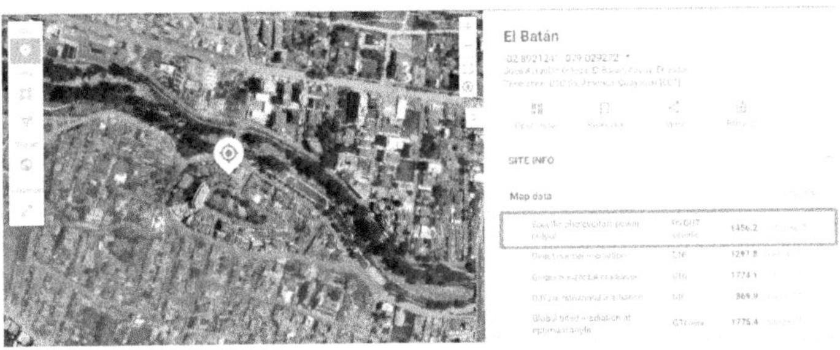

Figure 6.3 Annual representation of global solar irradiance levels in the research area – Cuenca, Ecuador.

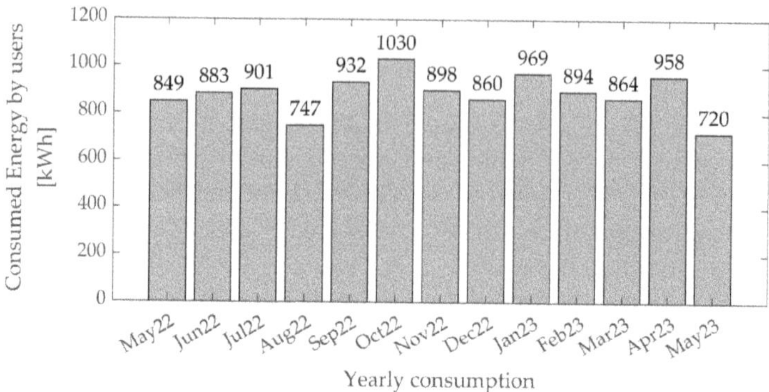

Figure 6.4 Annual energy consumption displayed by one of the subscribers.

of the additional 22 users, which were extracted from a publicly accessible information source.

Table 6.2 summarizes the data taken from the consumers connected to transformer 33443. This data includes the annual energy consumed, totaling 11.57 MWh, the estimated average installed power for the reference period, and the calculation of cost per kilowatt-hour ($/kWh).

6.4.4 Identification and validation of usable rooftop areas

To recognize and delineate usable areas on the roofs of the urban area, an orthophoto of the area of interest was acquired from the public municipality website [32], as depicted in Figure 6.5. This representation details each step followed to identify the roofs and define the areas available for the installation of PV panels. Figure 6.5a presents the original orthophoto of the district. Subsequently, through image processing, including color filtering and the use of grayscale, the roofs in the residential area were highlighted, as illustrated in Figure 6.5b. Next, a procedure in MATLAB was implemented to outline and generate polygons on the identified roofs. The georeferencing of the image facilitated the determination of the roof orientations (north, south, east, and west) and their corresponding areas, classifying them into four categories represented by different colors: north-facing roofs are represented in green, south facing in yellow, east facing in blue, and west facing in orange (see Figure 6.5c).

Furthermore, an analysis of the resulting image of the study district was conducted by manually selecting 23 households associated with the distribution mentioned above transformer. This process was carried out using the open-source GIS tool, QGIS. This allowed for verifying the roof areas and validating accuracy through OpenStreetMap, following the methodology described in Ref. [14].

Table 6.2 Consumer data for transformer 33443 in May 2023

Roof	Consumed energy (kWh)	Mean installed power (kW)	Energy billing ($)	Cost per kWh ($/kWh)
(a)	718	0.97	78.01	0.11
(b)	650	0.87	68.98	0.11
(c)	11	0.01	1.00	0.09
(d)	519	0.70	52.14	0.10
(e)	634	0.85	66.92	0.11
(f)	31	0.04	2.82	0.09
(g)	140	0.19	13.00	0.09
(h)	0	0.00	0.00	0.00
(i)	355	0.48	34.48	0.10
(j)	66	0.09	6.04	0.09
(k)	62	0.08	5.70	0.09
(l)	150	0.20	13.95	0.09
(m)	331	0.44	31.99	0.10
(n)	84	0.11	7.71	0.09
(o)	111	0.15	10.25	0.09
(p)	549	0.74	56.00	0.10
(q)	240	0.32	22.76	0.09
(r)	210	0.28	19.79	0.09
(s)	229	0.31	21.67	0.09
(t)	0	0.00	0.00	0.00
(u)	220	0.30	20.78	0.09
(v)	117	0.16	15.01	0.13
(w)	171	0.23	15.99	0.09

6.4.5 Selection of PV panel models

This study focuses on assessing a specific type of PV panel, the A-250P, whose main specifications are described in Table 6.3. The selection of this model is based on its availability in the Micro-Grid laboratory, which has a documented performance history over six years.

6.5 RESULTS AND DISCUSSION

6.5.1 Distribution of A-250P PV panels on rooftops in the urban district

Figure 6.6 depicts the outcomes of implementing the Roof-Solar algorithm to achieve an improved distribution of A-250P solar panels on rooftops. This algorithm provides georeferenced positions for the polygons comprising

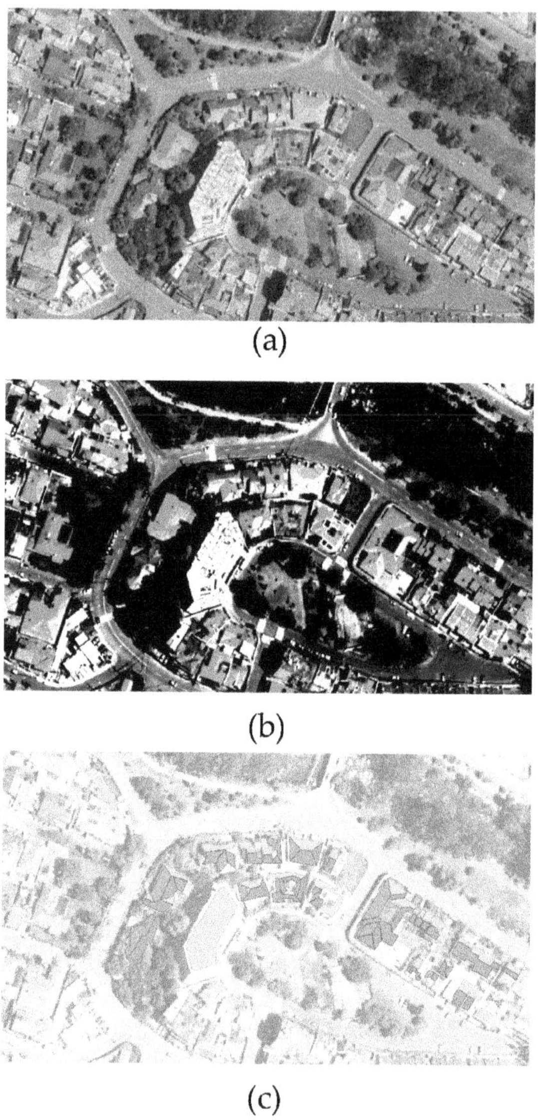

Figure 6.5 Roof localization through image processing. (a) Original aerial photograph of the study area. (b) Initial filtering to highlight available roofs (based on their shape and height). (c) Roof delineation and segmentation using georeferenced polygons.

Table 6.3 Main characteristics of the PV module considered in this study

PV module specification	A-250P
Efficiency (%)	15.35
Area (m²)	1.62
Dimensions (height, wide) (m)	1.645 × 0.99
Voltage (V)	29.53
Current (A)	8.45
Peak power output (W)	250
Peak power output density (W/m²)	154.32

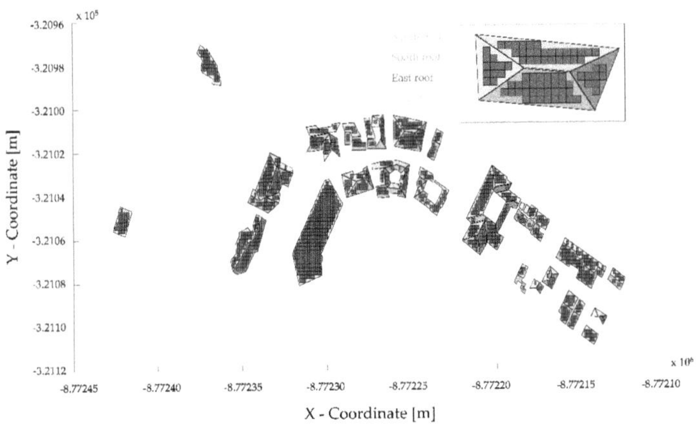

Figure 6.6 Distribution of A-250P solar panels on rooftops through the proposed application.

each solar panel array correctly placed on the rooftops. The solar panels are distributed in four orientations: yellow for north, orange for south, gray for east, and green for west. Consider the cardinal orientation of the rooftop panel contributes to obtaining more reliable numerical results when assessing the solar energy potential of each.

Figure 6.7 provides a detailed perspective of the results generated by the proposed algorithm. In this representation, each of the 23 analyzed roofs has been identified with a code to facilitate reference in the calculations presented in the subsequent sections. It is crucial to highlight that the Roof-Solar algorithm ensures that no PV panel extends beyond the roof polygon's area or overlaps with other panels, thus ensuring a practical distribution and optimal use of the available space.

Considering the capacity of each of the 1,372 panels arranged according to the algorithm, this layout achieves a PV solar capacity of 343 kW in the analyzed area. Table 6.4 provides a detailed synthesis, including the number of PV panels used per roof, the percentage of area utilized, and the total

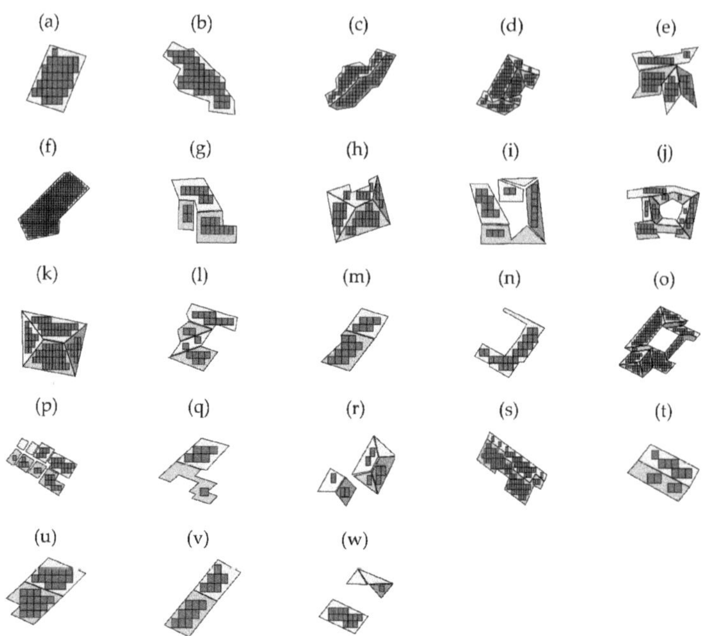

Figure 6.7 View of the 23 roofs identified through the algorithm and the arrangement of A-250P PV panels.

installed PV power. This detailed description underscores the effectiveness of the PV panel design in harnessing solar potential in the studied area.

Figure 6.8 presents the results obtained using the Roof-Solar algorithm for the A-250P PV panel. Figure 6.8a displays the number of PV panels that can be placed on each roof. It is evident from this figure that roofs identified as (f), (o), and (s) have a greater number of panels assigned by the proposed algorithm. Upon comparing this result with the orthophotos presented in Figure 6.7, it is observed that these roofs have larger surface areas and better contours for accommodating panels, which aligns with one of the primary objectives of the proposal. Conversely, the remaining roofs have configurations that hinder the accommodation of the commercial-sized panels considered in the study. Meanwhile, Figure 6.8b illustrates the percentage of roof area utilized to place PV panels. The results indicate that most of the considered roofs have a utilization level of over 50% of their physical surface. Improving this utilization percentage could be achieved by using smaller panel dimensions, albeit with a consequent reduction in the installed power capacity of the resulting array. However, since this research treats the urban district as a single aggregate generation entity, standardizing the size of individual panels is the focus of this study. Finally, Figure 6.8c highlights the total installed power capacity on the roofs. As expected,

Table 6.4 Results of the number of panels, percentage of area utilized, and installed power

Roof	Roof area (m²)	Number of PV solar panels (u)	Percentage of usable area used by PV panels (%)	Installed solar PV power (kW)
(a)	91	38	67.80	9.5
(b)	121	48	64.41	12
(c)	274	98	58.07	24.5
(d)	359	114	51.56	28.5
(e)	190	52	44.44	13
(f)	684	364	86.40	91
(g)	102	24	38.20	6
(h)	160	42	42.62	10.5
(i)	119	20	27.29	5
(j)	231	40	28.11	10
(k)	245	80	53.02	20
(l)	89	18	32.84	4.5
(m)	59	16	44.03	4
(n)	72	22	49.61	5.5
(o)	573	180	51.00	45
(p)	155	36	37.71	9
(q)	44	6	22.14	1.5
(r)	73	8	17.79	2
(s)	288	98	55.25	24.5
(t)	56	10	28.99	2.5
(u)	84	28	54.12	7
(v)	55	16	47.23	4
(w)	70	14	32.47	3.5
Total	4,194	1,372	45.00	343

the envelope of this histogram is directly related to the number of panels defined in Figure 6.8a. These three components of Figure 6.8 clearly visualize the differences between the two models and their impact on the number of panels and the total installed power capacity in the study area.

6.5.2 Annual net energy balance of residential PV systems

The proposed algorithm enhances the distribution of a finite number of commercial PV panels per roof, providing the immediate nominal capacity to be installed on each roof, as shown in Figure 6.8c. The proposed methodology calculates the annual net energy balance with this information and the consumption records of each user residing in the buildings. The results of this calculation are presented in Figure 6.9a, detailing the monthly

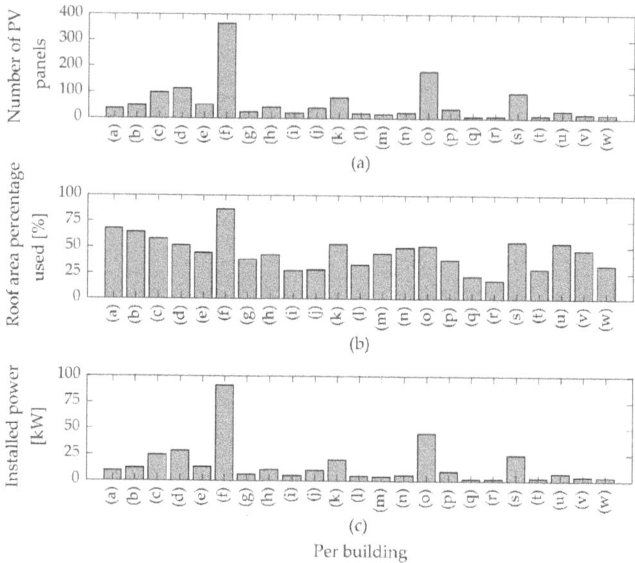

Figure 6.8 Quantitative results obtained through the implementation of the proposed approach for the case study. (a) Total number of assigned PV panels per roof. (b) Percentage of roof area utilized. (c) Installed nominal capacity per roof.

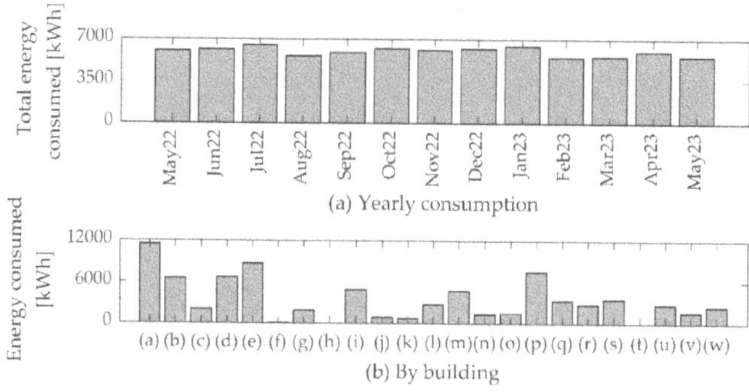

Figure 6.9 Projected energy balance for residences: (a) Monthly energy consumption for all buildings and (b) annual energy consumption per building.

annual consumption of the entire building complex within the studied district, while Figure 6.9b displays the total annual consumption per building. These data are then compared with the maximum energy that can be extracted from the installed PV panels in the study area, calculated based on historical irradiation data recorded at a nearby meteorological station.

The effective improvement of the roofs leads to an overproduction of energy, indicating the need to consider changing the initial transformer and the possibility of feeding excess energy back into the distribution grid for economic benefits. The algorithm yields an annual PV solar energy potential generated by the PV panels on the district's roofs under study that exceeds local demand, surpassing it by over four times the total electricity requirement, meeting the maximum energy demand for each building. The total nominal power of the PV panels in the proposed scenarios for the network of all buildings is 343 kW, with an average annual energy production of 431.72 MWh. However, since the system has been designed to extract the maximum amount of energy with the placement of PV panels, there is an average annual surplus of 420.15 MWh compared to the measured annual consumption of residential energy, which can be injected into the grid to supply additional households as shown in Figure 6.10.

Despite the encouraging results, it is essential to acknowledge that implementing large-scale PV systems in residential environments poses technical, economic, and regulatory challenges. Adequate policies and financing are needed to promote the adoption of PV energy, and it is crucial to involve the community and the government in the transition to sustainable and accessible solutions. The study also underscores the importance of considering aspects such as excess production stored in batteries, installation costs, net gains, and payback time as part of a more comprehensive analysis of the feasibility of large-scale solar projects in similar contexts.

In light of the results, the algorithm designed in this study demonstrates notable efficacy in assessing the solar potential on urban rooftops. As described in Figures 6.1 and 6.2, the calculation process commences with the identification of the urban district using the Geographic Information System software, QGIS, to georeference and categorize rooftops based on their orientations. Subsequently, the data is exported in ".shape" format for further processing in Matlab. Through a series of image filtering and

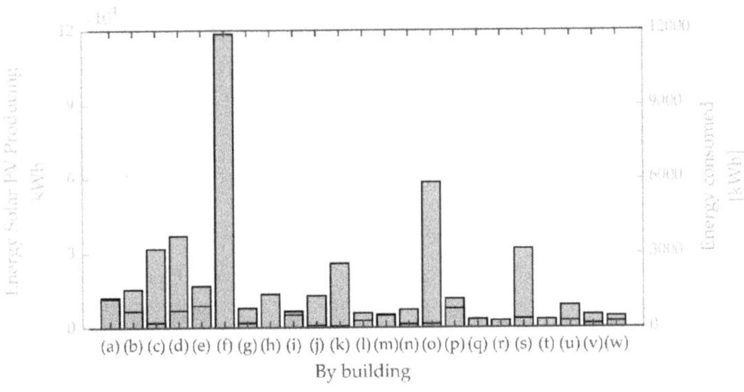

Figure 6.10 Potential excess energy with monthly building consumption in the case study.

segmentation processes, the algorithm delineates rooftop edges and identifies irregular polygons, incorporating additional information such as area, reference points, and names. This initial step, comprising approximately 65% of the total computational time, represents the most computationally intensive task.

Following the preprocessing stage, the "Roof-Solar" algorithm is deployed to address the placement of photovoltaic panels on each rooftop and orientation, ensuring compliance with constraints such as avoiding overlaps and staying within specified boundaries. This phase, representing 25% of the total computational effort, is characterized by its efficiency in swiftly identifying optimal panel locations.

Finally, leveraging existing energy parameters, the algorithm calculates the total installed power and potential energy generation, constituting the remaining 10% of computational effort. Notably, the computational burden is allocated such that the preprocessing stage consumes the majority of computational resources due to its complexity, while subsequent algorithmic processes are relatively swift, benefiting from the simplicity and efficiency of the implemented algorithm. This resource allocation underscores the computational efficiency of the proposed approach, which is easily implementable in various programming environments.

6.6 CONCLUSIONS

This chapter has successfully introduced the Roof-Solar method, designed to enhance the placement of PV panels in urban environments to increase solar energy production. The main conclusions include the demonstrated effectiveness of the method in enabling an efficient distribution of PV panels on urban rooftops, ensuring optimal energy production while avoiding overlaps and shadows. This globally applicable approach relies on globally available solar irradiation data and building polygons.

The study identifies significant potential for PV energy generation in the investigated area, reaching up to 343 kW, thereby highlighting the viability of PV energy as an important and practical source of electricity in urban environments. The successful implementation of PV systems on residential rooftops emerges as an effective strategy to reduce CO_2 emissions and address climate change, contributing to a transition toward a cleaner and more sustainable energy mix.

Furthermore, the study provides valuable insights for urban planners and policymakers on fostering widespread adoption of PV energy in urban settings. The exploration of surplus energy generation on the studied rooftops raises the possibility of injecting this excess into the electrical grid, with potential economic benefits and increased diversification of energy sources. Adjusting the model to address variable demand patterns, fluctuations in electricity prices, and other factors such as shading is suggested for future research.

In addition, if the research aims to achieve a net zero energy balance, it is imperative to incorporate technical limitations into the design process when sizing PV systems. By considering factors such as energy consumption patterns, storage capacity, and grid integration capabilities, designers can optimize the sizing of PV systems to align with the desired energy balance objectives. This approach ensures that the energy generated by the PV systems matches the energy consumed or stored, minimizing excess energy production and maximizing the overall efficiency and sustainability of the energy infrastructure.

Finally, beyond the performance of the algorithm extensively discussed in this work, the proposal faces some limitations, with one of the most significant being the issue of energy calculation, a characteristic of the final phase of the computation process. In the presented study, the input data for calculating the energy yield achieved with the distribution of panels on the rooftops defined in the preceding stages of the process corresponds to actual annual measurements obtained from nearby weather stations and an existing photovoltaic installation. Consequently, the energy results obtained reflect this particular case of study, obviating the need to consider factors such as seasonality, ambient temperature, shading effects, among others.

An opportunity for enhancing this component of the proposed algorithm would involve developing and incorporating a feature capable of importing meteorological data housed in global public databases and effectively considering the aforementioned aspects to achieve a more realistic energy estimation at any geographical point. Additionally, preprocessing meteorological data while considering meteorological and environmental constraints before introducing them into the algorithm is an alternative solution that warrants further investigation.

ACKNOWLEDGMENTS

Edisson Villa Ávila expresses his sincere gratitude for the opportunity to partially present the findings of his research, conducted as part of his doctoral studies in the Ph.D. program in Advances in Engineering of Sustainable Materials and Energies at the University of Jaen, Spain. Finally, the authors thank Universidad de Cuenca, Ecuador, for easing access to the facilities of the Micro-Grid Laboratory of the Centro Científico Tecnológico y de Investigación Balzay (CCTI-B), for allowing the use of its equipment, and for authorizing members of its staff the provision of technical support necessary to carry out the experiments described in this chapter.

Paul Arévalo thanks the Call for Grants for the Requalification of the Spanish University System for 2021-2023, Margarita Salas Grants for the training of young doctors awarded by the Ministry of Universities and financed by the European Union –Next Generation EU.

REFERENCES

1. Q. Chen *et al.*, "Remote sensing of photovoltaic scenarios: techniques, applications and future directions," *Appl Energy*, vol. 333, no. September 2022, p. 120579, 2023, doi: 10.1016/j.apenergy.2022.120579.
2. T. Hubinský, R. Hajtmanek, A. Šeligová, J. Legény, and R. Špaček, "Potentials and limits of photovoltaic systems integration in historic urban structures: the case study of monument reserve in bratislava, slovakia," *Sustainability (Switzerland)*, vol. 15, no. 3, 2023, doi: 10.3390/su15032299.
3. Q. Zhong, J. R. Nelson, D. Tong, and T. H. Grubesic, "A spatial optimization approach to increase the accuracy of rooftop solar energy assessments," *Appl Energy*, vol. 316, no. April, p. 119128, 2022, doi: 10.1016/j.apenergy.2022.119128.
4. J. Y. Han, Y. C. Chen, and S. Y. Li, "Utilising high-fidelity 3D building model for analysing the rooftop solar photovoltaic potential in urban areas," *Solar Energy*, vol. 235, no. February, pp. 187–199, 2022, doi: 10.1016/j.solener.2022.02.041.
5. V. Adjiski, G. Kaplan, and S. Mijalkovski, "Assessment of the solar energy potential of rooftops using LiDAR datasets and GIS based approach," *International Journal of Engineering and Geosciences*, vol. 8, no. 2, pp. 188–199, 2022, doi: 10.26833/ijeg.1112274.
6. X. Huang, K. Hayashi, T. Matsumoto, L. Tao, Y. Huang, and Y. Tomino, "Estimation of rooftop solar power potential by comparing solar radiation data and remote sensing data—a case study in Aichi, Japan," *Remote Sens (Basel)*, vol. 14, 1742, 2022, doi: 10.3390/rs14071742.
7. A. Alharbi, Z. Awwad, A. Habib, and O. de Weck, "Economical sizing and multi-azimuth layout optimization of grid-connected rooftop photovoltaic systems using mixed-integer programming," *Appl Energy*, vol. 335, no. October 2022, p. 120654, 2023, doi: 10.1016/j.apenergy.2023.120654.
8. Y. An, T. Chen, L. Shi, C. K. Heng, and J. Fan, "Solar energy potential using GIS-based urban residential environmental data: a case study of Shenzhen, China," *Sustain Cities Soc*, vol. 93, no. March, p. 104547, 2023, doi: 10.1016/j.scs.2023.104547.
9. X. Wang, X. Gao, and Y. Wu, "Comprehensive analysis of tropical rooftop PV project: a case study in nanning," *Heliyon*, vol. 9, no. 3, p. e14131, 2023, doi: 10.1016/j.heliyon.2023.e14131.
10. L. Yan *et al.*, "Estimation of urban-scale photovoltaic potential: a deep learning-based approach for constructing three-dimensional building models from optical remote sensing imagery imagery," *Sustain Cities Soc*, vol. 93, no. March, p. 104515, 2023, doi: 10.1016/j.scs.2023.104515.
11. X. Feng, T. Ma, Y. Yamaguchi, J. Peng, Y. Dai, and D. Ji, "Potential of residential building integrated photovoltaic systems in different regions of China," *Energy for Sustainable Development*, vol. 72, no. August 2022, pp. 19–32, 2023, doi: 10.1016/j.esd.2022.11.006.
12. K. Panicker, P. Anand, and A. George, "Assessment of building energy performance integrated with solar PV: towards a net zero energy residential campus in India," *Energy Build*, vol. 281, p. 112736, 2023, doi: 10.1016/j.enbuild.2022.112736.

13. D. Jurakuziev, S. Jumaboev, and M. Lee, "A framework to estimate generating capacities of PV systems using satellite imagery segmentation," *Eng Appl Artif Intell*, vol. 123, no. October 2022, p. 106186, 2023, doi: 10.1016/j. engappai.2023.106186.

14. U. Kafle, T. Anderson, and S. P. Lohani, "The potential for rooftop photovoltaic systems in Nepal," *Energies (Basel)*, vol. 16, no. 2, pp. 1–13, 2023, doi: 10.3390/en16020747.

15. H. Nasrallah, A. E. Samhat, Y. Shi, X. X. Zhu, G. Faour, and A. J. Ghandour, "Lebanon solar rooftop potential assessment using buildings segmentation from aerial images," *IEEE J Sel Top Appl Earth Obs Remote Sens*, vol. 15, pp. 4909–4918, 2022, doi: 10.1109/JSTARS.2022.3181446.

16. A. Pinna and L. Massidda, "A complete and high-resolution estimate of Sardinia's rooftop photovoltaic potential," *Applied Sciences (Switzerland)*, vol. 13, 7, 2023, doi: 10.3390/app13010007.

17. H. Ren, Z. Ma, A. B. Chan, and Y. Sun, "Optimal planning of municipal-scale distributed rooftop photovoltaic systems with maximized solar energy generation under constraints in high-density cities," *Energy*, vol. 263, no. PA, p. 125686, 2023, doi: 10.1016/j.energy.2022.125686.

18. D. H. C. Salim *et al.*, "Unveiling Fernando de Noronha Island's photovoltaic potential with unmanned aerial survey and irradiation modeling," *Appl Energy*, vol. 337, 120857, February, 2023, doi: 10.1016/j. apenergy.2023.120857.

19. T. Sun, M. Shan, X. Rong, and X. Yang, "Estimating the spatial distribution of solar photovoltaic power generation potential on different types of rural rooftops using a deep learning network applied to satellite images," *Appl Energy*, vol. 315, 119025, January, 2022, doi: 10.1016/j.apenergy.2022.119025.

20. M. Ernst, X. Liu, C. A. Asselineau, D. Chen, C. Huang, and A. Lennon, "Accurate modelling of the bifacial gain potential of rooftop solar photovoltaic systems," *Energy Convers Manag*, vol. 300, 117947, November 2023, 2024, doi: 10.1016/j.enconman.2023.117947.

21. G. Molnár, L. F. Cabeza, S. Chatterjee, and D. Ürge-Vorsatz, "Modelling the building-related photovoltaic power production potential in the light of the EU's Solar Rooftop Initiative," *Appl Energy*, vol. 360, 122708, November 2023, 2024, doi: 10.1016/j.apenergy.2024.122708.

22. Y. Yang *et al.*, "Whether rural rooftop photovoltaics can effectively fight the power consumption conflicts at the regional scale – a case study of Jiangsu Province," *Energy Build*, vol. 306, 113921, November 2023, 2024, doi: 10.1016/j.enbuild.2024.113921.

23. G. Li *et al.*, "A district-scale spatial distribution evaluation method of rooftop solar energy potential based on deep learning," *Sol Energy*, vol. 268, 112282, September 2023, 2024, doi: 10.1016/j.solener.2023.112282.

24. H. Jiang *et al.*, "Exploring the optimization of rooftop photovoltaic scale and spatial layout under curtailment constraints," *Energy*, vol. 293, 130721, February, 2024, doi: 10.1016/j.energy.2024.130721.

25. X. Zhu, Y. Lv, J. Bi, M. Jiang, Y. Su, and T. Du, "Techno-economic analysis of rooftop photovoltaic system under different scenarios in China University Campuses," Energies 16, 3123 2023. https://doi.org/10.3390/en16073123

26. M. F. M. Zublie, M. Hasanuzzaman, and N. A. Rahim, "Modeling, energy performance and economic analysis of rooftop solar photovoltaic system for

net energy metering scheme in Malaysia," *Energies (Basel)*, vol. 16, 723, 2023, doi: 10.3390/en16020723.

27. S. Lin *et al.*, "Accurate recognition of building rooftops and assessment of long-term carbon emission reduction from rooftop solar photovoltaic systems fusing GF-2 and multi-source data," *Remote Sens (Basel)*, vol. 14, 3144, 2022, doi: 10.3390/rs14133144.

28. A. Tamošiūnas, "Selecting rooftop solar photovoltaic modules by measuring their attractiveness by a categorical-based evaluation technique (MACBETH): the case of Lithuania," *Energies (Basel)*, vol. 16, 999, 2023, doi: 10.3390/en16072999.

29. J. L. Espinoza, L. G. Gonzalez, and R. Sempertegui, "Micro grid laboratory as a tool for research on non-conventional energy sources in Ecuador," *2017 IEEE International Autumn Meeting on Power, Electronics and Computing, ROPEC 2017*, vol. 2018–January, pp. 1–7, Jan. 2018, doi: 10.1109/ROPEC.2017.8261615.

30. "Global Solar Atlas." Accessed: Jan. 30, 2024. [Online]. Available: https://globalsolaratlas.info/map?c=-2.896753,-79.033585,11&m=site&a=-79.118729, -2.924526,-79.118729,-2.869664,-78.947754,-2.869664,-78.947754,-2.924526,-79.118729,-2.924526&s=-2.892124,-79.029272

31. "Geovisor Público." Accessed: May 19, 2023. [Online]. Available: https://geoportal2.centrosur.gob.ec/geoportal/apps/webappviewer/index.html?id=72a7304acd7a4df78866e70d8efc7c80

32. "Visor IDE Cuenca." Accessed: March 19, 2024. [Online]. Available: http://ide.cuenca.gob.ec/geoportal-web/viewer.jsf

A residual deep neural network–based noninvasive system for tomato crop health monitoring, disease type identification, and remedial measures

Anurag Chauhan, Altaf Alam,
Mohd Tauseef Khan, Anuj Kumar,
Ankit Bhatt, and Sanjay Kumar Maurya

ABBREVIATIONS

BRBFNN	radial basis function network
BST	bacterial spot of tomato
CNN	convolutional neural networks
DNN	deep neural networks
FLM	fungi leaf mold
FN	false negative
FP	false positive
FPR	false positive rate
GDP	gross domestic product
KNN	k-nearest neighbors
MSER	maximally stable extremal regions
NN	neural network
R-FCN	region-based fully convolutional network
ResNet	residual network
SIFT	scale-invariant feature transform
SSD	single shot multibox detector
SURF	speeded up robust features
SVM	support vector machine
TEB	tomato early blight
TLBM	tomato late blight mold
TMV	tomato mosaic virus
TN	true negative
TP	true positive
TPR	true positive rate
TSLS	tomato septoria leaf spot
TST	target spot of tomato
TTSSM	tomato two-spotted spider mite
TYLCV	tomato yellow leaf curl virus

DOI: 10.1201/9781003581246-7

7.1 INTRODUCTION

In the current scenario, smart agriculture monitoring systems using modern technologies have received greater attention to produce an adequate amount of food. Modern technologies such as artificial intelligence (AI), machine learning (ML), artificial neural network (ANN), big data, block chain, etc. are extensively used by researchers to solve the problems of the agriculture sector. Among different crops, tomato cultivation is prominent across the world as this crop offers a rich amount of nutrition. Therefore, this crop plays an important role in the global agricultural production.

In India, the agricultural sector has a share of almost 17% of the total GDP. Majorly four kinds of crops are cultivated in India: food grains crops, cash crops, plantation crops, and horticulture crops [1–3]. Major food grain crops are wheat, pulses, and rice, whereas cotton, sugarcane, oilseeds, and tobacco are major cash crops [4, 5]. Furthermore, tea, coffee, coconut, and rubber are major crop products of plantation crops and all vegetables and fruits come into the category of horticulture crops [6, 7].

Tomato under the horticulture category is the third major crop of India. Tomato plays an important role in the growth of the Indian economy, hence it is on the top priority of the government list of horticulture crops along with potatoes and onions. In India, the tomato crop is grown two times a year across the country, i.e. *kharif*/monsoon season (June to September) and *rabi*/spring season (October to February) [8]. Karnataka, Tamil Nadu, Andhra Pradesh, Madhya Pradesh, Bihar, Gujarat, and Uttar Pradesh are the main states of India growing tomatoes and thereby meet the tomato requirement of the country [9]. The production volume of processed tomatoes has significantly reached 154,300 metric tons and fresh tomato production reached 24.09 MT/ha in India during 2019–2020.

Food insecurity is a major problem faced by us [10]. Plant diseases are the major cause of food insecurity and the reduction of crop yields. Almost 16% of tomato crop yields are lost globally due to plant disease [11]. Although tomato crops can be cultivated on any moderately well-drained soil type, high-quality organic stuff is useful to both increase the productivity and reduce the chance of diseases. Certified seeds and plants must be used for good production and the tomato crop should not be cultivated on the same land more than once in three years for the conservation of organics in the agriculture field [12].

The rotation of crops helps to conserve the high quality of organic stuff in the field. Therefore, other crops are required to grow between the consecutive tomato crops grown. Corn is a good product for crop rotation, it supplies good organic stuff and enhances production. Despite using good quality seeds and organic stuff, disease in tomato plants may occur due to various pathogens. As already mentioned, plant diseases are the root cause of food insecurity and the reduction of crop yields. Tomato disease can be a serious problem unless taken care of promptly. Disease detection is an important task to cure the crop before it spreads all over the plants. Accordingly, this work is focused on the design and development of a non-invasive system to monitor the health and diseases of tomato crops along with the disease removal suggestion system.

7.1.1 Related work

The infected crop is an abnormal condition of the crop; it disturbs the crop growth and reduces the crop yields as well. Tomato crop diseases and the existing disease detection techniques have been extensively surveyed and discussed in this section. Several types of infection might occur in tomato crops that may cause serious loss in production. Some of them have been discussed in the tomato plant pathogens section. Almost 16% of the total crops all over the world are damaged due to plant infection, hence this is a serious issue among farmers and researchers to meet the food requirements of human beings [13, 14].

Researchers from eminent institutes all over the world have proposed various solutions to automatic detection and disease identification of tomato crops [15–18]. Various imaging technologies such as hyperspectral imaging [19], remote sensing [20], satellite imaging [21], thermal imaging [22], and visual imaging [23, 24] technologies have been utilized by researchers for crop condition monitoring and identification of infected plants. Image-based monitoring system utilizes various features, i.e., chromatic features [25], texture features [26], edge features (i.e., Canny, LoG, Differential, etc.) [27], geometrical features (i.e., Area, Minima, Maxima, Solidity, etc.) [28], multilevel features (i.e., SIFT, SURF, MSER, etc.) [29], and various ML techniques (i.e., NN, KNN, SVM, etc.) [30–32] for disease identification.

Feature extraction, suitable feature selection, and suitable algorithm adoption are complex, time-consuming, and labor-intensive tasks that lead the diagnosis system toward reducing the detection and identification accuracy [33]. Therefore, automatic feature extraction and selection might be useful to reduce the system complexity and get a highly accurate system. Convolutional neural network (CNN) architecture has been utilized by renowned researchers for automatic feature extractions and learning the pattern to identify particular objects in a sequence of images. Radial Basis Function Network (BRBFNN) [34], VGG-16 [35], ResNet-18 [36], ResNet-50 [37], ResNet-152 [38], and Faster R-CNN [39] are a few common deep neural network (DNN) architecture that has been utilized for crop disease detection, identification, and recognition tasks. CNN architecture consists of a DNN which has multiple hidden layers in its architecture that help to extract the feature from the image data automatically and give very promising results in detection and recognition.

Chouhan et al. [34] proposed a bacterial forging-based optimization CNN radial basis function network for plant disease detection and recognition. Arya and Singh [40] proposed a DNN-based disease identification in potato and mango leaves. A VGG-16 pretrained network has been utilized to estimate the harshness of the disease in the apple plant by Wang et al. [41]. LeNet DNN is utilized by Amara et al. [42, 43] to identify the disease in banana leaves. Kaur and Bhatia [44] utilized a pretrained residual network (ResNet) to classify the tomato diseases into seven classes. Rahman et al. [45] proposed a fully connected CNN architecture to classify the tomato leaves diseases into three diseases: bacterial spot of tomato (BST), tomato late blight mold (TLBM), and tomato septoria leaf spot (TSLS).

AlexNet and Squeeze Net architectures have been utilized by Durmuş et al. [46] for classifying ten diseases of tomato crops and getting an accuracy of 95.5%. Chakravarthy and Raman [47] utilized ResNet and Xception networks as classification networks and YOLO detector framework to detect and identify the tomato early blight (TEB) disease in tomato leaves. Antagonistic potential of plant growth–promoting rhizobacteria-based technique has been utilized by Attia et al. [48] to detect the TEB disease of tomato crop. Wang et al. [49] also proposed a system for the automatic identification of tomato disease types and infected areas on leaves based on faster R-CNN and mask R-CNN classifier, In addition, a DNN-based detector is also used for localizing the infected area in tomato leaves.

Abdulridha et al. [50] utilized hyperspectral imaging and unmanned aerial vehicle technology to identify the target spot of tomato and bacterial spot of tomato diseases in tomato leaves. Single shot multibox detector (SSD), region-based fully convolutional network (R-FCN) object detector network and faster region-based convolutional neural network (Faster R-CNN) have been used by Fuentes et al. [51] to detect the ten tomato crop diseases. A CNN deep neural network has been proposed by Agarwal et al. [52] for classification of ten diseases from tomato crop achieving an accuracy of 91.2%. Researchers also used different ML algorithms to monitor and identify diseases of crops [69–74].

The literature survey indicates that most of the proposed work were focused on the identification of a single disease or two to three diseases of tomatoes. Several other diseases could also affect the tomato crop and might damage it. Therefore, disease identification, disease classification, and disease detection of tomato crops are still an open issue for farmers and researchers all over the world. In this work, a system is proposed that is capable of diagnosing the tomato crop health and identifying the disease type in addition to suggesting the possible solution to remove the disease from the crop.

The proposed system identified nine different diseases of tomato: tomato early blight, tomato septoria leaf spot, fungi leaf mold, target spot of tomato, bacterial spot of tomato, tomato yellow leaf curl virus, tomato late blight mold, tomato mosaic virus, and tomato two-spotted spider mite. This work modified the fully connected layers of two ResNet-50 pretrained classifiers to fulfill the objectives. The first ResNet-50-based classifier is customized to monitor the health of tomato crops, i.e., healthy and unhealthy. The second ResNet-50 classifier has been customized to identify the disease type. Therefore, this work generated a rules-based system, based on tomato crop health, and identified disease types to suggest the solution and action to remove the disease from the crop.

7.1.2 Motivation

This work utilizes ResNet-50 DNN architecture to monitor the health and diseases of crops. The ResNet-50 classifier has been used by eminent researchers to develop the human health diagnosing system and other plant disease classification systems, achieving accurate and lucrative results. Sakib and colleagues [53] utilized ResNet-50 DNN architecture to detect the infection of COVID-19 disease in chest X-ray images. In Ref. [37], Sarwinda et al. utilize ResNet variants for colorectal cancer detection and achieved good accuracy. Mukti et al. [54] proposed a system that successfully identified 38 different diseases in plants with high accuracy of almost 99.8%. The above-discussed work and their accurate results motivate us to adopt the ResNet-50 DNN architecture to develop a system to monitor crop heath and to identify diseases. The following are the major contributions of the performed work:

1. Tomato leaf data collection and preparation (i.e., image data collection, image annotation, subimage generation) for developing the tomato crop health monitoring system.
2. Tomato leaf data collection and preparation for developing the tomato crop disease (i.e., TEB, TSLS, TST, FLM, BST, TLBM, TYLCV, TMV, and TTSSM) identification system.
3. The fully connected layer of ResNet-50 DNN architecture has been customized (i.e., optimizer selection, loss function selection, learning rate selection, batch size selection, number of epoch selection, etc.) to monitor the tomato crop health.
4. The fully connected layer of ResNet-50 DNN architecture has been customized (i.e., optimizer selection, loss function selection, learning rate selection, batch size selection, number of epoch selection, etc.) to identify the nine different disease types of tomato crop.
5. Generated a rule-based suggestion system based on the tomato crop health and identified disease types to suggest the solution and action for removing the disease from the crop.

This chapter is organized as follows: Section 7.2 formulates the tomato plant pathogens in detail. The material and methodology adopted to solve the crop monitoring and disease identification issue at hand are presented in Section 7.3. The quantitative result and discussion of the achieved consequence of the proposed system are analyzed and portrayed in Section 7.4. Finally, Section 7.5 concludes the chapter.

7.2 TOMATO PLANT PATHOGENS

The tomato cultivation in India is an intensive horticultural production due to the dynamic weather behavior. Infections are main restrictive feature for the vegetable that cause serious crop yield reduction which leads to severe economic losses. The researchers have observed throughout the study that the major pathogens for tomato plant infections are fungi, bacteria, mold, viruses, and mate [55–57]. In addition, these major pathogens are also cause of various subdivisions of disease, as shown in Figure 7.1. The pathogens of tomato crops are extensively described as follows:

7.2.1 Fungi

The fungi-related disease occurs on tomato crops due to natural opening on the plant, wounds caused by pruning, harvesting, hails, and insect, and fungi disease might occur due to mechanical damage to the crop. TEB, TSLS, and TST are some common types of fungi infections on the tomato crop which have been considered in this work for early detection through a deep neural network algorithm.

7.2.2 Bacteria

Bacterial pathogens in tomatoes are due to contaminated seeds and infected plants. After establishing in the crop, they spread through splashing rainwater runoff, use of infected equipment, and people working around the crop. Bacterial spot of tomato is a common disease in tomato caused by bacteria and has been considered in this work to identify automatically with the help of a developed algorithm.

7.2.3 Mold

Non-obligate pathogens that are the source of infection in a tomato plant are known as tomato leaf mold disease. The mold disease grows in crops

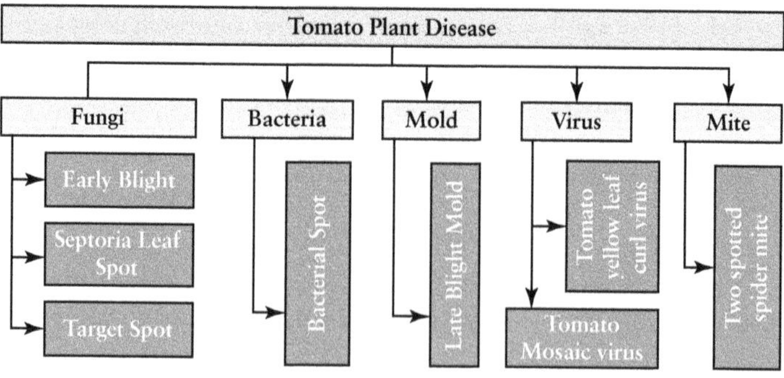

Figure 7.1 Tomato plant pathogens.

due to the excess humidity and cool environment. It affects the leaves, stems, and fruit of tomatoes as well. TLBM is a common infection that has been considered in this work for detection and identification.

7.2.4 Virus

The virus disease transmitted to the tomato crop by an insect vector that belongs to the family of Aleyrodidae and order Emiptera is the whitefly *Bemisia tabaci* (silver leaf whitefly). It affects the leaves and tomato fruit of crops. TYLCV and TMV are some common types of virus infections on the tomato crop that have been considered in this work for early detection and disease identification.

7.2.5 Mite

The mite disease occurred in tomato crops due to the existence of excess nitrogen and dry and dusty conditions which affect the leaves and stem of tomato crops. TTSSM is the disease that has been considered for identification and early disease detection.

The available literature concludes that bacterial infection can affect almost all parts of the plant and it can be cured by avoiding the use of infected seeds and proper sterilization of agricultural tools regularly. Virus infection can damage almost 90–100% of tomato crops and cannot be controlled with the help of fertilizer. Only proper treatment such as weed removal, disinfection of tools, disinfection of humans, and use of good quality seeds is useful in viral infection. Early recognition of these diseases might be helpful to the farmer in protecting the crop and the enhancement of its yield.

Therefore, various invasive and noninvasive techniques have been proposed by eminent researchers all over the world that work effectively in disease detection. This research work focuses on the development of a noninvasive tomato crop disease identification (NITCDI) system that utilizes a camera as a sensor to collect crop information and customized DNN architecture for crop health monitoring and disease type identification. The proposed NITCDI is extensively described in the material and method section.

7.3 MATERIAL AND METHODS OF PROPOSED NONINVASIVE TOMATO CROP DISEASE IDENTIFICATION SYSTEM

The workflow of the proposed NITCDI system is shown in Figure 7.2. As shown in the figure, the proposed system performed three different tasks in three different stages. In the first stage, the proposed system monitors the

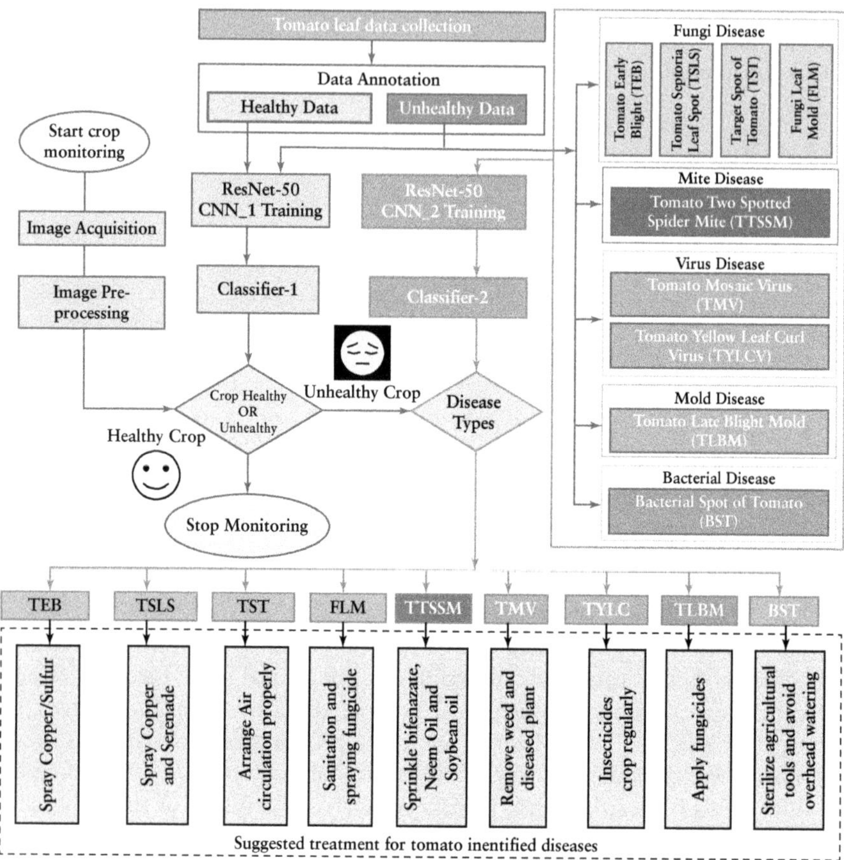

Figure 7.2 Schematic diagram of the proposed tomato health diagnosis and disease type identification system.

tomato crop health, i.e., crop is healthy or unhealthy. If the system output of the first stage is unhealthy crop, then the proposed system further identifies the type of disease, i.e., TEB, TSLS, TST, FLM, BST, TLBM, TYLCV, TMV, and TTSSM in the second stage. In addition, the proposed system also suggested the required solution to remove the disease from the crop.

7.3.1 Experimental setup

The proposed NITCDI is developed on a delicate processor that has an i7 Core processor with a 3.5 GHz processing speed, 8 GB internal memory (RAM), and graphics card specifications added as GTX 1650 TI NVIDIA with 4 GB inbuilt memory. The end-to-end experimental setup is developed and implemented through MATLAB 2021, a software platform.

7.3.2 Dataset used

Image captures details for the development and testing of the proposed NITCDI system. Images of tomato crops have been collected from Atarra town municipal area in the Indian state of Uttar Pradesh which is located at 25.28°N and 80.57°E. The average elevation of Atarra is 407 ft. (124 m) in India. The image capturing duration was fixed between morning 8 o'clock and evening 5 o'clock. While clicking the images for crop monitoring, the latitudes and longitudes of agricultural region were also fixed; the latitude was between 25°17′12.2676″ N and 25°28′46.6392″ N and the longitude was between 80°4′27.82″ E and 80°20′16.87″ E. A total of 11,000 images was prepared for creating the crop health monitoring system and another one was prepared for creating the disease identifier system.

7.3.2.1 Dataset used in the evolution of health monitoring system for tomato crop

To evolve a health monitoring system for tomato crop, this work utilized a total of 2,200 images, as shown in Table 7.1. Out of 2,200 images, 1,100 images were from healthy leaves and 1,100 images were from unhealthy leaves. Out of 1,100 unhealthy images, the unhealthy image dataset comprised 122 images of each type of diseased image leaf. Further, the dataset was subdivided into training and testing datasets in 80% and 20% ratios. Eighty percent of total data was used for training the crop health monitoring system, whereas 20% of data was used for testing the developed system. Figure 7.3a–j shows the sample image of both diseased and healthy leaves.

Table 7.1 Detail of tomato leaf image dataset utilized in crop health monitoring system

S. No.	Data type		Number of images
1	Healthy data		1,100
2	Unhealthy data	TEB	122
		TSLS	122
		TST	122
		FLM	122
		BST	122
		TLBM	122
		TYLCV	122
		TMV	123
		TTSSM	123
	Total images		2,200

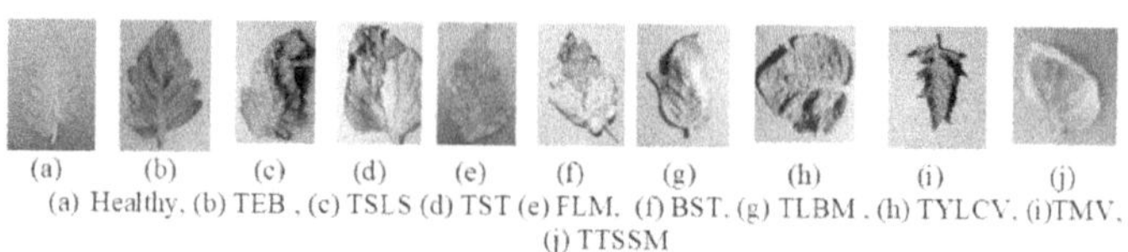

Figure 7.3 Sample of tomato leaves image dataset.

Table 7.2 Details of tomato leaf image dataset utilized in tomato crop disease identification

S. No.	Disease type	Infection type	Number of images
1	Tomato early blight (TEB)	Fungi disease	1,100
2	Tomato septoria leaf spot (TSLS)		1,100
3	Target spot of tomato (TST)		1,100
4	Fungi leaf mold (FLM)		1,100
5	Bacterial spot of tomato (BST)	Bacterial disease	1,100
6	Tomato late blight mold (TLBM)	Mold disease	1,100
7	Tomato yellow leaf curl virus (TYLCV)	Virus disease	1,100
8	Tomato mosaic virus (TMV)		1,100
9	Tomato two spotted spider mite (TTSSM)	Mite disease	1,100
Total images			**9,900**

7.3.2.2 Dataset used in the development of tomato crop disease identification system

In the development of the tomato crop disease identification system, this work utilizes a total of 9,900 images, as shown in Table 7.2. Out of 9,900 images, the dataset comprises 1,100 images from every nine types of tomato crop diseases. Further, the dataset was subdivided into training and testing sets. Eighty percent of the total data was used for training the disease identification system, whereas 20% of the data was used for testing the system.

7.3.3 Development of **NITCDI** system

The developed NITCDI system can diagnose the tomato crop health, i.e., healthy or unhealthy, and it can also identify the tomato crop disease type. Both the tasks (i.e., crop health monitoring and disease type identification) have been achieved by customizing the fully connected layers of two different ResNet-50 pretrained classifiers. ResNet is abbreviated for residual network, first introduced by He et al. [58]. ResNet allowed the training of a very deep neural network without vanishing the gradients. In deep network training, the gradient is used to backpropagate to the layers of earlier stages and repeated multiplication makes it very small; this issue is called the vanishing of gradient problem. It has been resolved in ResNet DNN by introducing the skip connection concept.

The traditional DNN (earlier to ResNet) was stacking multiple convolution layers for making the deep neural network, but ResNet proposed stacking multiple convolution layers and it also added the output of the convolution block to the original input. The addition of original input to the convolution output is known as a skip connection. This way ResNet alleviates the vanishing point of the gradient problem by adding a small path for the gradient to flow through the network, which permits the system to learn an identity function. This identity function ensures that the performance of higher-level layers will be as good as the lower-level layers. The identity mapping of ResNet is shown in Figure 7.4.

ResNet-50 networks comprise 50 layers of the residual network. ResNet-50 model combines the five stages of identity and five stages of convolution blocks to create the network architecture. In each stage of architecture, the convolution block consisted of three convolution layers and each identity block also consisted of three convolutional layers [59]. The architecture of the ResNet-50 DNN is shown in Figure 7.5. As shown in the figure, the training input image size of ResNet-50 is 224 × 224 × 3, while the first stage of the convolutional layer comprises 64 convolutional filters

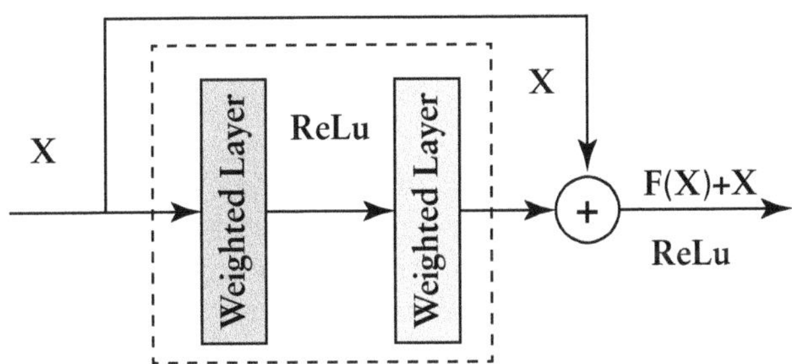

Figure 7.4 Identity mapping of ResNet architecture.

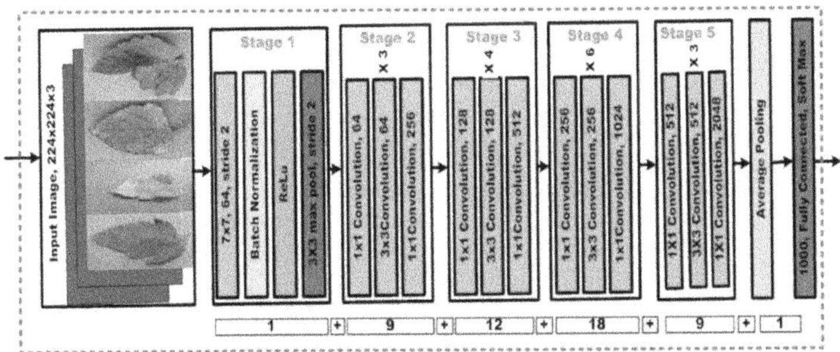

Figure 7.5 The architecture of ResNet-50 deep neural network model.

of 7×7 kernel size with stride 2 and zero padding. The ResNet-50 can extract almost 23 million trainable parameters from images which helps to enhance the system's accuracy.

7.3.3.1 Development of tomato crop health monitoring system

In this section, details of the customization of fully connected layers of ResNet-50 pretrained classifier for monitoring the tomato crop health have been discussed. In the development of the tomato health monitoring system, the fully connected layer (output layer) of the ResNet-50 pretrained model has been modified according to the desired output. The modified ResNet-50 architecture is trained with the created tomato crop dataset. Details of training parameters that have been utilized in the training of the crop health monitoring system are shown in Table 7.3. The health monitoring system training utilized an adaptive moment estimation (ADAM) optimizer, binary cross-entropy (BCE) loss function, and the initial learning rate of the system is 1×10^{-4}. The system used a maximum of five epochs to train the system and utilized a minimum batch size of 8. The weight learns factor of the model is 2 during the training, the bias learns factor of the system is also 2, and the health monitoring system training used Sigmoid activation function during the training.

- *Adaptive moment estimation (ADAM) optimizer*

ADAM optimizer engages a fusion of two gradient descent methodologies, i.e., gradient descent with momentum algorithm and the root mean square prop (RMSP) algorithm [60]. Gradient descent with momentum algorithm updates its weight as shown in Equations (1) and (2). The RMSP algorithm updates its weight as shown in Equations (3) and (4).

$$W_{(t+1)} = W_t - \propto m_t \tag{1}$$

$$m_t = \beta \times m_{t-1} + (1 - \beta) \times \frac{\partial L}{\partial W_t} \tag{2}$$

Table 7.3 Training parameter details of the proposed tomato crop health-monitoring system

Optimizer	Loss	Initial learning rate	Maximum epochs	Minimum batch size	Modified fully connected layer for disease recognition		
					Number of classes	Weight learn factor	Bias learn factor	Activation function
ADAM	Binary cross-entropy	1×10^{-4}	5	8	2	2	2	Sigmoid

Here m_t denotes the aggregate of the gradient at a particular time t, the initial value of m_t would be equal to 0, m_{t-1} represents the aggregate of the gradient at time $t-1$, W_t represents the weight at time t and W_{t+1} represents the weight at time $t + 1$, $\propto$ represents the learning rate of the model, β is the moving average parameters of optimizer – it would lie between constant and 0.9 range, ∂L is the derivative of the loss function. ∂W_t is derivative of weights at time t:

$$W_{t+1} = W_t - \frac{\alpha_t}{\sqrt{V_t + \epsilon}} \times \left[\frac{\partial L}{\partial W_t} \right] \tag{3}$$

$$V_t = \beta \times V_t + (1 - \beta)\lambda \times \left[\frac{\partial L}{\partial W_t} \right]^2 \tag{4}$$

Here W_t represents the weight at time t and $W_{(t+1)}$ represents the weight at time $t + 1$, $\propto$ represents the learning rate of the model, ∂L denotes the derivative of the loss function, ∂W_t is the derivative of weights at time t, V_t represents the sum of the square of past gradients, the initial value of V_t should be equal to 0, β is the moving average parameters of optimizer –it would lie between constant and 0.9 range, and ϵ is a small positive constant of the optimizer.

The ADAM optimizer has constructive quality of two methods (i.e., GDSM and RMSP) and creates upon them to develop a more optimized gradient descent. A new mathematical expression for m_t and V_t has been computed from gradient descent with momentum and RMSP which is shown in Equations (5) and (6), respectively.

$$m_t = \beta_1 m_t + (1 - \beta_1)\frac{\partial L}{\partial W_t} \tag{5}$$

$$v_t = \beta_2 v_{t-1} + (1 - \beta_2)\left[\frac{\partial L}{\partial W_t} \right]^2 \tag{6}$$

Here ϵ denotes a small positive constant to avoid the division by 0 error when v_t is greater than 0, β_1 and β_2 are the decay rates of an average of gradients, $\beta_1 = 0.9$ and $\beta_2 = 0.999$, and $\propto$ represents the step size parameters or learning rate (10^{-4}).

The initialized value for both m_t and V_t is 0, hence its tendency to be biased toward 0, and β_1 and β_2 are approximately equal to 1. The optimizer tries to fix this tendency of 0 biased by computing the bias-corrected m_t and V_t. Bias-corrected m_t and V_t are also achieved through controlling the weights by preventing them from converging at the global minima. The mathematical expression for bias-corrected m_t and V_t controlled weight is given by Equations (7) and (8):

$$\widehat{m_t} = \frac{m_t}{1-\beta_1^t} \quad \widehat{v_t} = \frac{v_t}{1-\beta_2^t} \tag{7}$$

$$W_{t+1} = W_t - \widehat{m_t}\,\frac{\alpha_t}{\sqrt{\widehat{v_t}+\in}} \tag{8}$$

Here $\widehat{m_t}$ is the bias-corrected aggregate of the gradient at time t and $\widehat{v_t}$ is the bias-corrected sum of the square of past gradients. Instead of simple m_t and V_t, the ADAM optimizer used this bias-corrected $\widehat{m_t}$ and $\widehat{v_t}$ for updating the weight, hence it gives higher performance.

- *Sigmoid activation function*

This classifier diagnoses whether the tomato is healthy or unhealthy, i.e., two-class classification problem. Therefore, the Sigmoid activation function has been used at the output layer of the model [61]. The mathematical expression for the Sigmoid function is shown in Equation (9):

$$f(x_i) = \frac{1}{\exp(x_j)} \tag{9}$$

7.3.3.2 Development of tomato crop disease identification system

Details of the development of the tomato crop disease type identification system have been discussed in this section. In the development of the tomato crop disease type identification system, the fully connected layer (output layer) of the ResNet-50 pretrained model has been modified according to the desired output and the trained network architecture again with the created tomato crop disease dataset. Details of the training parameters that have been utilized in the development of the tomato crop disease identification system are shown in Table 7.4.

Table 7.4 Training parameter details of the proposed tomato disease types identification system

| | | | | | | Modified fully connected layer for disease recognition | | |
Optimizer	Loss	Initial learning rate	Maximum epochs	Minimum batch size	Number of classes	Weight learn factor	Bias learn factor	Activation function
ADAM	Categorical cross-entropy	1×10^{-4}	5	8	10	10	10	Softmax

The crop disease type identification system training utilized an adaptive moment estimation optimizer, categorical cross-entropy (BCE) loss function, and the initial learning rate of the system is 1×10^{-4}. The system used a maximum of five epochs to train the system and utilized a minimum batch size of 8. The weight learns factor of the model is 10 during the training, the bias learns factor of the system is also 10, and the health monitoring system training used the SoftMax activation function during the training.

- *SoftMax activation function*

SoftMax activation function used probability distribution of features from features of the real number. The output range of the SoftMax function lies between 0 and 1 [62]. The mathematical expression for the SoftMax activation function is shown in Equation (10).

$$f\left(x_i\right) = \frac{\exp\left(x_i\right)}{\sum_j \exp\left(x_j\right)} \tag{10}$$

7.4 RESULTS AND DISCUSSIONS

This section represents the simulation details and obtained results of the proposed NITCDI system. Two ResNet-50 network architectures have been modified: the first one is modified for crop health monitoring, and the second one is modified for identifying the disease type if the crop is unhealthy. The complete tasks of system development (i.e., health diagnosis and disease type identification) were implemented on MATLAB 2021, a software using a delicate system that has an i7 Core processor with 3.5 GHz processing speed and 8 GB RAM. Further, the system had a GeForce GTX, 1650 TI NVIDIA graphics card with inbuilt 4 GB memory.

A dataset of 11,000 images of tomato leaves were collected which were used to monitor the health and identify the type of diseases. The details of the created dataset that has been used in the NITCDI system are given in Section 7.3.

Evaluation of the proposed NITCDI system is performed based on different parameters, i.e., accuracy, precision, F1 score, loss, true positive rate, and false positive rate, as provided in Table 7.5.

> *True positives (TP):* It is a condition when the system predicted healthy tomato crops as healthy tomato crops, which is considered the true positive.
> *True Negatives (TN):* It is a condition when the system predicted unhealthy tomato crops as unhealthy tomato crops, which is considered the true negative.

Table 7.5 Performance metrics mathematical expressions

S. No.	Name of parameter	Mathematical expression
1	True positive rate or recall	$TPR\% = \dfrac{TP}{TP + FN} \times 100$
2	False positive rate	$FPR\% = \dfrac{FP}{FP + TN} \times 100$
3	Accuracy	$ACC\% = \dfrac{TP + TN}{TP + TN + FP + FN} \times 100$
4	Precision	$Precision\% = \dfrac{TP}{TP + FP} \times 100$
5	F1 score	$F1\,Score = \dfrac{2 \times Precision \times Recall}{Precision + Recall}$
6	Loss	$Loss = (1 - Accuracy) \times 100$

False Positives (FP): It is a state when the system predicted unhealthy tomato crops as healthy tomato crops, which is considered the false positives.

False Negatives (FN): It is a case when the system predicted a healthy tomato crop as unhealthy tomato crop, which is considered the false negative.

7.4.1 Result of developed tomato crop health monitoring system

The developed tomato health crop monitor system has been evaluated based on the above-discussed parameters. The confusion matrix of the developed system is shown in Figure 7.6, i.e., information on TP, TN, FP, and FN prediction. The confusion matrix concludes that out of 1,100 positive samples and 1,100 negative samples, 1,095 samples were predicted as truly positive, 1,090 samples were predicted as truly negative, 15 samples were predicted as false positive, and 10 samples were predicted as false negative by the system. In addition, the performance parameters have been calculated with the predicted TP, TN, FP, and FN of the system, as provided in Table 7.6. Table 7.6 determines that the TPR of the proposed tomato crop health monitoring system is 99.09%, FPR is 1.3%, accuracy is 99.3%, precision is 98.64%, loss is 0.7% only, and F1 score is 98.86%.

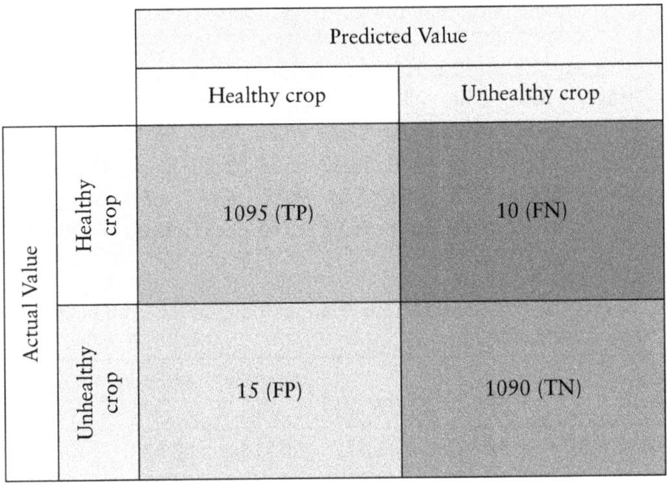

		Predicted Value	
		Healthy crop	Unhealthy crop
Actual Value	Healthy crop	1095 (TP)	10 (FN)
	Unhealthy crop	15 (FP)	1090 (TN)

Figure 7.6 Confusion metrics of the developed tomato crop health monitor.

Table 7.6   Results of the developed tomato crop health-monitoring system

TPR (%)	FPR (%)	Accuracy (%)	Precision (%)	Loss (%)	F1 score (%)
99.09	1.3	99.3	98.64	0.7	98.86

Table 7.7   Prediction of developed tomato crop disease type identification system

Disease prediction	TEB	TSL	TST	FLM	BST	TLBM	TYLCV	TMV	TTSSM	Total
Number of images	250	250	250	250	250	250	250	250	250	**2,250**
TP	242	240	241	243	239	240	241	244	240	**2,170**
FP	6	8	10	9	12	12	11	8	8	**84**
FN	8	10	9	7	11	10	9	6	10	**80**

7.4.2 Result of the developed tomato crop disease type identifier system

In this section, the performance of the developed disease type identifier has been evaluated and discussed. The crop health-monitoring system output predicted the crop health, i.e., healthy and unhealthy. In addition, the unhealthy crop is further processed for identification of the disease type with the help of the second modified ResNet-50 architecture. Evaluation of the diseased type identifier system has been performed based on TP, FP, FN, TPR, FPR, precision, and F1 score, as provided in Tables 7.7 and 7.8.

A total of 2,250 image samples have been analyzed for determining the performance of the developed system. Out of 2,250 total images, 250

Table 7.8 Results of the developed tomato crop disease type identification system

Disease prediction	TEB	TSL	TST	FLM	BST	TLBM	TYLCV	TMV	TTSSM	Total
Number of images	250	250	250	250	250	250	250	250	250	**2,250**
TPR (%)	96.80	96.00	96.40	97.20	95.60	96.00	96.40	97.60	96.00	**96.44**
Precision (%)	97.58	96.77	96.01	96.42	95.21	95.20	95.63	96.82	96.77	**96.27**
FPR (%)	2.41	3.22	3.98	3.57	4.78	4.76	4.36	3.17	3.10	**3.70**
F1 score	97.18	96.38	96.20	96.80	95.40	95.59	96.01	97.20	96.38	**96.34**

Table 7.9 Numerical results of various deep learning–based crop health-monitoring systems

Parameters Methods	TPR (%)	FPR (%)	Accuracy (%)	Precision (%)	Loss (%)	F1 score (%)
Brahimi et al. [63]	88.98	12.87	89.71	88.67	10.29	88.82
Brahimi et al. [63]	90.11	10.90	90.65	89.03	9.35	89.56
Barman et al. [64]	91.98	8.66	92.42	92.00	7.58	91.98
Maheswari et al. [65]	95.21	5.02	95.63	94.35	4.37	94.77
Proposed	**99.09**	**1.30**	**99.30**	**98.64**	**0.7**	**98.86**

images were from each class, and performance parameters were calculated. As shown in Table 7.7, out of 2,250 images, the proposed system identifies 2,170 diseases truly, whereas the system predicted only 84 diseases as false positive and 80 diseases as false negative.

The quantitative results of the evaluation parameters, i.e., TPR, precision, FPR, and F1 score, are provided in Table 7.9. The results of the disease type identification system for each disease, i.e., TEB, TSLS, TST, FLM, BST, TLBM, TYLCV, TMV, and TTSSM, have been evaluated and provided in Table 7.8. The overall TPR of the developed disease type identification system is 96.4%, precision is 96.27%, FPR is 3.70%, and F1 score is 96.34%.

7.4.3 Comparison of the developed monitoring system with existing **DNN** systems

In this section, a comparison of results of the developed system to monitor the crop health and identify the type of diseases with other existing deep learning–based state-of-the-art techniques is given in Tables 7.9 and 7.10, respectively. Brahimi et al. [63] analyzed the performance of existing pretrained DNN architecture, i.e., Google Net and Alex Net; this work compared the developed DNN system (i.e., diagnosis system for crop health and identification of diseases) with these two networks.

In addition, the developed crop health diagnosis and disease types identification system is also compared with Mobile Net [64] and ResNet-18 DNN [65] to check and evaluate the robustness and accuracy of the developed

Table 7.10 Numerical results of various deep learning–based tomato crop disease identification systems

Parameters methods	TPR (%)	FPR (%)	Accuracy (%)	Precision (%)	Loss (%)	F1 score (%)
Google Net	83.65	17.65	82.95	83.32	17.05	83.48
Alex Net	86.51	15.07	85.90	86.12	14.1	86.31
MobileNet	89.34	11.67	89.73	90.04	10.27	89.68
ResNet18	92.02	7.94	92.00	92.30	8.00	92.15
Proposed	**96.44**	**3.70**	**96.24**	**96.27**	**3.76**	**96.34**

system compared to the existing state-of-the-art DNN architecture. To achieve a fair comparison between the existing DNN system and the proposed system, this work utilized a similar dataset, hardware, and software platform to train other networks and proposed systems.

Table 7.10 shows the comparative result of the proposed health diagnosis system with the above-discussed state-of-the-art existing pretrained DNN. The Google Net DNN architecture achieved TPR of 88.98%, FPR of 12.87, accuracy of 89.71%, precision of 88.67%, loss of 10.29%, and F1 score of 88.82. The ResNet-18 DNN achieved TPR of 95.21%, FPR of 5.02%, accuracy of 95.63%, precision of 94.35%, loss of 4.37%, and F1 score of 94.77%. The proposed health-monitoring system achieved TPR of 99.09%, FPR of 1.30%, accuracy of 99.30%, precision of 98.64%, loss of 0.7%, and F1 score of 98.86%, as provided in Table 7.9.

However, the tomato health-monitoring results provided in Table 7.9 demonstrate that the proposed method gives better performance in comparison to the other existing deep learning systems. The proposed models achieved lucrative, robust, and accurate results; hence, they could be used as an efficient crop health diagnosis tool for monitoring the tomato crop health and its condition.

Table 7.10 shows the comparative result of the proposed disease types identification system with other existing state-of-the-art DNN network systems. In Table 7.10, Google Net achieved TPR of 83.65%, FPR of 17.65%, accuracy of 82.95%, precision of 83.32%, loss of 17.05%, and F1 score of 83.48%. ResNet-18 DNN achieved TPR of 92.02%, FPR of 7.94%, accuracy of 96.24%, precision of 96.27%, loss of 8%, and F1 score of 83.48%. The TPR of the proposed disease-type identification system is 96.44%, FPR is 3.70%, accuracy is 96.24%, precision is 96.27%, loss is 3.76%, and F1 score is 96.34%.

However, the proposed tomato crop disease identifier is more accurate, robust, and lucrative as compared to the existing DNN architecture. Farmers can leverage with proposed tomato crop health diagnosis and disease identification system, for identifying and removing the disease which leads to an increase in the crop yields and income of farmers as well.

7.4.4 Development of a rule-based system for tomato crop disease removal suggestion

The proposed health monitoring system diagnoses the tomato crop health, and disease type identification system identifies the disease in the crop. Furthermore, the work is extended and developed a rule-based system that suggests the remedy and action to remove the diseases from tomato crops. The details about the tomato crop disease and the required solution have been surveyed by eminent researchers and the report of Agricultural Society of India suggests the remedy and action to remove the tomato crop diseases

Table 7.11 A developed rule-based system for tomato crop disease removal

S. No.	Image sample	Health condition	Disease type	Solution
1		If the crop is healthy	And disease type is none	Not required
2		If the crop is unhealthy	And disease type is TEB	Spray copper/sulfur
3		If the crop is unhealthy	And disease type is TSLS	Spray copper and serenade
4		If the crop is unhealthy	And disease type is TST	Proper air circulation required
5		If the crop is unhealthy	And disease type is FLM	Sanitation and spray fungicide
6		If the crop is unhealthy	And disease type is TTSM	Sprinkle Bifenzate, neem oil, and soyabean oil
7		If the crop is unhealthy	And disease type is TMV	Remove weeds and disease plants
8		If the crop is unhealthy	And disease type is TYLC	Insecticides crop regularly
9		If the crop is unhealthy	And disease type is TLBM	Apply fungicide
10		If the crop is unhealthy	And disease type is BST	Sterilize agricultural tools and avoid overhead watering

[66, 67]. The developed rule-based system for applying the solution according to crop conditions is provided in Table 7.11.

The developed rule-based system suggests that if the crop is healthy and the disease type is none, then no action is needed. But if the crop is unhealthy and the developed disease type identification system identified the disease type, then the farmers can take action according to the proposed system suggested in Table 7.11. However, the developed rule-based system for disease removal might be helpful in detecting and identifying the accurate disease type as well as in finding a precise solution for each disease.

7.5 CONCLUSIONS

In this chapter, a noninvasive tomato crop disease identification system has been developed and tested to diagnose the tomato crop health and identification of the disease type. Further, a rule based system is used to obtain the solution for the disease. Image dataset has been collected from Atarra town in the Indian state of Uttar Pradesh.

In the chapter, output layers of two ResNet-50 pretrained DNN architecture has been modified for crop health monitoring and disease identification, respectively. The proposed tomato crop health monitoring system achieves TPR of 99.09%, FPR of 1.3%, accuracy of 99.3%, precision of 98.64%, loss of 0.7%, and F1 score of 98.86%. The proposed disease type identification system achieved TPR of 96.44%, FPR of 3.70%, accuracy of 96.24%, precision of 96.27%, loss of 3.76%, and F1 score 96.34%. Different performance parameters of the proposed system have been compared with the existing state-of-the-art DNN architecture. It has been concluded that the proposed system achieves more accurate and lucrative results, hence it could be beneficial for identifying the types of diseases in tomato crops, and the developed rule-based system would be helpful to overcome the identified disease.

Acknowledgment: Authors acknowledge the financial support from Ministry of Electronics and Information Technology (MeitY), Delhi, Government of India, under Administrative Approval Number: 26(6)2019-ESDA.

REFERENCES

1. A. Kumar, A.K. Padhee, S. Kumar, "How Indian agriculture should change after COVID-19," *Food Secur.*, Vol. 12 (4), pp. 837–840, 2020.
2. Government of India, National Food Security Act (NFSA), "Department of Food and Public Distribution, Ministry of Consumer Affairs," Food and Public Distirbution, New Delhi, India, 2013.

3. G.S. Bhalla, P. Hazell, "Foodgrain demand in India 2020, A Preliminary Exercise," *Economic and Political Weekly*, 1997.
4. A. Gulati, R. Meinzen-Dick, K.V. Raju, "Institutional reforms in Indian irrigation," International Food Policy Research Institute, Sage Publications, New Delhi, p. 322, 2005.
5. A. Gulati, S. Narayanan, *The Subsidy Syndrome in Indian Agriculture*, Oxford University Press, New Delhi, India, p. 297, 2003.
6. G.S. Kalkat, K.S. Pannu, K. Singh, P.S. Rangi, "Agricultural and Rural Development of Punjab: Transforming from Crisis to Growth, Punjab State Farmers Commission (PSFC)," Government of Punjab, 2006.
7. D.A. Kampman, "Water footprint of India: A study on water use in relation to consumption of agricultural goods in the Indian states," Masters Thesis. Enschede: University of Twente, 2007.
8. MoA & FW., "Horticultural statistics at a glance. Ministry of Agriculture and Farmers Welfare, Department of Agriculture, Cooperation and Farmers Welfare," *Horticulture Statistics Division. Government of India*, 2015.
9. F.A.O., "State of food and agriculture in Asia and the Pacific2006," FAO Regional Office for Asia and the Pacific, Bangkok, 2006.
10. A.P.K. Tai, M.V. Martin, C.L. Heald, "Threat to future global food security from climate change and ozone air pollution," *Nat Clim Change*, Vol. 4(9) pp. 817–821, 2014.
11. R.N. Strange, P.R. Scott, "Plant disease: a threat to global food security," *Annu Rev Phytopathol*, Vol. 43, pp.83–116, 2005.
12. H. Kibriya, R. Rafique, W. Ahmad, S. M. Adnan, "Tomato leaf disease detection using convolution neural network," *International Bhurban Conference on Applied Sciences and Technologies (IBCAST)*, pp. 346–351, 2021.
13. S.A. Miller, F.D. Beed, C.L. Harmon, "Plant disease diagnostic capabilities and networks," *Annu Rev Phytopathol*,Vol. 47, pp. 15–38, 2009.
14. M. B. Riley, M. R. Williamson, O. Maloy, "Plant disease diagnosis. The Plant Health Instructor," Vol. 10, 2016.
15. G. Wang, Y. Sun, J. Wang, " Automatic image based plant disease severity estimation using deep learning," *Comput Intell Neurosci*, 2917536, pp.1–8, 2017.
16. V. Singh, A.K. Misra, "Detection of plant leaf diseases using image segmentation and soft computing techniques," *Inf Process Agric*, Vol. 4(1), pp. 41–49, 2017.
17. K. Zhang, Q. Wu, A. Liu, X. Meng, "Can deep learning identify tomato leaf disease," *Adv Multimed*, 6710865, 1, pp. 1–10, 2018.
18. I.M. Hanssen, M. Lapidot, "Major tomato viruses in the mediterranean basin," *Adv Virus Res Acad Press*, Vol, 84, pp. 31–66, 2012.
19. K. Golhani, S.K. Balasundram, G. Vadamalai, B. Pradhan, "A review of neural networks in plant disease detection using hyperspectral data," *Inf Precis Agric*, Vol., 5 pp. 354–371, 2018.
20. M. Zhang, Z. Qin, X. Liu, "Remote sensed spectral imagery to detect late blight in field tomatoes," *Precision Agric*, Vol. 6, pp. 489–508, 2005.
21. C. Yang, "Remote sensing and precision agriculture technologies for crop disease detection and management with a practical application example," *Engineering*, Vol. 6, pp. 528–532, 2020.

22. W.J. Zhu, H. Chen, I. Ciechanowska, D. Spaner, "Application of infrared thermal imaging for the rapid diagnosis of crop disease," *IFAC-PapersOnLine*, Vol. 51 (17), pp. 424–430, 2018.

23. J. Lu, R. Ehsani, Y. Shi, Y. et al., "Detection of multi-tomato leaf diseases (late blight, target and bacterial spots) in different stages by using a spectral-based sensor," *Sci Rep*, Vol. 8, p. 2793, 2018.

24. C. Hao, F. Yong, Z. Dazhou, W. Chen, C. Zilong, "Study of shape feature of maize detection technology——based on image processing," *Agri Mech Res*, Vol. 6, pp. 49–52, 2015.

25. B. Luna-Benoso, J.C. Martínez-Perales, J. Cortés-Galicia, R. Flores-Carapia, V.M. Silva-García, "Detection of diseases in tomato leaves by color analysis," *Electronics*, Vol. 10, 1055, pp. 1–16, 2021.

26. C.S. Hlaing, S.M. Maung Zaw, "Tomato plant diseases classification using statistical texture feature and color feature," *IEEE/ACIS 17th International Conference on Computer and Information Science (ICIS)*, pp. 439–444, 2018.

27. S. Bankar, A. Dube, P. Kadam, S. Deokule, "Plant disease detection techniques using canny edge detection & color histogram in image processing," *Int J Comp Sci Inform Technol*, Vol. 5 (2), pp. 1165–1168, 2014.

28. A. Setya Budi, N. Merlina, M. Arie Hasan, D. Riana, S. Hadianti, "Classification of lycopersicon esculentum fruit based on color features with linear discriminant analysis (LDA) method," *Fourth International Conference on Informatics and Computing (ICIC)*, pp. 1–6, 2019.

29. D.G. Lowe, "Distinctive image features from scale-invariant keypoints," *Int J Comp Vision*, Vol. 60(2), pp. 91–110, 2004.

30. L. Tan, J. Lu, H. Jiang, "Tomato leaf diseases classification based on leaf images: a comparison between classical machine learning and deep learning methods," *Agric Eng*, Vol. 3, pp. 542–558, 2021.

31. K. Yamamoto, W. Guo, Y. Yoshioka, S. Ninomiya, "On plant detection of intact tomato fruits using image analysis and machine learning methods," *Sensors*, Vol. 14, pp. 12191–12206, 2014.

32. S. Budi, N. Merlina, M. Arie Hasan, D. Riana, S. Hadianti, "Classification of lycopersicon esculentum fruit based on color features with linear discriminant analysis (LDA) method", *Fourth International Conference on Informatics and Computing (ICIC)*, pp. 1–6, 2019.

33. J. Pretty, T. G. Benton, Z.P. Bharucha, L.V. Dicks, C.B. Flora, H.C.J. Godfray, D. Goulson, S. Hartley, N. Lampkin, C. Morris, "Global assessment of agricultural system redesign for sustainable intensification," *Nat Sustain*, Vol. 1, pp. 441–446, 2018.

34. S. S. Chouhan, A. Kaul, U. P. Singh, S. Jain, "Bacterial foraging optimization based radial basis function neural network (BRBFNN) for identification and classification of plant leaf diseases: An automatic approach towards plant pathology," *IEEE Access*, Vol. 6, pp. 8852–8863, 2018.

35. T.A. Salih, A.J. Ali, M.N. Ahmed, "Deep learning convolution neural network to detect and classify tomato plant leaf diseases," Open Access Library J, Vol. 7 (5), pp. 1–12, 2020.

36. R. Karthik, M. Hariharan, S. Anand, P. Mathikshara, A. Johnson, R. Menaka, "Attention embedded residual CNN for disease detection in tomato leaves," *Appl Soft Comput*, Vol. 86, pp. 105933, 2020.

37. D. Sarwinda R.H. Paradisaa, A. Bustamama, P. Anggiab, "Deep learning in image classification using residual network (ResNet) variants for detection of colorectal cancer," *Proc Comp Sci,* Vol. 179, pp. 423–431, 2021.
38. Y. Kawasaki, H. Uga, S. Kagiwada, H. Iyatomi, "Basic study of automated diagnosis of viral plant diseases using convolutional neural networks," *Basic Study of Automated Diagnosis of Viral Plant Diseases Using Convolutional Neural Networks,* Springer, pp. 638–645, 2015.
39. Q. Wang, F. Qi, "Tomato diseases recognition based on faster RCNN," *10th International Conference on Information Technology in Medicine and Education (ITME),* pp. 772–776, 2019.
40. S. Arya, R. Singh, "A comparative study of CNN and AlexNet for detection of disease in potato and mango leaf," *International Conference on Issues and Challenges in Intelligent Computing Techniques (ICICT),* pp. 1–6, 2019.
41. G. Wang, Y. Sun, J. Wang, "Automatic image-based plant disease severity estimation using deep learning," *Comput Intell Neurosci,* 2917536, 1, pp.1–10, 2017.
42. J. Amara, B. Bouaziz, A. Algergawy, "A deep learning-based approach for banana leaf diseases classification," *Datenbanksysteme für Business, Technologie und Web Workshopband,* pp. 79–87, 2017.
43. O. Ronneberger, P. Fischer, T. Brox, "U-net: Convolutional networks for biomedical image segmentation," *International Conference on Medical Image Computing and Computer-Assisted Intervention,* pp. 234–241, 2015.
44. M. Kaur, R. Bhatia, "Development of an improved tomato leaf disease detection and classification method," *IEEE Conference on Information and Communication Technology,* pp. 1–5, 2019.
45. M.A. Rahman, M.M. Islam, G.S. Mahdee, M.W.U. Kabir, "Improved segmentation approach for plant disease detection," *1st International Conference on Advances in Science, Engineering and Robotics Technology (ICASERT),* pp. 1–5, 2019.
46. H. Durmuş, E.O. Güneş, M. Kırcı, "Disease detection on the leaves of the tomato plants by using deep learning," *6th International Conference on Agro-Geoinformatics,* pp. 1–5, 2017.
47. A.S. Chakravarthy, S. Raman, "Early blight identification in tomato leaves using deep learning," *International Conference on Contemporary Computing and Applications (IC3A)* obacteri, pp. 154–158, 2020.
48. M.S. Attia, G.S. El-Sayyad, M.A. Elkodous, A.I. El-Batal, "The effective antagonistic potential of plant growth-promoting rhiza against Alternaria solani-causing early blight disease in tomato plant," *Scientia Horticulturae,* Vol. 266, p. 109289, 2020.
49. Q. Wang, F. Qi, M. Sun, J. Qu, J. Xue, "Identification of tomato disease types and detection of infected areas based on deep convolutional neural networks and object detection techniques," *Comput Intell Neurosci,* Vol. 2019, 9142753, pp.1–15, 2019.
50. J. Abdulridha, Y. Ampatzidis, S.C. Kakarla, P. Roberts, "Detection of target spot and bacterial spot diseases in tomato using UAV-based and benchtop-based hyperspectral imaging techniques," *Precision Agric,* Vol. 21, pp. 955–978, 2020.
51. A. Fuentes, S. Yoon, S.C. Kim, D.S. Park, "A robust deep-learning-based detector for real-time tomato plant diseases and pests recognition," *Sensors,* Vol. 17, p. 2022, 2017.

52. M. Agarwal, A. Singh, S. Arjaria, A. Sinha, S. Gupta, "ToLeD: tomato leaf disease detection using convolution neural network," *Procedia Comp Sci*, Vol. 167, pp. 293–301, 2020.

53. A. Seum, A. H. Raj, S. Sakib, T. Hossain, "A comparative study of CNN transfer learning classification algorithms with segmentation for COVID-19 detection from CT scan images," *11th International Conference on Electrical and Computer Engineering (ICECE)*, pp. 234–237 2020.

54. I.Z. Mukti, D. Biswas, "Transfer learning based plant diseases detection using ResNet50," *4th International Conference on Electrical Information and Communication Technology (EICT)*, pp. 1–6, 2019.

55. H.A. Atabay, "Deep residual learning for tomato plant leaf disease identification," *J Theor Appl Inf Technol*, Vol. 95, pp. 6800–6808, 2017.

56. G. Geetharamani, J. Arun Pandian, "Identification of plant leaf diseases using a nine-layer deep convolutional neural network," *Comput Electric Eng*, Vol. 76, pp.323–338, 2019.

57. K.G. Siva, K. Balasundram, G. Vadamala, B. Pradhan "A review of neural networks in plant disease detection using hyperspectral data," *Inform Proc Agri*, Vol. 5 (3), pp. 354–371, 2018.

58. K. He, X. Zhang, S. Ren, J. Sun, "Deep residual learning for image recognition," *Computer Vision and Pattern Recognition, arXiv:1512.03385*, pp. 1–12, 2015.

59. Y. Jia, E. Shelhamer, J. Donahue, S. Karayev, J. Long, R. Girshick, S. Guadarrama, T. Darrell, "Caffe: Convolutional architecture for fast feature embedding". *arXiv:1408.5093*, 2014

60. D.P. Kingma, J. Lei Ba, "Adam: A method for stochastic optimization," *arXiv:1412.6980* https://arxiv.org/pdf/1412.6980.pdf, 2015.

61. K.A. Toh, "Training a reciprocal-sigmoid classifier by feature scaling-space," *Mach Learn*, Vol. 65(1), pp. 273–308, 2006.

62. S. Maharjan, A. Alsadoon, P.W.C. Prasada, T. Al-Dalaina, O.H. Alsadoon, "A novel enhanced softmax loss function for brain tumour detection using deep learning," *J Neurosci Methods*, Vol. 330, pp. 1–20, 2020.

63. M. Brahimi, K. Boukhalfa, A. Moussaoui, "Deep learning for tomato diseases: classification and symptoms visualization," *Appl Artif Intell*, Vol. 31, no. 4, pp. 299–315, 2017.

64. U. Barman, R.D. Choudhury, D. Sahu, G.G. Barman, "Comparison of convolution neural networks for smartphone image based real time classification of citrus leaf disease," *Comput Electr Agri*, Vol. 177, pp. 1–9, 2020.

65. M. Prabhakar, R. Purushothaman, D.P. Awasthi, "Deep learning based assessment of disease severity for early blight in tomato crop," *Multimedia Tools Appl*, Vol. 79, pp. 28773–28784, 2020.

66. S. Panno, S. Davino, A.G. Caruso, S. Bertacca, A. Crnogorac, A. Mandi´c, E. Noris, S. Mati´c, "A review of the most common and economically important diseases that undermine the cultivation of tomato crop in the mediterranean basin," *Agronomy, Vol. 11, 2188,* pp. 1–41. 2021.

67. L.M. Suresh, et al., "Tomato disease - field guide," *Seminis Vegetable Seeds*, Vo. 1, pp. 1–168, 2017.

68. V.K. Singh, A.K. Singh, A. Kumar, "Disease management of tomato through PGPB: current trends and future perspective," *Biotech*, Vol.7, p.255, 2017.

69. K. Pankaja, V. Suma, "Plant leaf recognition and classification based on the whale optimization algorithm (WOA) and random forest (RF)," *J Inst Eng India Ser B*, Vol. 101, pp. 597–607, 2020.
70. M. Mishra, P. Choudhury, B. Pati, "Modified ride-NN optimizer for the IoT based plant disease detection," *J Ambient Intell Human Comput*, Vol. 12, pp. 691–703, 2021.
71. Y.A. Nanehkaran, D. Zhang, J. Chen, et al. "Recognition of plant leaf diseases based on computer vision," *J Ambient Intell Human Comput*, Vol. 1, 2020.
72. L. Selvam, P. Kavitha, "Classification of ladies finger plant leaf using deep learning," *J Ambient Intell Human Comput*, Vol. 1, 2020.
73. F. Sun, X. Wang, R. Zhang, "Task scheduling system for UAV operations in agricultural plant protection environment," *J Ambient Intell Human Comput*, Vol. 1, 2020.
74. Z. Zhang, X. Wang, X. Wang, et al. "A simulation-based approach for plant layout design and production planning," *J Ambient Intell Human Comput*, Vol. 10, pp. 1217–1230, 2019.

Chapter 8

Advancements in healthcare

Machine learning applications for quick disease diagnostics and individualized treatment

Vinay Anand and Himanshu Sharma

ABBREVIATIONS

CT scan: computed tomography scan
EDA: exploratory data analysis
EHR: electronic health record
KNN: k-nearest neighbors
ML: machine learning
MRI: magnetic resonance imaging
SVC: support vector classifier
SVM: support vector machine
XG Boost: extreme gradient boost

8.1 INTRODUCTION

Machine learning (ML) is rapidly changing the healthcare industry and playing a key role in disease diagnosis. Its capacity to process extensive volumes of intricate data, encompassing medical images, electronic health records (EHRs), and genetic data, leads to swifter, more precise, and less intrusive diagnostic outcomes. This research examines the current status of ML in disease diagnosis and outlines its applications, advantages, challenges, and future directions. Utilizing ML algorithms, medical image analysis involves the examination of medical images, including X-rays, CT scans, and MRIs, to identify abnormalities with a high level of precision. It is particularly effective in detecting cancer, heart disease, and neurological disorders. For example, deep learning models have excelled at detecting breast cancer, in some cases even outperforming expert radiologists.

EHR analysis: ML can extract information from the EHR, including patient demographics, symptoms, lab tests, and medical history. This information can be used to predict disease risk, customize treatment plans, and identify patients at high risk of complications. For example, ML algorithms can predict the onset of sepsis or heart failure with high accuracy, allowing early intervention and better outcomes.

DOI: 10.1201/9781003581246-8

Genetic analysis: ML can analyze genetic data to identify patients with inherited diseases or susceptibility to certain diseases. This can lead to earlier diagnosis, preventive measures, and individualized treatments. For example, ML has been used to identify genetic mutations associated with Alzheimer's disease and other neurodegenerative conditions.

Better accuracy and early detection with ML algorithms: ML algorithms possess the capability to scrutinize intricate data patterns that might go unnoticed by conventional methods, resulting in enhanced precision in diagnoses and the early detection of diseases. Consequently, this has the potential to greatly enhance patient outcomes and diminish mortality rates. Effective diagnosis and early intervention can minimize the need for expensive tests and treatments, resulting in overall savings for healthcare systems. ML can customize treatment plans according to the unique characteristics of individual patients, taking into account factors such as their genetic makeup, medical history, and other relevant information. This can lead to more effective treatment and improve the patient experience. The human genome contains information that affects a person's susceptibility to disease, response to therapy, and overall health. ML models can scrutinize genetic data, pinpointing particular genetic markers or mutations linked to specific diseases. By understanding the patient and their genetic makeup, healthcare professionals can tailor treatment plans to target underlying causes or predispositions and optimize treatment outcomes. The patient's medical history including past illnesses, treatments, medications, and the hospital stay may be evaluated and this finding were researched [1]. ML algorithms can analyze extensive sets of historical patient data, discerning patterns and correlations within the information. This analysis helps healthcare providers anticipate potential challenges, understand the patient and their unique health condition, and make informed decisions about current and future treatment options. In addition to genetics and medical history, ML models can account for many other factors, such as lifestyle, environmental influences, socioeconomic status, and patient-reported outcomes. Combining this diverse set of data points allows for a more holistic understanding of the patient. ML algorithms can unravel the complex relationships between these factors, providing a comprehensive view that goes beyond traditional medical assessments. ML algorithms demonstrate exceptional proficiency in recognizing intricate patterns and connections within extensive datasets. In the realm of personalized medicine, these models possess the capability to discern optimal treatment alternatives by analyzing a distinctive amalgamation of genetic markers, medical history, and other significant factors. This adaptation of treatment plans allows health professionals to choose interventions that are more likely to be successful for a given patient, thus optimizing the effectiveness of treatment. By tailoring treatment plans to individual patients, the probability of treatment success increases. ML models can predict how patients are likely to respond to certain medications or treatments, allowing the most effective and least harmful interventions to be selected. This not only improves treatment results but also reduces the probability of side effects or ineffective treatment, ultimately improving the overall effectiveness of medical procedures. Personalized care plans improve the patient experience in several ways. Patients are more likely to respond positively to treatment tailored to their unique needs and characteristics. In addition, patients receive faster relief and fewer side effects, reducing the trial-and-error process of finding effective treatments. This personalized approach builds empowerment and trust in the healthcare system, which improves the overall patient experience.ML algorithms have some challenges, The effectiveness of ML algorithms hinges on the quality of their training. If there's bias in the practical data used for training, it can result in biased algorithms that uphold and reinforce existing disparities in healthcare [2]. Therefore, ensuring high-quality and unbiased information is crucial for the ethical and responsible use of ML in healthcare. ML models can be complex and opaque, making their decision-making process difficult. This can raise concerns about transparency and accountability, especially for important medical decisions. Continued Development of Advanced ML Algorithms: Further exploration and advancement are essential to enhance the precision, interpretability, and applicability of ML models in medical settings. The primary emphasis of this investigation revolves around the contribution of ML models and their precision in the early detection of diverse diseases. ML has emerged as a transformative influence in healthcare, particularly in the realm of disease diagnosis. This research assesses the current landscape of ML in disease diagnosis, delving into its applications, benefits, challenges, and prospective trajectories [3]. The research concentrates on three principal domains: medical image analysis, EHR scrutiny, and genetic analysis. The capacity to scrutinize diverse datasets, encompassing medical images, patient records, and genetic data, facilitates swifter, more precise, and less invasive disease diagnosis.

8.2 LITERATURE REVIEW

Relevant research articles are reviewed and organized in Table 8.1 containing the titles and authors of various research articles analyzed for this research.

8.3 RESEARCH PROBLEM

The research discussed focuses on ML-based disease diagnostics, specifically for breast cancer, heart failure, and diabetes. While the research methodology is comprehensive and aims to address key healthcare challenges, there are some potential problems and considerations to be aware of:

Table 8.1 Relevant research article reviewed

Sl. No.	Methodology	Findings	Related research
1	Meta-analysis	ML algorithms enhance breast cancer detection Accuracy	[4]
2	Cross-validation	XG Boost outperforms other algorithms in heart failure diagnostic	[5]
3	Cohort study	Lifestyle factors influencing diabetes risk	[6]
4	Systematic review	Meta-analysis indicates ML algorithms enhance breast cancer detection accuracy compared to traditional methods.	[7]
5	Cross-validation and comparative study	XG Boost outperforms logistic regression and SVM in heart failure diagnostics with an average test score of 93.78%	[8]
6	Longitudinal cohort study	Unhealthy lifestyle choices, such as prolonged periods of inactivity and an unbalanced diet, markedly elevate the likelihood of acquiring type 2 diabetes. Top of Form	[9]
7	Ethical analysis	Identified challenges and biases in ML for healthcare, emphasizing transparency and accountability issues	[10]
8	Qualitative analysis	Challenges and biases in ML for health care	[11]

1. *Data bias and representativeness:* One of the critical issues in ML-based diagnostics is the potential bias in the training data. If the dataset used for training the ML models is not representative of the diverse patient population, it can lead to biased predictions and inaccurate diagnostic outcomes. Ensuring data diversity and inclusivity is essential to mitigate this problem.

2. *Algorithm selection and performance:* While the research employs a variety of ML algorithms (e.g., logistic regression, SVM, naïve Bayes, random forest, KNN) for disease diagnostics, the selection and performance evaluation of these algorithms should be carefully conducted. Different algorithms may perform differently depending on the dataset characteristics, and improper algorithm selection could affect the accuracy and reliability of diagnostic results and the algorithm corrects residual motion and significantly decreases artifacts. Increasing the gating window width reduces image noise and enhances vessel contrast [12].

3. *Interpretability and explainability:* ML models, especially complex ones like random forest and SVM, can lack interpretability and explainability, making it challenging to understand how the models arrive at their predictions. This lack of transparency can be a concern in healthcare settings where stakeholders require insights into the decision-making process of the models.

4. *Ethical and regulatory considerations:* ML-based diagnostic tools in healthcare must adhere to ethical guidelines and regulatory standards to ensure patient privacy, data security, and responsible use of AI technologies. Issues such as algorithmic fairness, transparency, and accountability need to be addressed to build trust and confidence in ML-driven diagnostic systems.

5. *Real-world implementation challenges:* Transitioning from research prototypes to real-world clinical settings poses practical challenges such as data integration, system interoperability, scalability, and usability. Ensuring seamless integration of ML-based diagnostic tools into existing healthcare infrastructure is crucial for successful deployment and adoption.

8.4 BREAST CANCER DIAGNOSTICS

A breast comprises three main elements: lobules, ducts, and connective tissue. The lobules act as glands, producing milk, and the ducts serve as conduits, transporting the milk to the nipple. The connective tissue, which encompasses and offers structural support to these components, is comprised of both fibrous and fatty tissues. The anatomy of the human breast labels key structures like lobules, lobes, ducts, fat, the nipple, and the areola.

1. *Lobules:* These are the milk-producing glands in the breast. Each breast typically contains 15–lobes, and within each lobe are many smaller lobules, connected to ducts. Lobules are a crucial part of the mammary glandular system, and during lactation, they produce milk which is carried to the nipple through the ducts.

2. *Lobes:* The breast is divided into multiple lobes, and each lobe contains multiple lobules. The lobes are organized in a radial pattern within the breast, which is essential for its function during milk production.

3. *Ducts:* Milk ducts are thin tubes that carry milk from the lobules to the nipple. The ducts converge at the nipple, and during breastfeeding, milk flows through these ducts. The network of ducts is crucial for the milk ejection process.

4. *Fat:* Fat tissue surrounds the lobes and ducts, giving the breast its shape and size. The amount of fat within the breast varies between individuals, affecting breast size, but does not influence milk production directly. The fat tissue provides structural support.

5. *Nipple and areola:* The nipple is the external opening where the ducts terminate, allowing milk to be delivered during lactation. The areola is the pigmented skin surrounding the nipple and contains glands that help lubricate the nipple during breastfeeding. The darker pigmentation of the areola and its glands play a role in breastfeeding by helping the infant locate the nipple and maintaining skin moisture.

6. *Breast structure overview:* Each breast is composed of glandular tissue (lobules and ducts), fibrous connective tissue, and adipose (fat) tissue. The proportion of glandular to fat tissue varies across individuals and changes throughout life, especially during menstruation, pregnancy, and menopause.

7. *Breast anatomy in clinical context:* Understanding breast anatomy is essential in the context of diagnosing and treating breast diseases such as breast cancer. Mammary glands, ducts, and lobular structures are often areas where pathological changes occur, making accurate anatomical representation crucial for medical education.

Breast cancer is identified by the uncontrollable proliferation of cells in the breast as in Figure 8.1. There are different forms of breast cancer, and the particular type is defined by the cells within the breast that undergo malignant changes.

The majority of breast cancers typically originate in either the ducts or lobules. Cancerous cells from the breast have the potential to spread beyond their original location through blood vessels and lymph vessels. Also, automatic retinal blood vessel segmentation and classification aids in identifying various conditions like glaucoma, high blood pressure, macular

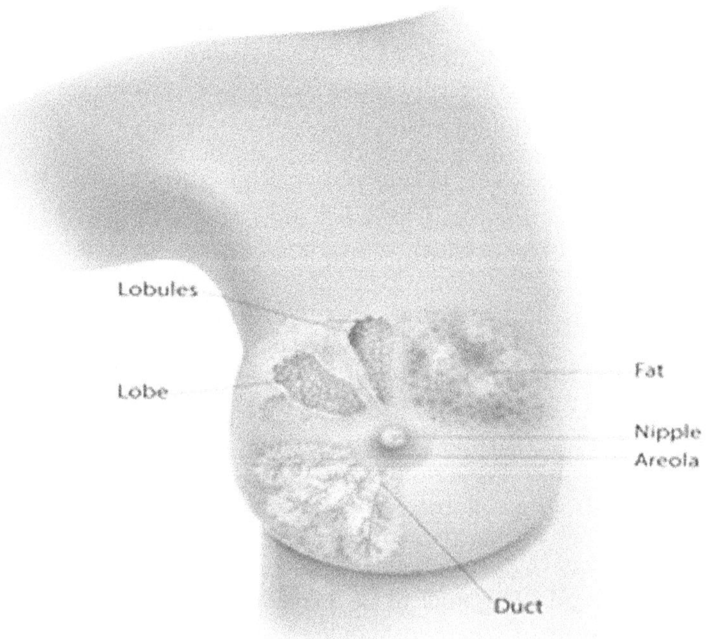

Figure 8.1 Breast diagram shows the area of the lobules, lobe, duct, areola, nipple, and fat.

degeneration, diabetes, and more [13]. When breast cancer extends to other parts of the body, it is referred to as having metastasized.

The two main types of breast cancer are discussed below:

1. *Invasive ductal carcinoma*
   Cancerous cells initiate within the ducts and extend beyond them into additional regions of the breast tissue. These invasive cells have the potential to metastasize and spread to other parts of the body.
2. *Invasive lobular carcinoma*
   Cancer cells originate in the lobules and disseminate from the lobules to nearby breast tissues. These invasive cancer cells also can spread to other parts of the body.

Various methods are employed for screening breast cancer, even though elastic registration techniques prove significantly more effective than rigid methods in aligning dynamic abdominal CT scans across the entire field of view [14], with mammograms being a prominent tool for early detection in young women. A mammogram is essentially an X-ray of the breast. For many women, mammograms serve as the most effective means of detecting breast cancer early, making it easier to treat before it becomes palpable or causes symptoms. Regular mammograms contribute to lowering the risk of mortality from breast cancer. Currently, a mammogram stands as the primary method for detecting breast cancer in most women of screening age.

Undergoing a mammogram is often uncomfortable for the majority of women, with some finding it to be a painful experience. However, the procedure is brief, and the discomfort subsides quickly [15]. The level of discomfort can vary based on factors such as the technologist's expertise, breast size, and the extent of compression needed. Sensitivity may be heightened if the mammogram is performed around the time of menstruation [16]. A specialized doctor, known as a radiologist, will analyze the X-ray to identify early indications of breast cancer or other potential issues.

Breast magnetic resonance imaging (MRI) employs magnetic fields and radio waves to capture images of the breast. This technique is employed in conjunction with mammograms to screen women with a high risk of developing breast cancer. However, it is not recommended for women at an average risk since breast MRIs can exhibit abnormalities even in the absence of cancer [16].

However, the use of ML algorithms will be the easier and more comfortable solution when compared to mammograms or MRIs. To assess the precision of different ML models, the dataset is obtained from the AI for Social Good: Women Coders' Bootcamp, a program organized by Artificial Intelligence for Development in partnership with UNDP Nepal. This dataset is employed to evaluate the accuracy of diverse ML models [17].

8.4.1 Data preprocessing

Data preprocessing plays a pivotal role in ML, encompassing activities such as cleaning, transforming, and structuring raw data to prepare it for model training. The effective pre-processing of data significantly improves the performance and accuracy of ML models. In the provided dataset, there are 568 entries and 6 columns, specifically labeled as mean radius, mean texture, mean perimeter, mean area, mean smoothness, and diagnosis.

8.4.2 Data visualization

For the dataset as in Figure 8.2, the distribution of diagnosis consists of 357 benign and 212 malignant cases.

Data visualization is essential in ML for various reasons where it acts as a potent instrument for comprehending, explaining, and conveying intricate patterns and insights embedded in datasets. An example of such visualization is the heat map, a graphical portrayal of data in which individual values are depicted using different colors [18]. In the realm of ML and data analysis, a frequently employed visualization tool is the heat map, which is utilized to display the correlation matrix of a dataset. This type of heatmap offers a rapid and instinctive means of comprehending the connections among various variables within the dataset (Figures 8.3 and 8.4).

8.4.2.1 Diagnosis using logistic regression

Logistic regression stands out as a widely employed classification algorithm within the realm of ML. Its efficacy becomes evident when dealing with a

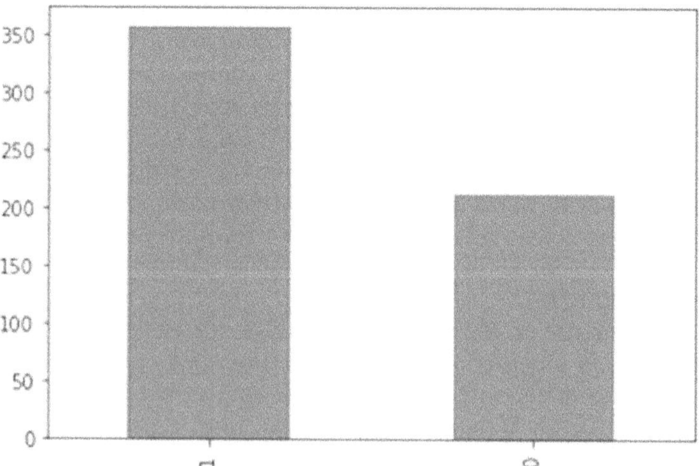

Figure 8.2 Distribution of diagnosis in the dataset.

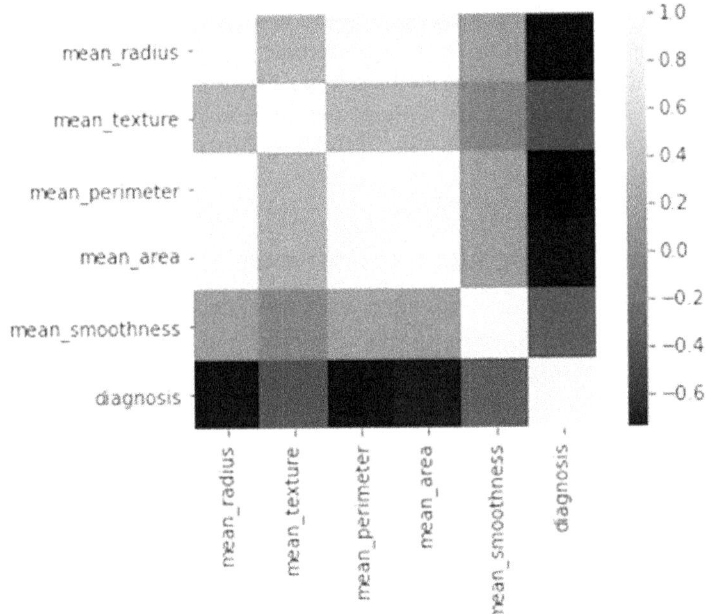

Figure 8.3 Graphical representation of variables.

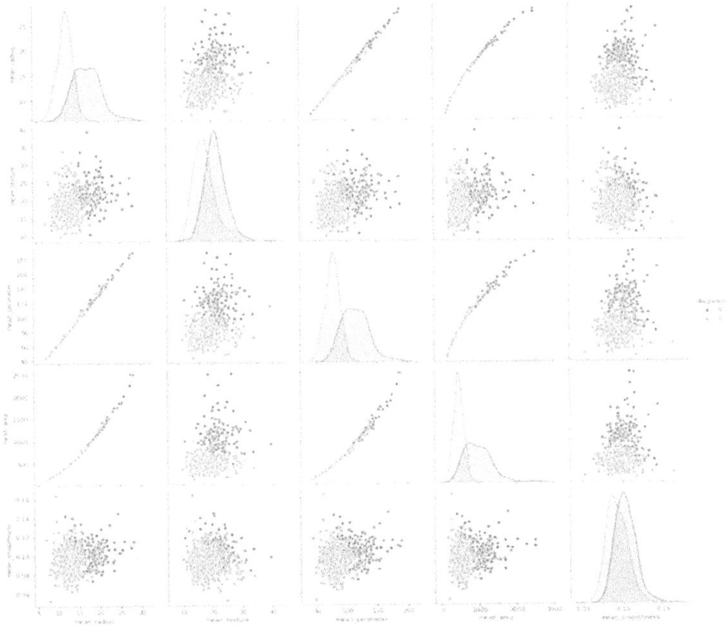

Figure 8.4 Graphical representation of all parameters.

binary dependent variable, where there exist only two potential outcomes. The primary function of Logistic regression involves modeling the likelihood that a specific instance pertains to a particular category (Figure 8.5).

The logistic regression score is calculated using the sci-kit learn library and for more accurate analysis, the average test score is calculated. Interestingly, the score obtained is 0.9210526315789473 and the average precision-recall score is close to 0.90.

8.4.2.2 Diagnosis using support vector machine (SVM) classification

SVM represents a category of supervised ML algorithms that are highly effective in disease diagnosis. These algorithms demonstrate exceptional efficacy when dealing with intricate and nonlinear datasets. SVMs are adept at efficiently managing high-dimensional data, a crucial aspect in disease diagnostics where a plethora of features such as biomarkers, genetic data, and imaging features are often present. The utility of SVMs extends to aiding in the selection of pertinent features or transforming data into a more suitable representation. While SVMs are inherently designed as binary classifiers, their adaptability allows for the handling of multiclass problems through techniques like one-versus-one or one-versus-all. In the realm of disease diagnostics, where outcomes often boil down to binary decisions such as the presence or absence of a disease, SVMs can be effectively employed for binary classification tasks (Figure 8.6).

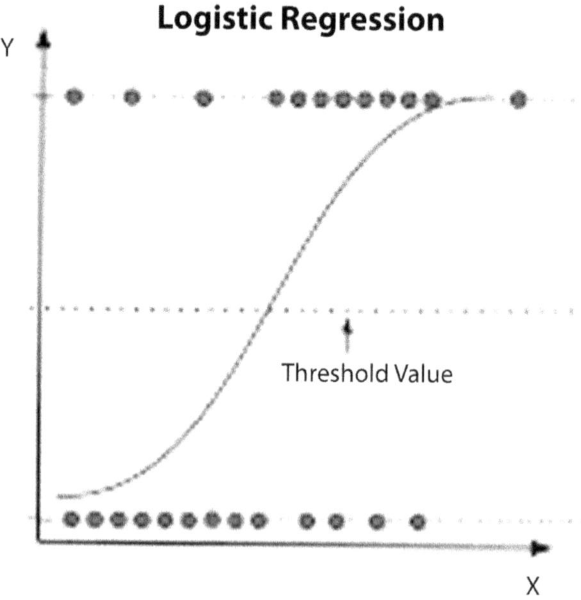

Figure 8.5 Logistic regression classification.

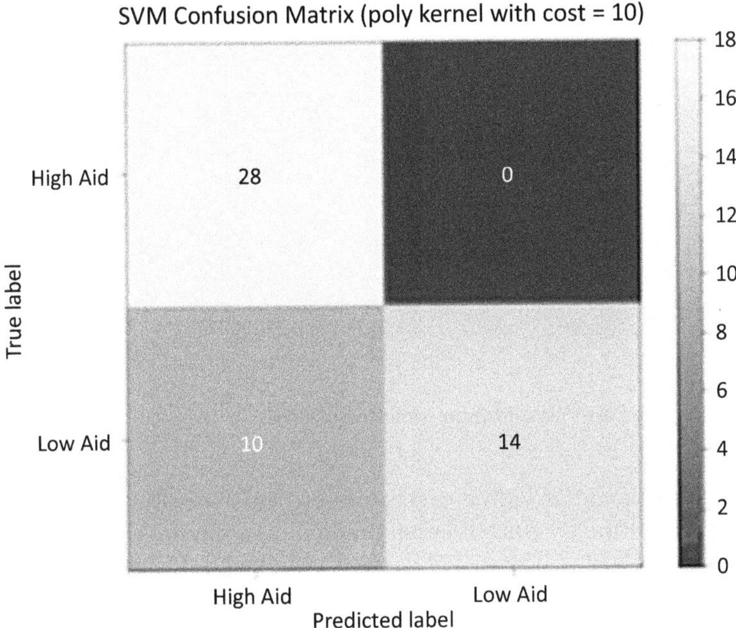

Figure 8.6 SVM classification confusion matrix.

The score obtained for SVM is 0.9035087719298246 and the average precision-recall score is close to 0.88.

8.4.2.3 Diagnosis using naïve Bayes classification

Utilizing the naïve Bayes classification for diagnosis involves applying a probabilistic ML algorithm frequently employed in tasks such as disease diagnostics. This algorithm, based on Bayes' theorem, assumes conditional independence of features given the class label, a simplification commonly known as the "naïve" assumption. It is crucial to recognize that while naïve Bayes is straightforward and computationally efficient, it relies on the assumption of feature independence, which may not be valid in all real-world situations. Nevertheless, even with its simplicity, naïve Bayes can exhibit effective performance, particularly when dealing with a substantial amount of data (Figure 8.7).

The average test score is calculated. Interestingly, the score obtained is 0.9035087719298246 and the average precision-recall score is close to 0.88.

8.4.2.4 Diagnosis using random forest classification

The random forest classification stands out as a robust ensemble learning algorithm suitable for the classification of diseases. In disease classification,

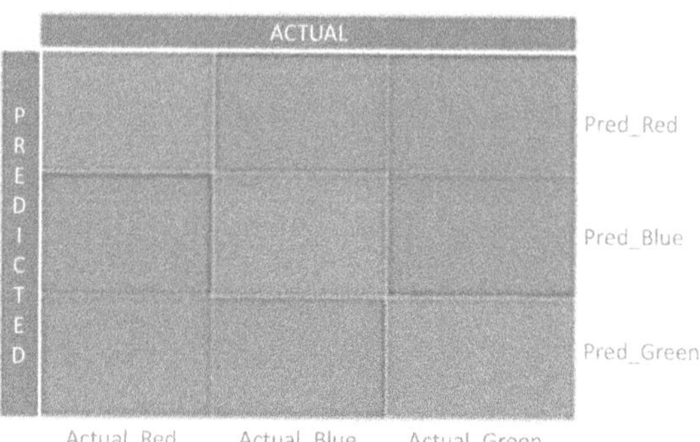

Figure 8.7 Naïve Bayes classification confusion matrix.

ensemble methods, including random forest, enhance overall performance and generalization by consolidating predictions from multiple models. The average test score is calculated. The score obtained is 0.9210526315789473 and the average precision-recall score is close to 0.91 (Figure 8.8).

8.4.2.5 Diagnosis using k-nearest neighbor (KNN) classification

The KNN algorithm serves as a straightforward and intuitive method for medical diagnosis. Its fundamental concept involves classifying a data point by considering the predominant class among its KNN within the feature space. KNN's effectiveness depends on selecting an appropriate distance metric and the value assigned to k. It performs admirably when the decision boundary is relatively smooth, and the dataset is not excessively large. However, challenges may arise, particularly in high-dimensional feature spaces, where KNN might be susceptible to the curse of dimensionality.

Additionally, KNN does not learn explicit patterns during training, and predictions can be sensitive to noise in the data. Despite its simplicity, KNN can be effective for disease classification, especially in situations where the dataset is not very large, and there is a need for an interpretable model. The average test score is calculated. The score obtained is 0.8947368421052632 and the average precision-recall score is close to 0.87, which is classified as in Figure 8.9.

8.5 COMPARISON OF TECHNIQUES

The presentation of diverse ML methods in disease diagnosis includes an assessment and comparison of the effectiveness of various algorithms specifically in predicting or diagnosing breast cancer. The results are summarized below:

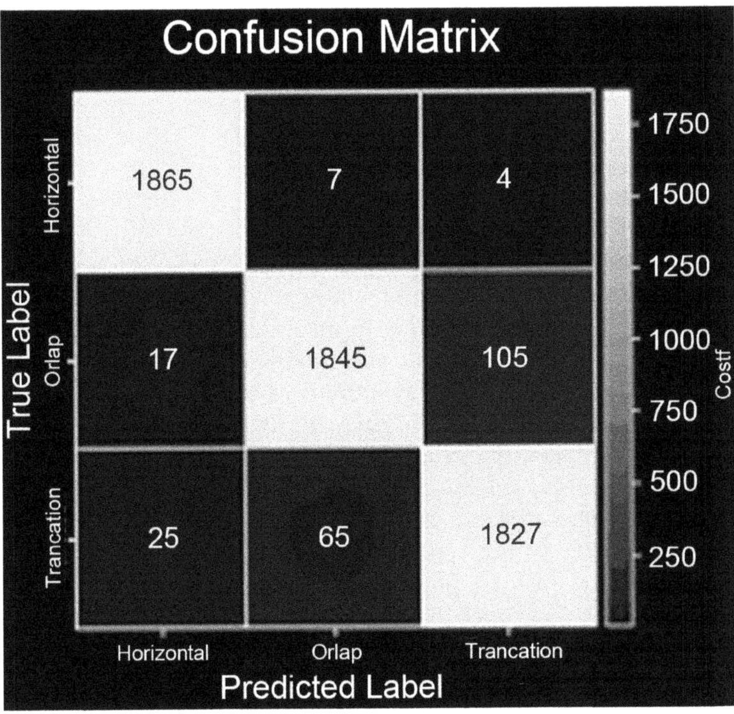

Figure 8.8 Random forest classification confusion matrix.

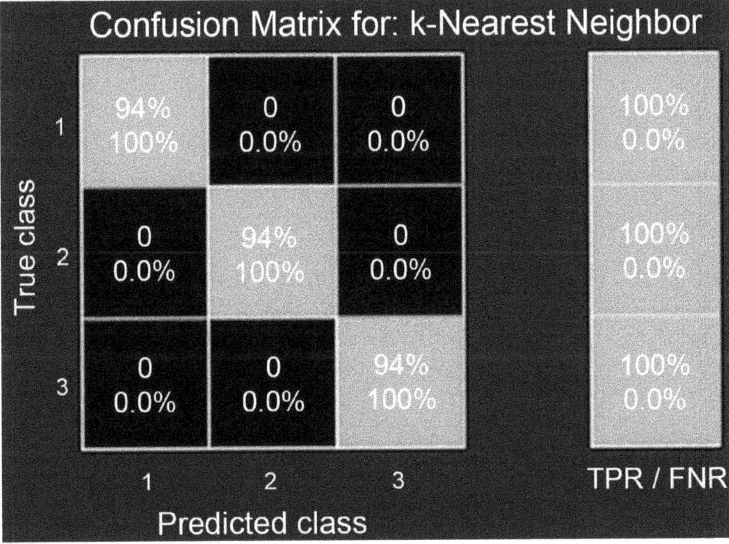

Figure 8.9 KNN Classification confusion matrix.

- Logistic regression score: 0.921053
- SVM score: 0.903509
- Naïve Bayes score: 0.903509
- Random forest Score: 0.921053
- KNN score: 0.894737

8.6 HEART FAILURE DIAGNOSTICS

Heart failure is a medical condition characterized by the heart's inability to efficiently pump blood, resulting in insufficient blood flow to meet the body's requirements. This chronic and progressive ailment can stem from various underlying heart issues. Symptoms of heart failure vary and commonly include shortness of breath, particularly during physical exertion or when lying down, persistent coughing or wheezing, fluid retention leading to swelling in the legs, ankles, and abdomen, fatigue and weakness, and a rapid or irregular heartbeat.

Diagnosing heart failure typically involves a comprehensive approach, combining medical history, physical examination, imaging tests such as echocardiography, and other diagnostic procedures. The treatment objective is to manage symptoms, enhance the quality of life, and address the root causes. Common treatment approaches encompass medications like diuretics, beta-blockers, and ACE inhibitors, lifestyle modifications involving diet and exercise, and occasionally surgical interventions such as heart valve repair or heart transplant.

Individuals diagnosed with heart failure should closely collaborate with healthcare professionals to effectively manage the condition and adopt a heart-healthy lifestyle. Early identification and appropriate management play a crucial role in improving outcomes and enhancing the quality of life for those dealing with heart failure.

Here also the early detection of heart failure is attempted using different ML algorithms and the accuracy is analyzed using different scores.

As heart failure depends on very factors/attributes, first the analysis is done to find the correlation between the heart failure probability and various factors like age, sex, smoking habits, BP, etc., and then to carry out the analysis in the proper manner, certain attributes were removed where the correlation was comparatively low Figure 8.10 shows distribution chart whereas Table 8.2 shows data information to describe the data set used.

8.7 DATA INFORMATION

8.7.1 Exploratory data analysis

Exploratory data analysis (EDA) stands as a pivotal stage in the data analysis workflow, with a significant focus on comprehending the distribution

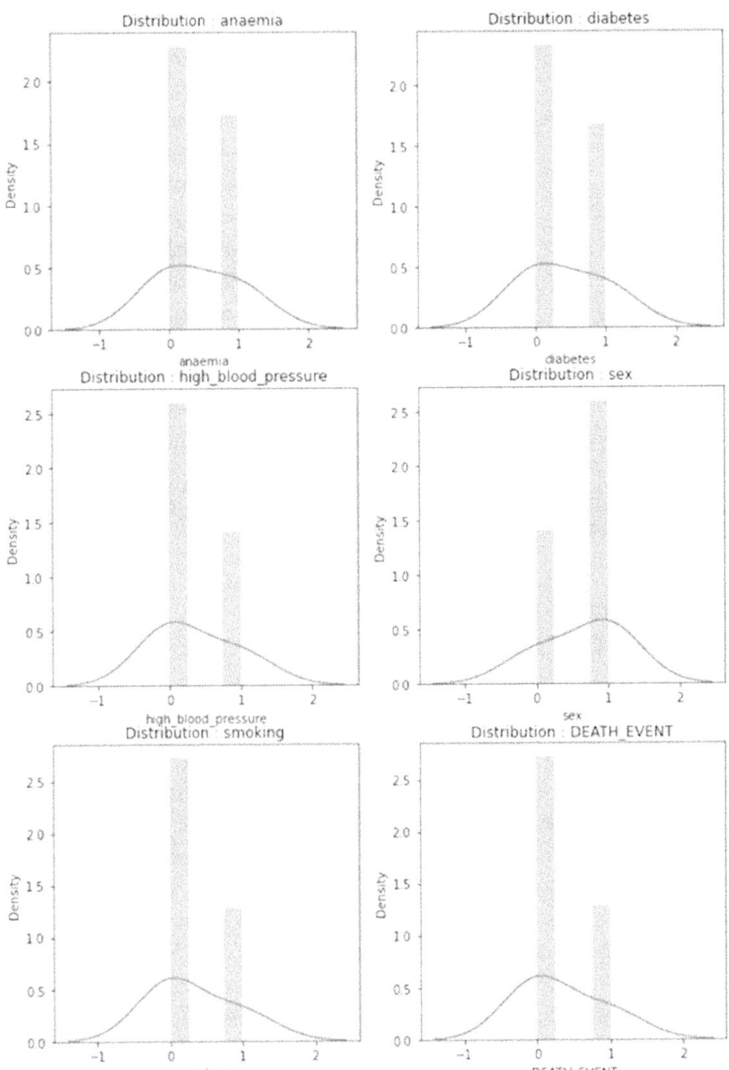

Figure 8.10 Distribution chart.

of categorical features (Figure 8.10). These features encapsulate qualitative methodology data reported in the detailed analysis of a newborn's skull dimensions in early infancy, aiming to identify growth patterns in healthy newborns [20].

Distribution of features: The distribution of different features which is shown in Figure 8.11 is important to analyze, which provides the insight data to know the in-depth effect.

Table 8.2   Data Information

Age	With 299 non-null values of type float 64
Anemia	With 299 non-null values of type int 64
Creatinine phosphokinase	With 299 non-null values of type int 64
Diabetes	With 299 non-null values of type int64
Ejection fraction	With 299 non-null values of type int64
High blood pressure	With 299 non-null values of type int64
Platelets	With 299 non-null values of type float64
Serum creatinine	With 299 non-null values of type float64
Serum sodium	With 299 non-null values of type int64
Sex	With 299 non-null values of type int64
Smoking	With 299 non-null values of type int64
Death event	With 299 non-null values of type int64
Time	With 299 non-null values of type int64

G. Kotronoulas *Et Al.*, "An Overview Of The Fundamentals Of Data Management, Analysis, And Interpretation In Quantitative Research," *Semin. Oncol. Nurs.*, Vol. 39, No. 2, Pp. 1–9, 2023, Doi: 10.1016/J.soncn.2023.151398. [19]

8.8 ML ALGORITHM FOR HEART FAILURE DETECTION

8.8.1 XG Boost classifier

XG Boost, an acronym for e Xtreme Gradient Boosting, stands out as a robust and widely adopted ML algorithm falling within the gradient boosting frameworks category. Engineered for efficiency, scalability, and adaptability, XG Boost finds extensive applications in both classification and regression tasks, demonstrating considerable success in numerous machine-learning competitions. In this context, an evaluation was conducted to verify the algorithm's performance, resulting in a score of 0.9378, while the average precision-recall score approached 0.83. Subsequently, the average test score was determined, yielding a score of 0.8887, with a comparable average precision-recall score of approximately 0.83 (Figures 8.12 and 8.13).

8.9 LOGISTIC REGRESSION

8.10 SUPPORT VECTOR CLASSIFIER

A support vector classifier (SVC) as shown in Figure 8.14 falls under the category of supervised ML models within the SVM family. SVMs are robust algorithms employed for both classification and regression tasks. In classification scenarios, the model endeavors to identify a hyperplane that effectively divides the data into distinct classes. The average test score,

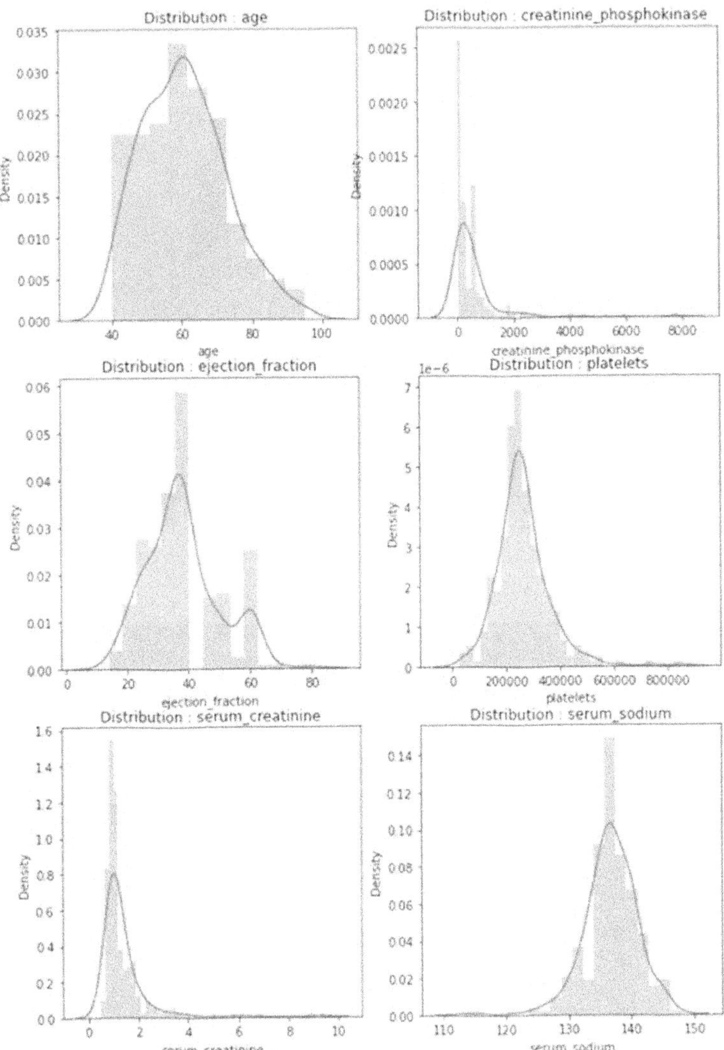

Figure 8.11 Distribution features.

calculated in this context, is 0.8627, and the average precision-recall score closely approximates.

8.11 DIABETES DIAGNOSIS

Diabetes has emerged as a significant health issue in India, leading to the country being recognized as the global "diabetes capital." The prevalence

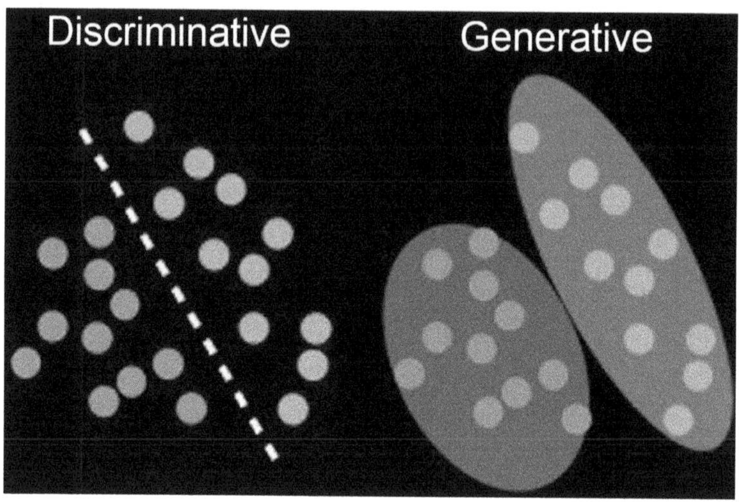

Figure 8.12 Classifier.

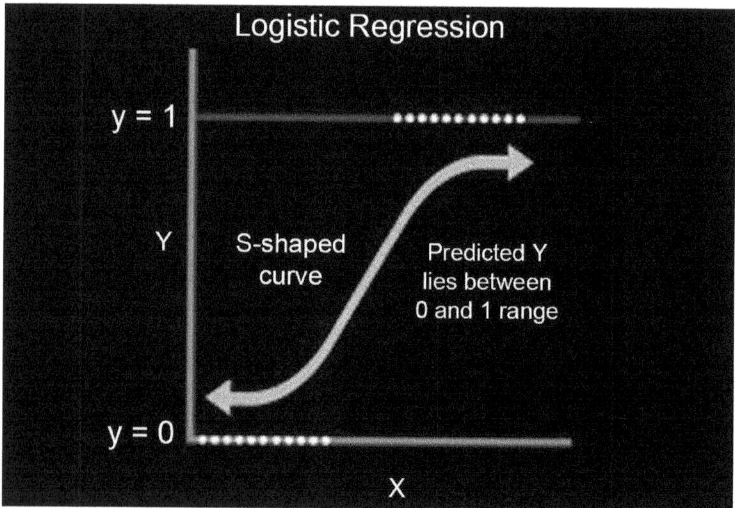

Figure 8.13 Logistic regression.

of both type 1 and type 2 diabetes is notable, with type 2 diabetes being the predominant form, constituting more than 90% of diagnosed cases. This rise in diabetes prevalence is attributed to factors such as urbanization, sedentary lifestyles, poor dietary habits, genetic predisposition, and an aging population. Complications of diabetes, including cardiovascular diseases, neuropathy, nephropathy, retinopathy, and diabetic foot complications,

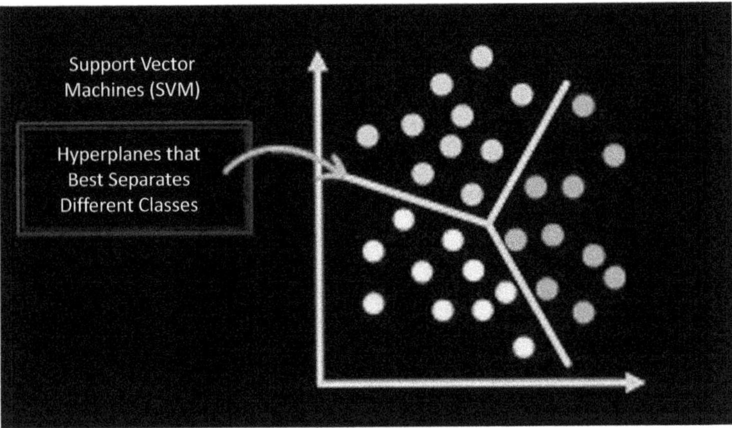

Figure 8.14 Support vector classifier.

Table 8.3 Algorithmic chart for regression and classifier

S. No.	ML algorithm	Cross-validation score	ROC AUC score
1	XGB classifier	92.51%	86.33%
2	Logistic regression	88.87%	83.39%
3	Support vector classifier	86.27%	79.68%

further exacerbate the health burden. Despite efforts by the government to implement awareness and management programs, challenges persist due to limited access to healthcare, lack of awareness, and affordability issues. Research and innovation are ongoing to better understand the disease and develop cost-effective solutions tailored to the Indian context. Non-governmental organizations (NGOs), healthcare entities, and community associations are pivotal contributors to the prevention and management endeavors by conducting awareness initiatives, providing screening services, and offering assistance to individuals with diabetes and their families. Addressing the diabetes epidemic in India requires a comprehensive approach involving various stakeholders to promote prevention and early detection, the suggested approach facilitated the creation of a comprehensive feature set, enabling ML models to detect vulnerable plaque more precisely, even when dealing with low-quality and limited-scale datasets [21].

In brief, diabetes poses a significant public health challenge in India due to its high prevalence and associated complications. Tackling this challenge necessitates a collaborative, multi-sectoral approach involving healthcare providers, policymakers, communities, and individuals. To effectively address diabetes, the following key factors should be taken into account.

1. *Genetics*
   The role of family history and genetic predisposition is substantial in diabetes development, with individuals having a family history facing a higher risk.
2. *Lifestyle factors*
   Sedentary lifestyles, lack of physical activity, and unhealthy dietary habits, including high consumption of processed foods, sugary beverages, and saturated fats, contribute to an increased risk of type 2 diabetes.
3. *Obesity*
   Excess body weight, particularly central obesity (abdominal fat), is strongly associated with insulin resistance and the onset of type 2 diabetes.
4. *Physical activity*
   Regular physical activity improves insulin sensitivity, helps maintain a healthy weight, and lowers blood sugar levels, thereby reducing the risk of diabetes and its complications.
5. *Age and ethnicity*
   Age is a non-modifiable risk factor, with diabetes risk increasing with age. Certain ethnic groups, such as South Asians, Hispanics, African Americans, and Native Americans, are predisposed to a higher risk of diabetes.
6. *Gestational diabetes*
   Women who experience gestational diabetes during pregnancy and their offspring are at an elevated risk of developing type 2 diabetes later in life.
7. *Medical conditions*
   Certain medical conditions, such as polycystic ovary syndrome (PCOS), hypertension, dyslipidemia, and cardiovascular disease, are associated with an increased risk of diabetes.
8. *Stress*
   Chronic stress can lead to elevated levels of cortisol and adrenaline, impairing insulin sensitivity and contributing to diabetes development.
9. *Environmental factors*
   Factors like pollution, exposure to endocrine-disrupting chemicals, and socioeconomic conditions can influence the risk of diabetes.
10. *Sleep*
    Poor sleep quality, sleep deprivation, and disorders like obstructive sleep apnea are linked to insulin resistance and an increased risk of diabetes.
11. *Smoking and alcohol*
    Smoking and excessive alcohol consumption are associated with an increased risk of type 2 diabetes and can worsen diabetes-related complications.

Understanding and considering these factors are crucial for implementing effective prevention, early detection, and management strategies to alleviate the burden of diabetes and its associated complications.

8.12 COMPARATIVE ANALYSIS OF THE PERFORMANCES OF ML ALGORITHMS IN DISEASE DIAGNOSTICS

In disease diagnostics, ML algorithms offer several advantages over traditional diagnostic methods.

1. *Accuracy*: ML algorithms often outperform traditional methods in terms of accuracy, especially in complex data analysis tasks such as medical imaging. For example, in breast cancer diagnostics, ML-based medical imaging analysis demonstrates higher accuracy in identifying abnormalities compared to traditional visual interpretation methods.
2. *Early detection*: ML algorithms excel in early disease detection due to their ability to analyze large datasets and detect subtle patterns that may not be easily discernible by human observers. This early detection can lead to timely interventions and improved patient outcomes.
3. *Personalization:* ML algorithms can personalize treatment plans by analyzing EHRs and genetic data to identify individual risk factors and tailor interventions accordingly. This personalized approach is more effective than generic treatment strategies used in traditional diagnostic methods.
4. *Efficiency*: ML algorithms can process vast amounts of data rapidly, leading to faster diagnostic results compared to traditional methods. This efficiency is crucial in healthcare settings where timely diagnosis and treatment are critical.
5. *Scalability:* ML algorithms are scalable and can handle large volumes of data, making them suitable for population-level screenings and analyses. Traditional methods may struggle to scale effectively, especially in scenarios with extensive data requirements.
6. *Adaptability:* ML algorithms can adapt and improve over time as they learn from new data and feedback. This adaptability allows for continuous refinement and enhancement of diagnostic accuracy, which is not achievable with static traditional methods.

While ML algorithms offer significant advantages, it is essential to acknowledge some challenges, such as biased algorithms, model interpretability issues, and the need for robust validation and regulatory oversight. However, the added value of ML in disease diagnostics, particularly in terms of accuracy, early detection, personalization, efficiency, scalability, and adaptability, underscores its pivotal role in advancing healthcare practices.

8.13 BIASED ALGORITHMS AS A CHALLENGE IN ML-BASED DIAGNOSTICS

Acknowledging biased algorithms as a challenge in ML-based diagnostics is crucial, and implementing bias mitigation strategies is essential to ensure fairness and equity in healthcare. Here are some key strategies and considerations that enrich the discussion on addressing bias in ML-based diagnostics:

1. *Data preprocessing*: Begin by carefully preprocessing the dataset to identify and mitigate biases. This includes detecting and handling imbalanced class distributions, addressing missing or biased data points, and ensuring representative demographic and disease stage diversity in the dataset. Techniques like oversampling, undersampling, and synthetic data generation can be used to balance the dataset.
2. *Feature selection:* Choose features thoughtfully to avoid introducing biases. Conduct feature importance analysis to understand which features significantly impact the model's predictions and ensure that these features are relevant, non-discriminatory, and representative of diverse patient populations.
3. *Fairness-aware algorithms:* Utilize fairness-aware ML algorithms that explicitly consider fairness constraints during model training and decision-making. Techniques such as fairness constraints, adversarial training, and bias detection algorithms can help mitigate biases and promote fairness in predictions, especially regarding sensitive attributes like race, gender, or socioeconomic status.
4. *Bias audits and monitoring:* Conduct bias audits and ongoing monitoring of model performance to detect and mitigate biases in real-time. Implement feedback loops that continuously evaluate the model's predictions across different demographic groups and disease stages, ensuring that disparities are identified and addressed promptly.
5. *Interpretability and transparency:* Prioritize model interpretability and transparency to understand how the model makes decisions and identify potential sources of bias. Explainable AI techniques, such as feature importance visualization, decision tree explanations, and model-agnostic interpretability methods, help stakeholders understand the underlying factors influencing predictions and identify biased patterns.
6. *Diverse stakeholder involvement:* Involve diverse stakeholders, including healthcare professionals, ethicists, patients, and community representatives, in the development and evaluation of ML models. Incorporate their perspectives, feedback, and domain knowledge to ensure that the model's predictions are fair, ethical, and aligned with healthcare best practices.
7. *Ethical guidelines and regulations:* Adhere to ethical guidelines, regulations, and industry standards related to fairness, privacy, and data protection in healthcare ML. Collaborate with regulatory bodies, such as the FDA or relevant healthcare authorities, to ensure compliance with ethical principles and legal requirements in deploying ML-based diagnostic tools.

By implementing these bias mitigation strategies and fostering a culture of fairness, transparency, and inclusivity in ML-based diagnostics, healthcare practitioners can mitigate biases, promote equitable healthcare outcomes, and build trust in AI-driven decision-making processes.

8.14 METHODOLOGY

The primary emphasis of this research centers on diagnosing breast cancer, heart failure, and diabetes. Specifically, regarding breast cancer, the study employs ML algorithms to assess different models using a dataset obtained from the AI for Social Good: Women Coders Bootcamp. The data is pre-processed and visualized to improve model performance and interpretation. For the diagnosis of heart failure, the study uses EDA to understand the relationship between the probability of heart failure and various factors. The ML algorithms are used to detect heart failure using XG Boost, logistic regression, SVM, naïve Bayes, random forest, and KNN algorithms and their accuracy is assessed by cross-validation, ROC AUC score. As for the diagnosis of diabetes, the study uses various risk factors from genetics, lifestyle factors, environmental influences, and other relevant sources. The study highlights the need for a holistic, multidisciplinary approach to tackle the diabetes epidemic in India. We can understand the flowchart explanation given below:

(a) *Breast cancer diagnostics*: The study utilizes ML algorithms to examine a dataset containing 569 records and six columns, encompassing parameters such as mean radius, mean texture, mean perimeter, mean area, mean smoothness, and diagnosis. Different ML models, such as logistic regression, SVM, naïve Bayes, Random Forest, and KNN, are employed to assess the accuracy of the analysis.

(b) *Heart failure diagnostics*: The study analyzes the correlation between heart failure probability and various factors such as age, sex, smoking habits, and blood pressure. ML algorithms, such as XG Boost, logistic regression, and SVC, are utilized for the identification of heart failure, with their precision evaluated.

(c) *Diabetes diagnostics*: The research addresses the significant public health challenge of diabetes in India. Factors influencing diabetes, including genetics, lifestyle, obesity, physical activity, age, gestational diabetes, medical conditions, stress, environmental factors, sleep, smoking, and alcohol, are considered. The study explores ML applications for diabetes detection and evaluates the performance of XG Boost, logistic regression, and SVC.

(d) *Principle findings:* The ML algorithms demonstrate promising accuracy in disease diagnostics, with specific findings highlighted for each medical condition. For breast cancer, logistic regression and Random

Forest exhibit high accuracy. In heart failure detection, XG Boost outperforms other algorithms, while logistic regression and SVC show comparable performance in diabetes diagnosis.

(e) *Implications of findings:* The use of ML in disease diagnostics offers improved accuracy, early detection, and personalized treatment plans. However, challenges such as biased algorithms and model complexity need careful consideration for ethical and responsible use in healthcare.

The rationale behind selecting specific ML algorithms in healthcare research enhances understanding by showing how different algorithms complement each other. Here is how this rationale can be elucidated:

1. *Logistic regression:* This algorithm is chosen for its simplicity and interpretability, making it suitable for initial analyses and establishing baseline performance in disease diagnostics. It provides insights into linear relationships between input variables and outcomes, aiding in understanding basic patterns in the data.
2. *SVM:* This is selected for its ability to handle nonlinear data and complex decision boundaries. It complements logistic regression by capturing more intricate relationships in the data, especially in scenarios where classes are not easily separable. SVM contributes to improving diagnostic accuracy by considering higher-dimensional feature spaces.
3. *Naïve Bayes*: Naïve Bayes is advantageous for its computational efficiency and robustness to noisy data. It works well with relatively small datasets and is particularly useful in handling categorical or text-based data in healthcare, such as medical records or textual descriptions of symptoms. Naïve Bayes provides a probabilistic framework for decision-making, aiding in risk prediction and classification tasks.
4. *Random forest*: Random forest is chosen for its ensemble learning capability, which combines multiple decision trees to improve accuracy and reduce overfitting. It excels in handling high-dimensional data and capturing complex interactions among variables. Random forest contributes to the robustness of the model by reducing variance and enhancing generalizability across diverse patient populations.
5. *KNN:* This is selected for its simplicity and non-parametric nature, making it suitable for pattern recognition tasks and similarity-based classification. It complements other algorithms by considering local patterns and nearest neighbors in the feature space, providing additional insights into data clusters and patient similarities.

By integrating these diverse ML algorithms, the research methodology leverages their respective strengths to address different aspects of disease diagnostics comprehensively. Logistic regression establishes baseline performance, SVM handles complex data structures, naïve Bayes deals with categorical data, Random Forest reduces overfitting, and KNN considers local patterns. This collective approach enhances the overall model's accuracy, robustness, and interpretability, aligning with the research objectives of accurate disease detection and personalized treatment recommendations in healthcare.

8.15 CONCLUSION

The study suggests that disease diagnostics greatly benefit from the essential contribution of ML algorithms, especially in the early detection phase. It highlights the importance of ongoing research and development efforts aimed at improving the precision, interpretability, and applicability of ML models in medical contexts. The results emphasize the considerable potential of ML to transform healthcare practices and enhance the overall well-being of patients. Future research directions in the advancement of ML-based disease diagnostics explore fairness-aware algorithms, enhance data diversity, ensure interpretability, engage diverse stakeholders, and develop real-time bias monitoring for equitable and accurate ML-based disease diagnostics.

8.16 HIGHLIGHTS

- *Early disease detection:* The proposed ML-based technique can facilitate early detection of breast cancer, heart failure, and diabetes, leading to timely interventions and improved patient outcomes.
- *Personalized treatment plans:* By analyzing diverse data factors, including genetics, lifestyle, and environmental influences, the technique enables personalized treatment plans tailored to individual patient needs, enhancing healthcare efficacy.

REFERENCES

1. A. Belderrar and A. Hazzab, "Real-time estimation of hospital discharge using fuzzy radial basis function network and electronic health record data," *Int. J. Med. Eng. Inform.*, vol. 13, no. 1, pp. 75–83, 2021, doi: 10.1504/IJMEI.2021.111870.
2. A. Esteva *et al.*, "Dermatologist-level classification of skin cancer with deep neural networks," *Nature*, vol. 542, no. 7639, pp. 115–118, 2017, doi: 10.1038/nature21056.
3. G. Litjens *et al.*, "A survey on deep learning in medical image analysis," *Med. Image Anal.*, vol. 42, no. December 2012, pp. 60–88, 2017, doi: 10.1016/j.media.2017.07.005.

4. J. F. AlSamhori *et al.*, "Artificial intelligence for breast cancer: Implications for diagnosis and management," *J. Med. Surgery, Public Heal.*, vol. 3, no. April, p. 100120, 2024, doi: 10.1016/j.glmedi.2024.100120.

5. S. Patidar, D. Kumar, and D. Rukwal, "Comparative Analysis of Machine Learning Algorithms for Heart Disease Prediction," *Adv. Transdiscipl. Eng.*, vol. 27, pp. 64–69, 2022, doi: 10.3233/ATDE220723.

6. D. Mozaffarian, A. Kamineni, Carnethon M, and E. Al, "Lifestyle risk factors and new-onset DM in older adults," *Arch Intern Med*, vol. 169, no. 8, pp. 798–807, 2009, doi: 10.1001/archinternmed.2009.21.Lifestyle.

7. J. S. Ahn *et al.*, *Artificial Intelligence in Breast Cancer Diagnosis and Personalized Medicine*, vol. 26, no. 5. 2023. doi: 10.4048/jbc.2023.26.e45.

8. G. Nahler, "Observational Study," *Dict. Pharm. Med.*, vol. 34, no. May, pp. 125–125, 2009, doi: 10.1007/978-3-211-89836-9_951.

9. U. Galicia-garcia, A. Benito-vicente, S. Jebari, and A. Larrea-sebal, "Costus ignus: Insulin plant and it's preparations as remedial approach for diabetes mellitus," *Int. J. Mol. Sci.*, pp. 1–34, 2020.

10. S. Kushwaha, R. Srivastava, H. Vats, and P. Khanna, "Machine learning in healthcare," *Mach. Learn. Soc. Improv. Mod. Prog.*, no. 6, pp. 50–70, 2022, doi: 10.4018/978-1-6684-4045-2.ch003.

11. S. Kushwaha, R. Srivastava, H. Vats, and P. Khanna, "Machine learning in healthcare," *Mach. Learn. Soc. Improv. Mod. Prog.*, pp. 50–70, 2022, doi: 10.4018/978-1-6684-4045-2.ch003.

12. S. Li, J. C. Nunes, C. Toumoulin, and L. Luo, "3D Coronary Artery Reconstruction by 2D Motion Compensation Based on Mutual Information," *Irbm*, vol. 39, no. 1, pp. 69–82, 2018, doi: 10.1016/j.irbm.2017.11.005.

13. K. Balasubramanian and N. P. Ananthamoorthy, "Robust retinal blood vessel segmentation using convolutional neural network and support vector machine," *J. Ambient Intell. Humaniz. Comput.*, vol. 12, no. 3, pp. 3559–3569, 2021, doi: 10.1007/s12652-019-01559-w.

14. M. E. Cohen *et al.*, "Quantitative evaluation of rigid and elastic registrations for abdominal perfusion imaging with X-ray computed tomography," *Irbm*, vol. 34, no. 4–5, pp. 283–286, 2013, doi: 10.1016/j.irbm.2013.07.007.

15. J. H. Chen and S. M. Asch, "Machine Learning and Prediction in Medicine —Beyond the Peak of Inflated Expectations," *N. Engl. J. Med.*, vol. 376, no. 26, pp. 2507–2509, 2017, doi: 10.1056/nejmp1702071.

16. V. Gulshan *et al.*, "Development and validation of a deep learning algorithm for detection of diabetic retinopathy in retinal fundus photographs," *JAMA – J. Am. Med. Assoc.*, vol. 316, no. 22, pp. 2402–2410, 2016, doi: 10.1001/jama.2016.17216.

17. J. Lee *et al.*, "2017-Lee-Deep Learning in Medical Imaging_ Gen," *Korean J. Radiol.*, vol. 18, no. 4, pp. 570–584, 2017.

18. R. Miotto, L. Li, B. A. Kidd, and J. T. Dudley, "Deep Patient: An Unsupervised Representation to Predict the Future of Patients from the Electronic Health Records," *Sci. Rep.*, vol. 6, no. January, pp. 1–10, 2016, doi: 10.1038/srep26094.

19. G. Kotronoulas *et al.*, "An Overview of the Fundamentals of Data Management, Analysis, and Interpretation in Quantitative Research," *Semin. Oncol. Nurs.*, vol. 39, no. 2, pp. 1–9, 2023, doi: 10.1016/j.soncn.2023.151398.

20. M. Mohtasebi *et al.*, "Modeling of Neonatal Skull Development Using Computed Tomography Images To cite this version: HAL Id: hal-03602674 Modeling of Neonatal Skull Development Using Computed Tomography Images," 2023.

21. X. Xu *et al.*, "Multi-Feature Fusion Method for Identifying Carotid Artery Vulnerable Plaque," *Irbm*, vol. 43, no. 4, pp. 272–278, 2022, doi: 10.1016/j.irbm.2021.07.004.

Chapter 9

IoT-based home automation using NodeMCU

Sourav Diwania, Sumit Sharma, Salim, and Varun Gupta

9.1 INTRODUCTION

Internet of Things (IoT) is a concept where each device is assigned an IP address that makes it identifiable over the Internet. It started as the "Internet of Computers." Research studies have forecast an explosive growth in the number of "things" or devices that will be connected to the internet. The resulting network is called the IoT [1]. The recent developments in technology that permit the use of Bluetooth and Wi-Fi have enabled different devices to have the capabilities of connecting. Using a Wi-Fi shield to act as a micro web server for the Arduino eliminates the need for wired connections between the Arduino board and a computer which reduces cost and enables it to work as a stand-alone device. The Wi-Fi shield needs a connection to the internet from a wireless router or wireless hotspot and this would act as the gateway for the Arduino to communicate with the internet. With this in mind, an internet-based home automation system for remote control of home appliances is designed. The real-world appliances are being prepared with intellect and computing capability so that they can configure themselves accordingly. Sensors attached to embedded devices along with the low-power wireless connectivity can facilitate remote monitoring and control of the devices. This forms an integral component of the IoT network. In contrast to the majority of household automation structures now on the market, the anticipated device is scalable, meaning a single server may oversee several hardware interface modules as long as it is connected to a Wi-Fi network. Certain systems provide a broad range of home automation techniques, such as security and energy control components. In terms of scalability and flexibility, the suggested system performs better than the home computerization structures that are available for purchase. IoT also helps in transferring data from sensors through wireless networks, achieving recognition and informational exchange in open computing networks [2]. Things that we are using in our daily lives are becoming smart with the current technologies but it isn't sufficient until we connect them to act with the dynamic environment in addition to making their inter-network, that is machine-to-machine communication.

DOI: 10.1201/9781003581246-9

Objects like electronic devices, software, sensors, actuators, home appliances, and vehicles are connected to a wireless network. The IoT is considered a wireless network of these objects and they can exchange data through lightweight protocols like MQTT and CoAP [3]. By eliminating the need for unstable connections between the Arduino board and laptop, Wi-Fi serves as a micro internet server for the Arduino, lowering costs and enabling it to function as a stand-alone device. A wireless network or hotspot connection is required for the Wi-Fi to function as a gateway for the Arduino to communicate with the internet. A network-based home automation system is meant to operate remote household equipment. Once this IoT-based activity is completed successfully, you will be able to operate the appliances in your house, such as the refrigerator, TV, door lock, fan, lightbulb, and so on. Use your smartphone to access content from anywhere in the world. There are many types of radio modules, out of which GSM, 3G, Wi-Fi, Bluetooth, and ZigBee [4] are common. However, owing to the surging number of Wi-Fi hotspots and range sufficient to perform the required control and monitoring, Wi-Fi is chosen as the mode of communication in the prototype, and the devices are controlled through the Blynk App implemented using ESP8266 [5]. Using wireless internet access, an ESP8266 module will wirelessly get instructions from your smartphone. To carry out this task, we need a utility to be activated for the challenge. This application will encode the instructions into a smartphone and send it to ESP8266. Although there are other applications available, we will choose the one that meets our needs and provides us with easy access to "Blynk" these products. In addition, it is accessible on all platforms, including IOS and Android. The Blynk app may help you develop creative projects [6]. Blynk is a Platform with iOS and Android apps to control Arduino, Raspberry Pi, and the like over the internet. It is a digital dashboard where you can build a graphic interface for your project by simply dragging and dropping widgets. It is really simple to set everything up and you will start tinkering in less than 5 minutes. Blynk is not tied to some specific board or shield. Instead, it supports the hardware of your choice. Whether your Arduino or Raspberry Pi is linked to the internet over Wi-Fi, Ethernet, or this new ESP8266 chip, Blynk will get you online and ready for the Internet of Your Things. Through the utilization of the MQTT (Message Queuing Telemetry Transport) protocol, effective communication and data exchange between devices are achieved, ensuring there is real-time responsiveness and reliability. Additionally, the research tackles challenges such as interoperability, scalability, and data security that are present in IoT deployments. Strategies to handle these challenges are discussed, such as adopting standardized communication protocols, having a modular system architecture, and using encryption techniques to protect sensitive information. The effectiveness of the suggested IoT-based home automation system is tested through practical experimentation, measuring parameters like energy consumption, response time, and user satisfaction.

The findings reveal the system's ability to streamline household operations, optimize resource utilization, and boost overall comfort and security. The project's main focus is on implementing NodeMCU, a potent and reasonably priced development board that is entirely dependent on the ESP8266 Wi-Fi module [7]. NodeMCU is the brain of our home automation system, allowing various devices and appliances to communicate verbally and connect effortlessly. The concept aims to provide homeowners with the capacity to remotely control and observe their living spaces, promoting a more favorable comfort level, energy economy, and average well-being.

The second section of this chapter is about system design, which discusses system design and block diagram of the whole process, the third section is about the relays which are used in this project, and the fourth section is about NodeMCU. Its specification and circuit diagram explain the whole process and how our project works. The fifth and sixth sections are about the software we use in our project and the app from which we can control home appliances from any place; in the seventh section, there is a flowchart in which the working of our project is given.

9.2 SYSTEM DESIGN

9.2.1 *Description of block diagram*

A better job to manage the devices in a timely and organized manner is home automation that is based on a real-time clock. Wi-Fi allows users to control the devices from many locations. The gadgets or household equipment's operational parameters may be recorded by the RTC using EEPROM technology. The task is essentially an idea to bring automation into the home or business. A smartphone app will be used to operate every device in the home. For organized operation, the home or business equipment will be interfaced with a centralized microcontroller NodeMCU. In Figure 9.1, the controller's built-in RTC and EEPROM generation may be turned on for the task. To get the management instructions from the Wi-Fi protect fabric (Wi-Fi hotspot), the controller is interfaced with Wi-Fi devices [1]. Devices including temperature sensors, fuel sensors, motion sensors, light switches, electrical plugs, and others were integrated with the suggested home control structures to demonstrate the viability and efficacy of this machine. It makes use of an IP-connected integrated micro-net server in the NodeMCU microcontroller to provide remote access to and management of devices and appliances. These devices may be controlled by Bluetooth Android-based Smart smartphone bundles or online apps [2]. In comparison to comparable structures, the future device allows protocol connectivity to display and control the home switching capabilities without the need for a dedicated server PC. The operator will be satisfied with the mobile app provided by Wi-Fi being enough. The user must swap out the control button provided in

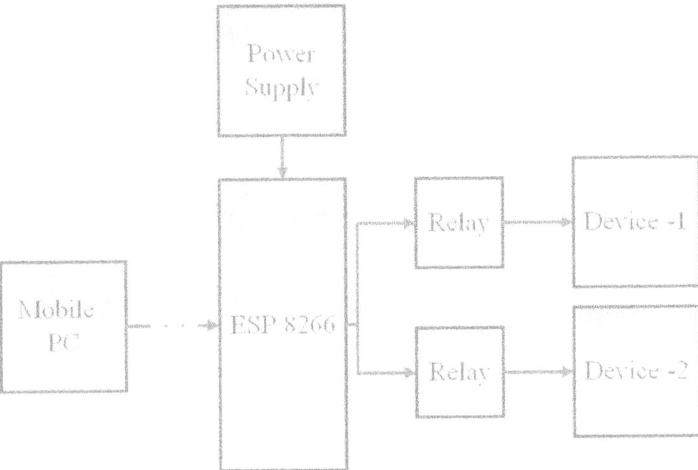

Figure 9.1 Block diagram of system.

the app to turn the light on or off. Thereafter, the data will be transferred to the microcontroller's Wi-Fi. The microcontroller activates via RTC and EEPROM as soon as the request is received, allowing movement to be performed by the request. All other appliances may be operated in the same way.

9.2.2 Display

It is electromagnetic switching, but the relay itself is nothing. Relays let one circuit switch out for another when they are apart. Relays are used in low-voltage circuits to indicate when a gadget needs high voltage to operate and to switch it off [8]. For example, a lamp run by 230 V AC mains may be powered by a 5 V delivery linked to the relay. Relays are available in a variety of operating voltage configurations, including 6 V, 9 V, 12 V, and 24 V. Typically, a relay is divided into two parts: an input and an output. There is no input component, but there is a coil that creates a magnetic field when a little input voltage is applied to it. The first normally closed (NC), second usually opened (NO), and third not uncommon (COM) contactors make up this relay. Electrical devices in homes may be turned on or off by using the correct contactor combinations.

9.2.3 Flow diagram

This flowchart describes the process of controlling a hardware device using a NodeMCU and a relay module, likely triggered by user input within an application in Figure 9.2. The steps involved in the study are as follows:

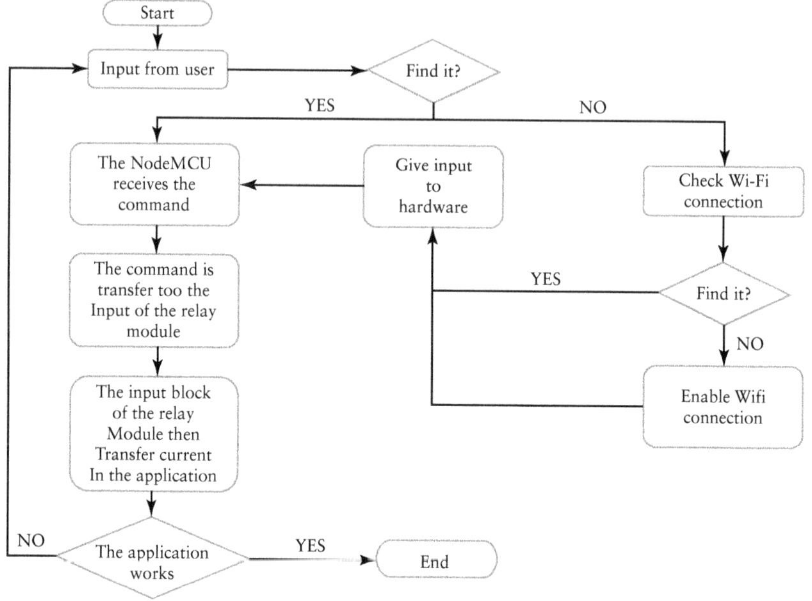

Figure 9.2 Working diagram of IoT-based system.

- Start: The process begins.
- Input from the user: The system receives a command from the user.
- Find it? (Decision): The system checks if the command is recognized or valid.
  - YES: The process proceeds.
  - NO: The system prompts the user to provide valid input.
- The NodeMCU receives the command: The validated command is sent to the NodeMCU (a microcontroller board) [9].
- The command is transferred to the input of the relay module: The command is forwarded to the relay module, which controls an electrical switch.
- Check Wi-Fi connection: The system verifies if there is an active Wi-Fi connection.
  - YES: The process continues.
  - NO: The system attempts to enable a Wi-Fi connection.
- The input block of the relay module then transfers current in the application: The relay module, based on the command, acts as a switch to control the flow of electrical current, likely turning a device on or off.
- Does the application work? (Decision): The system checks if the application is functioning as intended based on the executed command.
  - YES: The process ends successfully.
  - NO: The system may indicate an issue or prompt the user for further action (not visualized in this chart).

9.2.4 NODEMCU

The ESP8266 microcontroller is integrated into a broad circuit board. An integrated USB port on the board is already fixed, with tension built up inside the chip. Physical reset button, standard-sized GPIO (General Purpose Input Output) pins that can be plugged into a breadboard, LED lights, and a Wi-Fi antenna. Its CPU, known as the L106 32-bit RISC microprocessor core, is based on the Ten silica Xtensa Diamond Standard 106 Micro and operates at 80 MHz. Its memory comprises 16 kB of system data RAM, 32 kB of training cache RAM, and 80 kB of user data RAM. It is equipped with built-in IEEE 802.11 b/g/n Wi-Fi modules [10]. The call of a microcontroller created by Expressive Systems is the ESP8266. The ESP8266 is a self-contained Wi-Fi networking solution in and of itself. It can also be used to create employer self-contained packages and functions as a bridge from a microcontroller to Wi-Fi [11, 12]. The specification of the proposed study has been presented in Table 9.1.

In Figure 9.3, a 5 V input is given to NodeMCU (main microcontroller) and four relays which are working as a switch. We can give 5 V to NodeMCU and relay separately or by supplying 5 V in parallel so that it can work as a bus. Two plugs are connected to the AC mains since the LED bulb and fan work on AC voltage at 50 Hz frequency. When we give the command to NodeMCU via the mobile app through Wi-Fi to turn on a light or a fan, the NodeMCU sends signals to the relays and then they allow current to flow through the fan and LED bulb, thus creating an app-based switching control system. All the home appliances will be controlled by a mobile app. The appliances in the industry or home will be interfaced with a centralized microcontroller NodeMCU for systematic working. This will activate the RTC and EEPROM built-in technology inside the controller [13]. To accept

Table 9.1 Complete specifications of NodeMCU

Specification	Values
Voltage	3.3 V
Wi-Fi Direct (P2P)	Soft-AP
Current consumption	10 µA–170 mA
Flash memory attachable	16 MB max (512K normal)
Integrated protocol stack	TCP/IP
Processor	Ten silica L106 32-bit
Processor speed	80–160 MHz
RAM	32K + 80K
GPIOs	17 (multiplexed with other functions)
Analog to digital	1 input with 1,024 step resolution +19.5 dBm output power 802.11b mode 802.11 support b/g/n
Maximum concurrent TCP connection	5

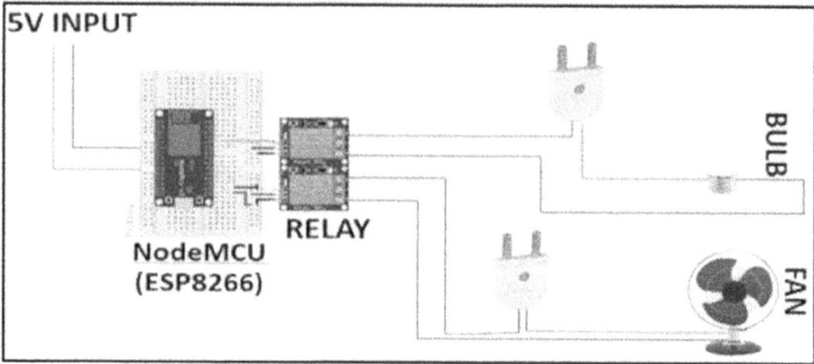

Figure 9.3 Circuit diagram of the proposed study.

controlling orders from the Wi-Fi shield (Wi-Fi hotspot), the controller additionally interfaces with Wi-Fi. An app for mobile devices with Wi-Fi built in will be given to the operator. The control button in the app must be switched on for the operator to turn the light on or off.

9.2.5 BLYNK DESCRIPTION

Blynk evolved into an IoT design. In addition to doing many other amazing things, it may display sensor data, store and visualize records, and remotely modify hardware.

9.2.5.1 Working and establishment of Blynk

Blynk evolved into an IoT design. It has remote hardware control capabilities.
The platform is made up of three essential parts:
Blynk App: With the help of the many widgets we provide, you may design amazing user interfaces for your jobs.
Blynk Server: This device is responsible for all communications between the hardware and the smartphone. You may run your device locally or use our Blynk Cloud. The accessible resources may also be deployed on a Raspberry Pi and can easily transfer thousands of devices at once.
Blynk Libraries provide for communication with the server and handle all incoming and outgoing instructions for all well-known hardware structures [14].

9.2.5.2 Registration of new Blynk account after downloading the Blynk app

Step 1: If you already have an account, this one is different from the ones used for the Blynk Forums.
Step 2: Start by establishing a new project: Following a successful account login, begin by creating a new project.

Step 3: Select the hardware.

Step 4: Auth Token which is a special code is required to link your smartphone and hardware [15].

9.2.6 MICROPYTHON

MicroPython is a lightweight Python 3 implementation optimized for microcontrollers and embedded systems. It can be used to program NodeMCU, providing a Python environment for development. Before you can use MicroPython on NodeMCU, you need to flash the MicroPython firmware onto the NodeMCU board. You can find the MicroPython firmware for ESP8266 on the official MicroPython website. Once the firmware is flashed, you can connect to the NodeMCU using a serial communication tool, such as PuTTY or minicom [16]. This allows you to interact with the NodeMCU using the MicroPython REPL (Read-Eval-Print Loop) over the serial connection. With MicroPython installed, you can write and run Python code directly on the NodeMCU. You can use a text editor or an integrated development environment (IDE) to write your Python scripts and then transfer them to the NodeMCU using tools like ampy or Web REPL. Figure 9.4 presents the complete flowchart of the working of the complete structure.

Start: The procedure is indicated by the start point in the flowchart above.

System initialization: Configure the NodeMCU and attach it to the necessary parts to begin the IoT-based home automation system.

Connect to Wi-Fi/hotspot: To create a communication connection with the internet, the NodeMCU joins the nearby Wi-Fi network.

Start Blynk: Start the NodeMCU's Blynk application. IoT application creation is made simple and rapid with the help of the Blynk platform.

Authenticate device: To enable communication between the NodeMCU and the Blynk app, authenticate the NodeMCU with the Blynk server.

Initialize sensors: Set up any sensors that are attached to the NodeMCU, such as those for motion.

User involvement: See if the Blynk app allows for any user involvement.

Control actuators: Utilizing sensor data and human inputs and regulating actuators to do tasks like regulating thermostats and turning on and off lights, among other things.

9.3 RESULT AND DISCUSSION

By integrating NodeMCU, a cost-effective Wi-Fi-enabled microcontroller, with the Blynk IoT platform, we create a seamless system for managing home appliances and security devices. This project demonstrates the powerful potential of combining IoT technology with home automation [17,

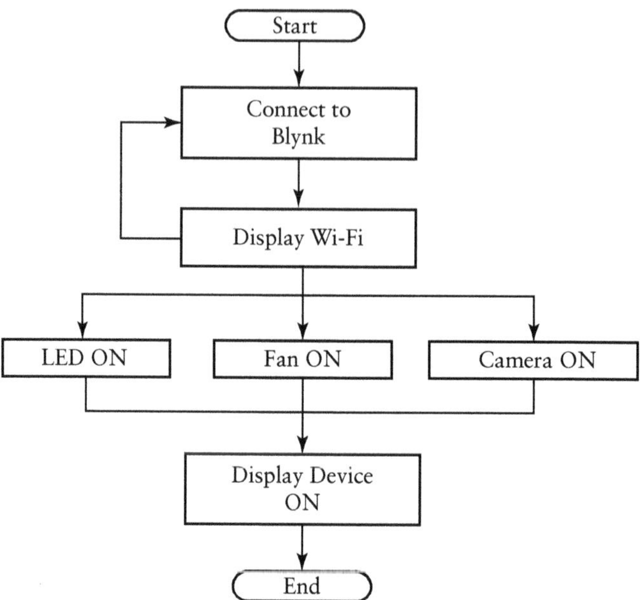

Figure 9.4 Flowchart of the proposed structure.

18]. By utilizing NodeMCU and the Blynk platform, we create an integrated system that offers enhanced convenience, energy efficiency, and security for homeowners. The use of affordable and versatile components like NodeMCU makes this approach accessible and practical for a wide range of users, highlighting the future of smart home technology. Figure 9.5 represents the home automation structure using NodeMCU and Blynk.

The Blynk app allows you to create custom web dashboards for IoT projects shown in Figure 9.6. A variety of widgets are used for dashboards, including switches, sliders, graphs, and gauges. The goal is to develop a web dashboard using Blynk for home automation that controls devices like lights, fans, and monitors security through a thermal camera. Blynk is an IoT platform that allows users to control and monitor hardware remotely using a smartphone app shown in Figure 9.7. The dashboard in the image provides controls for a fan, light, and camera. This project uses NodeMCU (ESP8266) to interface with Blynk, allowing users to manage their home devices remotely. Figure 9.8 represents the hardware implementation of the proposed system.

This project lets you remotely control two appliances (like lights, DC fans, or even an AC appliance through a special socket) using your smartphone. The NodeMCU controller acts as the brains, receiving commands from the Blynk app you install on your phone. The Blynk app provides a user-friendly interface with buttons or sliders to turn on/off appliances connected to relays controlled by the NodeMCU. When you interact with

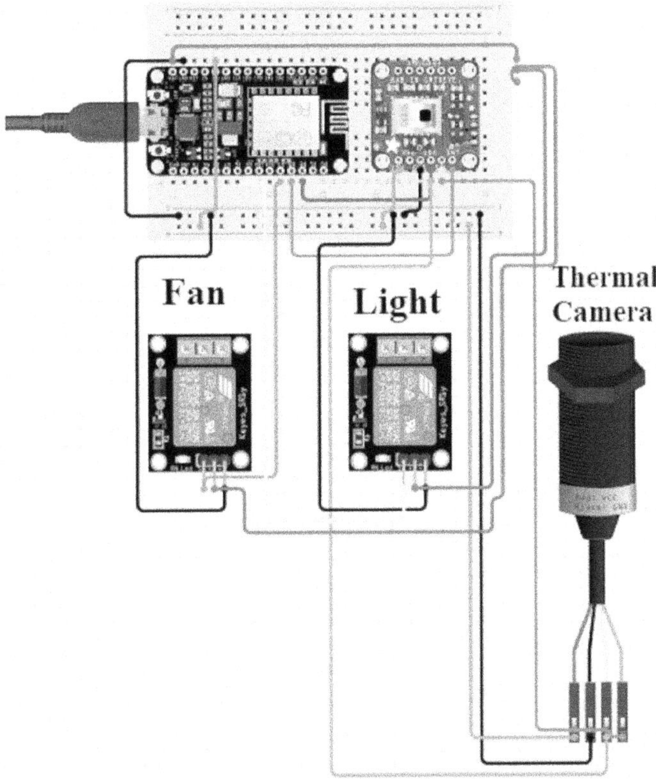

Figure 9.5 Home automation using NodeMCU and Blynk.

the app, it sends a signal to the NodeMCU, which then activates the corresponding relay, ultimately controlling the power to your appliances.

9.4 CONCLUSION

The high cost of home automation is a major barrier to its acceptance. This chapter reviews existing systems, which often require additional network devices like hubs and increasing costs. The use of NodeMCU and IoT platforms can make these devices more cost-effective. This approach offers remote control convenience via a web page or application, making it platform independent and easily operable without specific operating systems. Overall, this system promises optimal results. The NodeMCU ESP8266 module enables the design of a smart home system controlled via a web application. It can manage lighting, fans, temperature, and more, providing convenience and energy savings. This project demonstrates a cost-effective

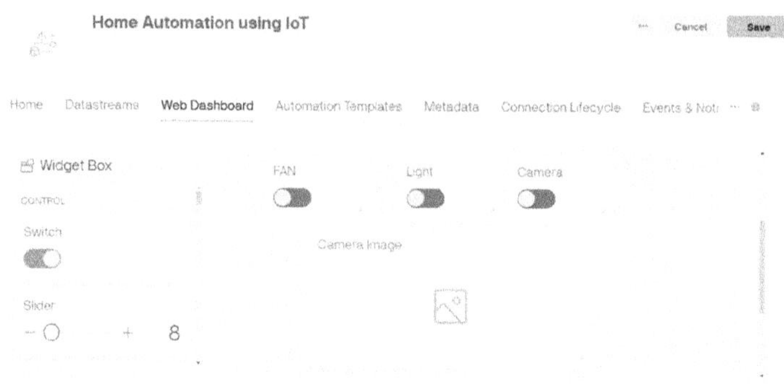

Figure 9.6 Home automation web dashboard in Blynk.

Figure 9.7 Home automation mobile dashboard in Blynk.

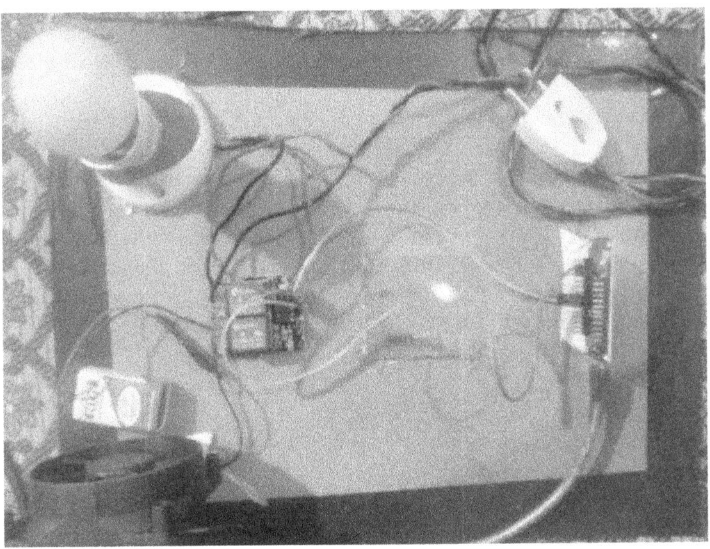

Figure 9.8 Hardware implementation of proposed system.

home automation system using the NodeMCU Board and various sensors, controlled remotely by an Android smartphone. It allows for the control and monitoring of household appliances like security lights, TVs, and air conditioners. The system is flexible and uses the Blynk app for control, with notifications sent to users.

REFERENCES

1. M. Murphy (2011). *Beginning Android 3*, Après. ISBN-13 (pbk): 978-1-4302-3297-1. ISBN-13 (electronic): 978-1-4302-3298-8.
2. Addison-Wesley (2011). *Android Wireless Application Development*, 2nd edn. ISBN-13: 978-0-321-74301-5, ISBN-10: 0-321-74301-6.
3. Wikipedia (2009). Home automation. http://en.Wikipedia.org/wiki/Home automation. Retrieved on 20/5/2018.
4. M. Bates (2006). *Interfacing PIC Microcontrollers Embedded Design by Interactive Simulation*. Newness, London.
5. S. Alam, M. M. R. Chowdhury, and J. Noll (2010). "Sena as: An event-driven sensor virtualization approach for internet of things cloud," in *Networked Embedded Systems for Enterprise Applications (NESEA), 2010 IEEE International Conference*.
6. Z. Shelby, K. Hartke, and C. Bormann (2013) Constrained application protocol (Coop), Core Working Group Internet-Draft.
7. D. Norris, Smart home automation based on IOT and android technology, M. Abivandhana, K. Divya, D. Gayathri, R. Ruhin Kouser, Student, Assistant Professor, Department of CSE Kingston Engineering College, Katpadi, Vellore, India.

8. K. Baraka, M. Ghobril, S. Malek, R. Kanj, and A. Kayssi (2013). "Low-cost Arduino/android-based energy-efficient home automation system with smart task scheduling," in *Fifth International Conference on Computational Intelligence*.

9. Giuseppe D'Aniello, Matteo Gaeta,Francesco Orciuoli, Giuseppe Sansonetti and Francesca Sorgente (2014). "Star-city: semantic traffic analytics and reasoning for city," in *Proceedings of the 19th International Conference on Intelligent Users Interfaces*.

10. Luis Sanchez, José Antonio Galache, Veronica Gutierrez, José Manuel Hernandez, Jesús Bernat, and Alex Gluhak (2011). "SmartSantander: The meeting point between future internet research and experimentation and the smart cities," in *Future Network & Mobile Summit*.

11. Blynk, "https://www.blynk.cc/," Blynk, 2017. [Online]. Available: https://www.blynk.cc/.

12. T. Song, R. Li, B. Mei, J. Yu, X. Xing, and X. Cheng (2017). "A privacy preserving communication protocol for IoT applications in smart homes", *IEEE Internet of Things Journal,* vol. 4, no. 6, 2017.

13. J. Chhabra and P. Gupta (2016). *IoT Based Smart Home Design Using Power and Security Management.*

14. P. D. Ranjith Balakrishnan (2015). *IoT Based Monitoring and Control System for Home Automation.*

15. R. K. Kodali, V. Jain, S. Bose, and L. Boppana (2016). *IoT Based Smart Security and Home Automation System.*

16. T. Churasia and P. Kumar Jain (2019). Enhance smart home automation system based on internet of things", in *Proceedings of the Third International Conference on I-SMAC (IoT in Social, Mobile, Analytics and Cloud) (I-SMAC 2019) IEEE Xplore Part Number: CFP19OSV-ART, ISBN: 978-1-7281-4365-1.*

17. W. A. Jabbar, M. Hayyan Alsibai, N. S. S. Amran, and S. K. Mahayadin (2018). *Design and Implementation of IoT- Based Automation System for Smart Home.*

18. N. Vikram, K. S. Harish, M. S. Nihaal, R. Umesh, and S. A. Ashok Kumar (2017). A low cost home automation system using Wi-Fi based wireless sensor network incorporating internet of things", *2017 IEEE 7th International Advance Computing Conference.*

Chapter 10

Stabilization and synchronization of Chen–Lee chaotic system using sliding mode control approach

Pallav, Himesh Handa, and Sumit Sharma

10.1 INTRODUCTION

The scientific community has shown considerable interest in investigating methods to stabilize chaotic systems during the last two decades [1, 2]. Chaotic systems, recognized for their high sensitivity to initial conditions, have emerged as a prominent and extensively studied research domain. Significantly, in 1990, Pecora and Carroll made a groundbreaking contribution by pioneering the concept of complete synchronization for two identical chaotic systems [3]. This innovative approach opened new avenues for understanding and controlling chaotic dynamics, laying the foundation for further developments in the arena of chaos theory and non-linear dynamics. The concept of complete synchronization has since become a crucial tool in the study and application of chaotic systems, offering insights into their behaviour and potential avenues for practical utilization in various scientific and technological domains. Following the revolutionary efforts of Pecora and Carroll, the investigation into synchronization phenomena within chaotic systems has broadened, encompassing a range of diverse synchronization types. Researchers have delved into various synchronization patterns, each shedding light on distinct aspects of the dynamics within chaotic systems. This investigation explores diverse synchronization scenarios, from complete state mirroring to more nuanced interactions involving selective state matching (generalized and partial synchronization), phase alignment, and specific scaling relationships (projective synchronization) [4–8]. These synchronization phenomena have found practical applications in various domains, including secure communication, system identification, neural networks, brain activity modelling, non-linear system optimization, and pattern recognition [9–14]. In recent times, researchers have proposed several control methods for chaos synchronization. These include various control approaches to achieve synchronization, including feedback control (linear and non-linear), contraction theory, backstepping design, impulse control, sliding mode control (SMC), active control, and adaptive methods. Additionally, specific techniques like time-delay feedback, pushback control, and feedback linearization are investigated for their application in

DOI: 10.1201/9781003581246-10

chaos synchronization [15–25]. These advancements align with the foundational principles of chaotic control theory, illustrating ongoing efforts to leverage chaos synchronization for diverse applications and addressing complex dynamics in non-linear systems.

SMC, a robust non-linear control approach, offers a direct and effective method to handle chaos in non-linear systems, demonstrating its success in controlling chaotic systems and proving its adaptability to complex and unpredictable dynamics. Extensive literature exists on various SMC schemes, showcasing their applications across diverse fields. The essence of SMC resides in developing control principles which guide system states to reach and maintain the switching surface. The dynamic performance of a SMC system is influenced by the chosen sliding surface, determining the shift in the control structure. Linear hyperplanes commonly serve as sliding surfaces in this context. Within the framework of SMC, a discontinuous control rule compels the system's state trajectories to adhere to a designated switching surface.

One of the key setups in chaos synchronization research involves two coupled systems, where the goal is to achieve identical chaotic behaviour between them, often referred to as the "drive" and "driven" systems. The process of synchronization is instigated by the drive system, acting as the initiating influence, while the response system replicates the behaviour of the drive system. The fundamental thing of this synchronization setup lies in the careful design of control methodologies, with the objective of ensuring that the states of the driven system closely track and synchronize with those of the master system. This intricate coordination enables a harmonized and correlated dynamic evolution between the two chaotic systems, facilitating the attainment of synchronization in the overall system behaviour.

Chen and Lee introduced the Chen–Lee system in 2004 with the aim of implementing anti-control of chaos in rigid body motion [26]. This system is one of more than 30 of its sort that exhibits autonomous dynamics in three dimensions. Since Lorenz first proposed the idea in 1963 [27], other researchers have discovered a number of systems with related features. The Chen–Lee system is distinct among the other recently discovered models that may generate a complicated chaotic attractor, placing it within the larger class of such dynamic systems.

In the current era, various control techniques are being employed to achieve synchronization in such systems. The literature often features the utilization of numerous controllers for synchronizing chaotic systems, with the intriguing observation that the quantity of state variables in the system aligns with the quantity of controllers employed. Alternatively, achieving synchronization in two chaotic systems may involve using two or more controllers, depending on the specific context. When devising a control strategy, it is pivotal to lessen the quantity of control inputs, aiming for simplicity and straightforwardness in the control scheme. As a result, the authors concluded that investigating chaotic synchronization using minimum number of control inputs is beneficial. Implementing control strategies in real-world systems that rely on minimum number of control inputs not only simplifies the process but also enhances understanding and facilitates effective execution.

This study is dedicated to overcoming the challenge of devising a control scheme that achieves both stabilization and synchronization of Chen–Lee (drive–response) chaotic systems while employing a minimal number of control inputs. The research integrates Lyapunov stability concept with a SMC approach to attain this objective. Leveraging a proportional-integral (PI) sliding surface, the design guarantees closed-loop error system stability even during switching transients. The anticipated controller instigates sliding motion, effectively achieving synchronization in Chen–Lee chaotic systems configured in a drive–response setup. The chapter wraps up by substantiating the efficacy of the anticipated strategy via numerical simulations.

The organization of this chapter unfolds in the following manner: Section 10.2 provides an explanation of the Chen–Lee chaotic system, while Section 10.3 introduces the anticipated design for switching surface and stabilizing controllers. Proceeding to Section 10.4, the chapter elaborates on the formulation of sliding surfaces and SMCs, addressing the synchronization challenge within the Chen–Lee chaotic systems. Section 10.5 provides numerical simulation results, highlighting the efficiency of the anticipated controller. The finishing notes are outlined in Section 10.6.

10.2 INTRODUCTION OF CHEN–LEE CHAOTIC SYSTEM

In 2004, Chen and Lee introduced the Chen–Lee system, a novel chaotic system. Unlike the Lorenz system's focus on thermal convection, the Chen–Lee system tackles a different physical phenomenon, providing a mathematical representation of gyro motion with feedback control within its equations. This unique characteristic ensures that no physical meanings are overlooked in its representation. Additionally, the Chen–Lee system is noted for its commendable symmetrical behaviour, particularly in the right-hand side coordinate system $(x - y - z)$. The mathematical model of the system is comprised of the following non-linear differential equations [26]:

$$\begin{cases} \dot{x}_1 = -x_2 x_3 + a x_1 \\ \dot{x}_2 = x_1 x_3 + b x_2 \\ \dot{x}_3 = \frac{1}{3} x_1 x_2 + c x_3 \end{cases} \tag{10.1}$$

With the state variables denoted as x_1, x_2, and x_3, the system exhibits chaotic dynamics when its parameters a, b, and c take on the values 5, –10, and –3.8, as illustrated in Figure 10.1.

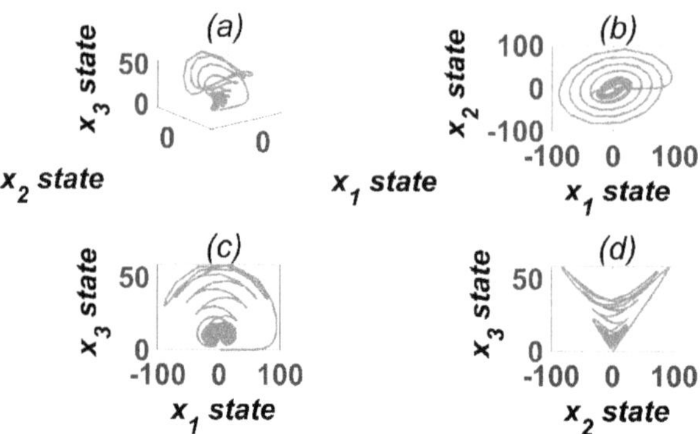

Figure 10.1 The dynamics of the Chen–Lee chaotic system exhibiting chaotic behaviour.

10.3 STABILIZATION OF CHEN–LEE CHAOTIC SYSTEM

Let us frame the dynamics of the Chen–Lee chaotic system in the following manner:

$$\begin{cases} \dot{x}_1 = -x_2x_3 + ax_1 + u \\ \dot{x}_2 = x_1x_3 + bx_2 \\ \dot{x}_3 = \dfrac{1}{3}x_1x_2 + cx_3 \end{cases} \tag{10.2}$$

Here u represents the controller to be formulated to stabilize the system defined in equation (10.2). The introduction of controller term is done through a meticulous analysis of the system dynamics employing the sliding mode technique.

To achieve stability in the sliding mode and to drive the system in equation (10.2) to equilibrium (zero state), the controller design follows a two-step approach. The first step involves selecting a suitable switching surface, which is critical for ensuring stability within the sliding mode. This surface is then used in the subsequent step, where a SMC law is designed to enforce the sliding mode ($s = 0$). Recognizing the importance of surface selection for stability, a PI switching surface function is defined as follows:

$$s = x_1 + \int_0^t (k_1x_1 + \frac{4}{3}x_2x_3)d\tau \tag{10.3}$$

Here the positive constant k_1 is explicitly determined by the designer. As per the findings by Utkin [28], when the system operates in the sliding mode, the following equation is valid:

$$\dot{s} = 0 \tag{10.4}$$

By utilizing equations (10.3) and (10.4), the sliding mode dynamics can be derived as follows:

$$\dot{s} = \dot{x}_1 + k_1 x_1 + \frac{4}{3} x_2 x_3 \tag{10.5}$$

From equation (10.5), one can express:

$$\dot{x}_1 = -k_1 x_1 - \frac{4}{3} x_2 x_3 \tag{10.6}$$

Hence, based on equation (10.6), the equivalent sliding mode dynamics for the system can be derived in the following manner:

$$\begin{cases} \dot{x}_1 = -k_1 x_1 - \dfrac{4}{3} x_2 x_3 \\[2mm] \dot{x}_2 = x_1 x_3 + b x_2 \\[2mm] \dot{x}_3 = \dfrac{1}{3} x_1 x_2 + c x_3 \end{cases} \tag{10.7}$$

Leveraging the Lyapunov stability principle, a specific Lyapunov function is constructed to assess the stability of the sliding mode dynamics described in equation (10.7):

$$V = \frac{1}{2}\left(x_1^2 + x_2^2 + x_3^2\right) \tag{10.8}$$

The stability of the system can be analysed by examining the time derivative of the Lyapunov function described previously, represented as follows:

$$\begin{aligned} \dot{V} &= x_1 \dot{x}_1 + x_2 \dot{x}_2 + x_3 \dot{x}_3 \\ &= x_1\left[-k_1 x_1 - \frac{4}{3} x_2 x_3\right] + x_2\left[x_1 x_3 + b x_2\right] + x_3\left[\frac{1}{3} x_1 x_2 + c x_3\right] \\ &= -k_1 x_1^2 + b x_2^2 + c x_3^2 \end{aligned}$$

In this expression, the design parameter k_1, representing a positive number, is to be selected by the designer, while b and c are the system parameters, both of which are negative.

$$\text{Hence, } \dot{V} \le 0 \tag{10.9}$$

The Lyapunov stability principle asserts that sliding motion on a sliding manifold is stable, ensuring that $\lim\limits_{t \to \infty} \| x_i(t) \| = 0$ for $i = 1, 2, 3.$

Establishing an SMC approach is essential to guide system trajectories onto the sliding surface $s = 0$. This achievement is realized through the identification of the appropriate sliding surface (10.3), as explained earlier.

The subsequent controller is introduced to guarantee the initiation of sliding motion.

$$u = \left(-k_1 - a\right)x_1 - \frac{1}{3}x_2x_3 - \psi\left(\text{sign}\left(s\right)\right), \; \psi > 0 \tag{10.10}$$

In the subsequent theorem, the outlined controller structure is employed to ensure the presence of sliding motion and, consequently, to guarantee stabilization.

Theorem 10.1: When the system dynamics outlined in equation (10.2) are governed by the controller specified in equation (10.10), then the system trajectories converge onto the sliding surface $s = 0$ and fulfil the condition $\lim_{t \to \infty} \| x_i\left(t\right) \| = 0$ for $i = 1, 2, 3$.

Proof: Introduce a suitable Lyapunov function as follows:

$$V = \frac{1}{2}s^2 \tag{10.11}$$

Now, by computing the time derivative of equation (10.11) and incorporating equations (10.2), (10.5), and (10.10), the expression obtained is as follows:

$$\dot{V} = s\dot{s}$$

$$= s\left[\dot{x}_1 + k_1x_1 + \frac{4}{3}x_2x_3\right]$$

$$= s\left[-x_2x_3 + ax_1 + u + k_1x_1 + \frac{4}{3}x_2x_3\right]$$

$$\leq -\psi|s| \tag{10.12}$$

When ψ is chosen in a way that $\psi > 0$, then $\dot{V} \leq 0$, thus confirming that the reaching condition is continuously satisfied. Therefore, the proof is convincingly established.

10.4 SYNCHRONIZATION OF TWO CHEN–LEE CHAOTIC SYSTEMS

If the system specified in equation (10.2) is designated as the driver system, then the response system can similarly be characterized as specified below:

$$\begin{cases} \dot{y}_1 = -y_2 y_3 + a y_1 + u_1 \\[4pt] \dot{y}_2 = y_1 y_3 + b y_2 + u_2 \\[4pt] \dot{y}_3 = \dfrac{1}{3} y_1 y_2 + c y_3 \end{cases} \tag{10.13}$$

Here u_1 and u_2 represent the control inputs which are to be intended for synchronizing the systems defined in equations (10.1) and (10.13).

Let the error dynamics between equations (10.13) and (10.1) be expressed as follows:

$e_i = y_i - x_i$ for $i = 1 - 3$

The computation of error dynamics can be carried out as follows:

$$\begin{cases} \dot{e}_1 = -y_2 e_3 - y_3 e_2 + e_2 e_3 + a e_1 + u_1 \\[4pt] \dot{e}_2 = y_1 e_3 + y_3 e_1 - e_1 e_3 + b e_2 + u_2 \\[4pt] \dot{e}_3 = \dfrac{1}{3} y_1 e_2 + \dfrac{1}{3} y_2 e_1 - \dfrac{1}{3} e_1 e_2 + c e_3 \end{cases} \tag{10.14}$$

The two functions representing the PI switching surface intended for synchronization can be described as follows:

$$\begin{cases} s_1 = e_1 + \displaystyle\int_0^t \left(\beta_1 e_1 + \dfrac{1}{3} y_2 e_3 - \dfrac{1}{3} e_2 e_3\right) d\tau \\[10pt] s_2 = e_2 + \displaystyle\int_0^t \left(\beta_2 e_2 + \dfrac{1}{3} y_1 e_3\right) d\tau \end{cases} \tag{10.15}$$

Here the positive constant β_1 and β_2 are explicitly determined by the designer. As per the findings by Utkin [28], if the system goes in the sliding mode, the subsequent equation remains valid:

$$\begin{cases} \dot{s}_1 = 0 \\[4pt] \dot{s}_2 = 0 \end{cases} \tag{10.16}$$

By utilizing equations (10.15) and (10.16), the sliding mode dynamics may be derived as follows:

$$\begin{cases} \dot{s}_1 = \dot{e}_1 + \beta_1 e_1 + \dfrac{1}{3} y_2 e_3 - \dfrac{1}{3} e_2 e_3 \\[6pt] \dot{s}_2 = \dot{e}_2 + \beta_2 e_2 + \dfrac{1}{3} y_1 e_3 \end{cases} \tag{10.17}$$

From equation (10.17), one can express:

$$\begin{cases} \dot{e}_1 = -\vartheta_1 e_1 - \dfrac{1}{3} y_2 e_3 + \dfrac{1}{3} e_2 e_3 \\[2mm] \dot{e}_2 = -\vartheta_2 e_2 - \dfrac{1}{3} y_1 e_3 \end{cases} \tag{10.18}$$

Hence, based on equation (10.18), the system's corresponding sliding mode dynamics may be derived in the following way:

$$\begin{cases} \dot{e}_1 = -\vartheta_1 e_1 - \dfrac{1}{3} y_2 e_3 + \dfrac{1}{3} e_2 e_3 \\[2mm] \dot{e}_2 = -\vartheta_2 e_2 - \dfrac{1}{3} y_1 e_3 \\[2mm] \dot{e}_3 = \dfrac{1}{3} y_1 e_2 + \dfrac{1}{3} y_2 e_1 - \dfrac{1}{3} e_1 e_2 + c e_3 \end{cases} \tag{10.19}$$

Grounded in the Lyapunov stability principle, the choice of the Lyapunov function (presented below) facilitates the evaluation of the sliding mode dynamics'' stability, as described in equation (10.19):

$$V = \frac{1}{2}\left(e_1^2 + e_2^2 + e_3^2\right) \tag{10.20}$$

The following representation of the time derivative of the previously defined Lyapunov function serves as the basis for assessing the system's stability.

$$\dot{V} = e_1 \dot{e}_1 + e_2 \dot{e}_2 + e_3 \dot{e}_3$$

$$= e_1\left[-\vartheta_1 e_1 - \frac{1}{3} y_2 e_3 + \frac{1}{3} e_2 e_3\right] + e_2\left[-\vartheta_2 e_2 - \frac{1}{3} y_1 e_3\right]$$

$$+ e_3\left[\frac{1}{3} y_1 e_2 + \frac{1}{3} y_2 e_1 - \frac{1}{3} e_1 e_2 + c e_3\right]$$

$$= -\vartheta_1 e_1^2 - \vartheta_2 e_2^2 + c e_3^2$$

In this expression, ϑ_1 and ϑ_2 act as adjustable parameters, offering the designer control over its functionality. These parameters are characterized as positive numbers. On the other hand, the variable c represents a parameter associated with the system, and it is specified as a negative value.

Hence, $\dot{V} \le 0$ $\hspace{4cm}$ (10.21)

The Lyapunov stability principle assures that a system confined to a sliding manifold exhibits stable sliding motion, ensuring that $\lim_{t \to \infty} \| e_i(t) \| = 0$ for $i = 1, 2, 3$.

The implementation of a SMC strategy plays a crucial role in driving the system's trajectories towards the desired sliding surface $s_i = 0$. This achievement is realized through the identification of the appropriate sliding surface (10.17), as explained earlier.

The subsequent controllers are introduced to guarantee the initiation of sliding motion:

$$\begin{cases} u_1 = \left(-8_1 - a\right)e_1 + \dfrac{2}{3}y_2e_3 + y_3e_2 - \dfrac{2}{3}e_2e_3 - \psi\left(\text{sign}\left(s_1\right)\right), & \psi > 0 \\[4mm] u_2 = \left(-8_2 - b\right)e_2 - \dfrac{4}{3}y_1e_3 - y_3e_1 - e_1e_3 - \psi\left(\text{sign}\left(s_2\right)\right), & \psi > 0 \end{cases}$$ $\hspace{1cm}$ (10.22)

In Theorems 10.2 and 10.3, the outlined controller structure is employed to ensure the presence of sliding motion and, consequently, to guarantee synchronization.

Theorem 10.2: If the error dynamics outlined in equation (10.14) are governed by the control input u_1 specified in equation (10.22), then the system trajectories converge to the sliding surface $s_i = 0$ and fulfil the condition $\lim_{t \to \infty} \| e_i(t) \| = 0$ for $i = 1, 2, 3$.

Proof: Introduce a suitable Lyapunov function as follows:

$$V_1 = \frac{1}{2}s_1^{\;2}$$ $\hspace{4cm}$ (10.23)

Now, by computing the time derivative of equation (10.23) and utilizing equations (10.14), (10.17), and (10.22), the expression obtained is as follows:

$$\dot{V}_1 = s_1 \dot{s}_1$$

$$= s_1\left[\dot{e}_1 + 8_1e_1 + \frac{1}{3}y_2e_3 - \frac{1}{3}e_2e_3\right]$$

$$= s_1\left[-y_2e_3 - y_3e_2 + e_2e_3 + ae_1 + u_1 + 8_1e_1 + \frac{1}{3}y_2e_3 - \frac{1}{3}e_2e_3\right]$$

$$\le -\psi\left|s_1\right|$$ $\hspace{4cm}$ (10.24)

When ψ is chosen such that $\psi > 0$, then $\dot{V_1} \leq 0$, thus confirming that the reaching requirement is continuously satisfied. Therefore, the proof is convincingly established.

Theorem 10.3: If the error dynamics outlined in equation (10.14) are governed by the controller u_2 defined in equation (10.22), then the system trajectories converge to the sliding surface $s_i = 0$ and fulfil the condition $\lim_{t \to \infty} \| e_i(t) \| = 0$ for $i = 1, 2, 3$.

Proof: Introduce a suitable Lyapunov function as follows:

$$V_2 = \frac{1}{2} s_2^2 \tag{10.25}$$

Now, by computing the time derivative of equation (10.25) and incorporating equations (10.14), (10.17), and (10.22), the expression obtained is as follows:

$$\dot{V_2} = s_2 \dot{s_2}$$

$$= s_2 \left[\dot{e_2} + 8_2 e_2 + \frac{1}{3} y_1 e_3 \right]$$

$$= s_2 \left[y_1 e_3 + y_3 e_1 - e_1 e_3 + b e_2 + u_2 + 8_2 e_2 + \frac{1}{3} y_1 e_3 \right]$$

$$\leq -\psi |s_2| \tag{10.26}$$

When ψ is chosen such that $\psi > 0$, then $\dot{V_2} \leq 0$, thus confirming that the reaching requirement is continuously satisfied. Therefore, the proof is convincingly established.

10.5 SIMULATION RESULTS AND DISCUSSION

To assess the effectiveness of the proposed control tactic in achieving stabilization and synchronization, numerical simulations are performed on the Chen–Lee chaotic system within this segment. The simulations are implemented using MATLAB's ODE 45 solver with a step size of 0.01, allowing the simulation to extend over a duration of 6 seconds. The chosen system parameters for these simulations are $a = 5$, $b = -10$, and $c = -3.8$.

In this segment of the simulation, stabilization is attained through the utilization of the controller specified in equation (10.10). The following initial conditions are considered for the system states: $x_1(0) = 0.2$, $x_2(0) = 0.2$,

$x_3(0) = 0.2$. The control gain is chosen as follows: $k_1 = 10$. Figure 10.1a represents the 3D phase trajectory of Chen–Lee chaotic system and Figure 10.1b–d represents the phase portrait of Chen–Lee chaotic system in two dimensions with respect to different axes. Figure 10.2a–c represents the state responses of the variables in the Chen–Lee chaotic system when the control input is not introduced. Figure 10.3a–c demonstrates the state responses of the variables in the Chen–Lee chaotic system when the controller is activated. The introduction of the control signal results in the evident convergence of system states to zero, demonstrating the effectiveness of the control strategy in controlling the states of the Chen–Lee chaotic system.

Within this phase of simulation, the achievement of synchronization is realized by employing the controller outlined in equation (10.22). The initial conditions for both the drive system and the driven system are set

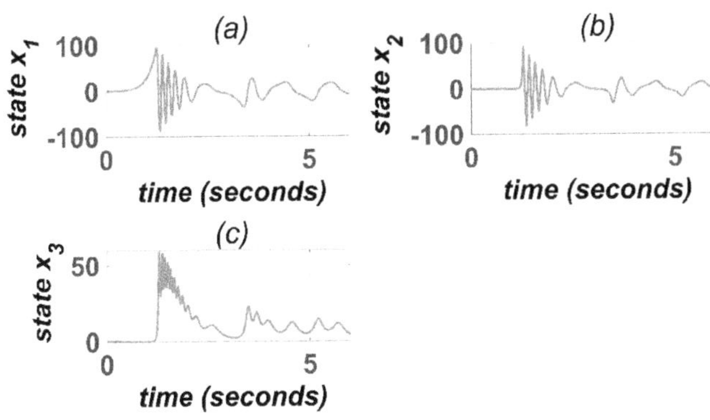

Figure 10.2 Time evolution of states in the absence of controller activation.

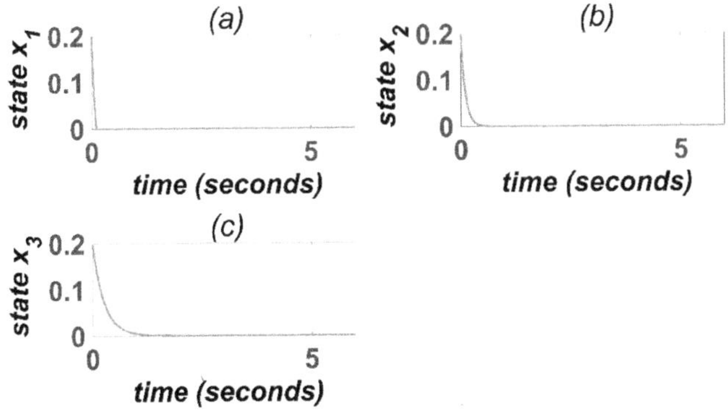

Figure 10.3 Time response of states when controller is activated.

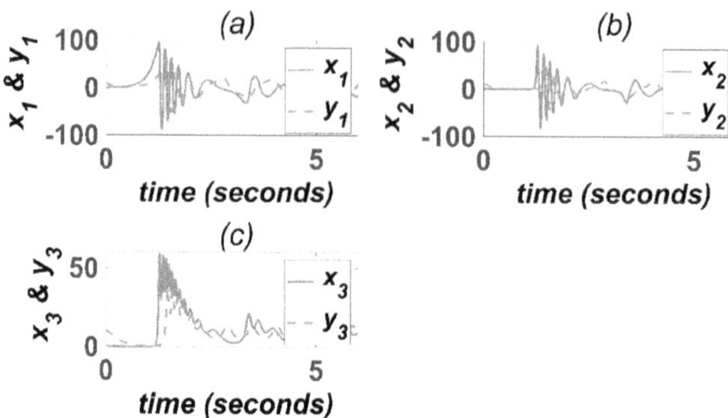

Figure 10.4 Time response of the corresponding states of master and slave system when controller is not activated.

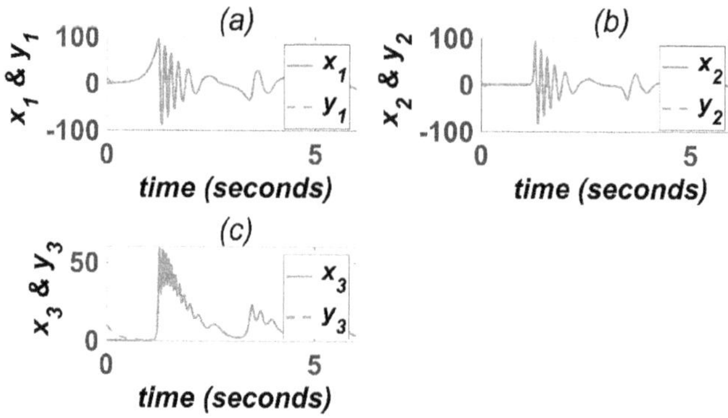

Figure 10.5 Time response of the corresponding states of master and slave system when controller is activated.

as follows: $x_1(0) = 0.20$, $x_2(0) = 0.20$, $x_3(0) = 0.20$, and $y_1(0) = 10.00$, $y_2(0) = 10.00$, $y_3(0) = 10.00$. The selection of control gains is outlined as follows: $\delta_1 = 10$ and $\delta_2 = 20$. Figure 10.4a–c illustrates the respective states of the drive and response Chen–Lee chaotic systems when control signals remain inactive. Observing the figures, it is evident that in the absence of the triggered control signals, the states of the drive and driven systems do not exhibit synchronization. Figure 10.5a–c presents the variation of individual states in the drive and driven systems, highlighting the impact of applied control signals. This graphical representation distinctly illustrates the synchronization, with both systems mirroring each other's states. The time-domain behaviour of synchronization error for the drive–response system, measured before controller implementation, is presented in Figure 10.6a–c. The figure clearly demonstrates that the error dynamics exhibit a lack of convergence towards zero. Figure 10.7a–c depicts the dynamics of synchronization error variation upon the initiation of the control signal. The figure vividly depicts the rapid convergence of synchronization errors to zero, highlighting the remarkable efficiency of the implemented synchronization method. Figure 10.8 demonstrates how sliding surfaces evolve over time, clearly showing that they eventually converge within a finite time frame. Figure 10.9 illustrates the evolution of control inputs over time, specifically designed to minimize errors throughout the synchronization process.

Chaotic systems, exemplified by the Chen–Lee system in the chapter, exhibit intricate behaviours that artificial intelligence (AI) and machine

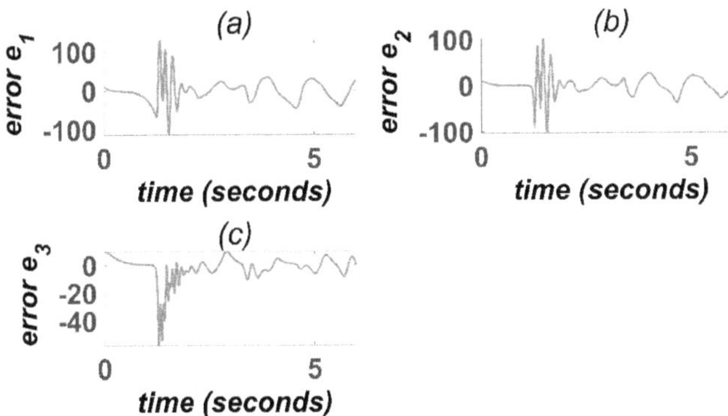

Figure 10.6 Time response of the synchronization errors between master and slave system when controller is not activated.

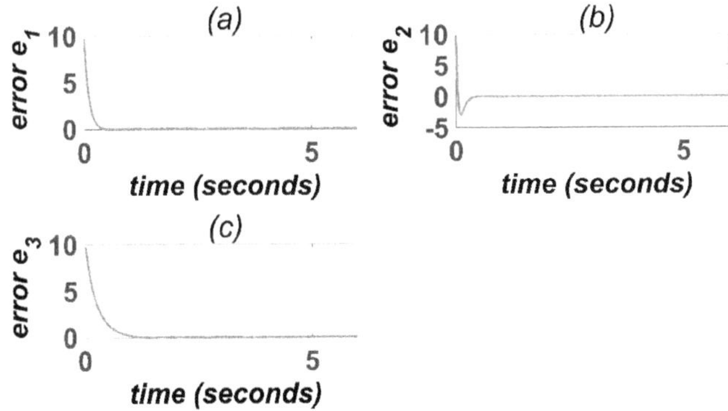

Figure 10.7 Time response of the synchronization errors between master and slave system when controller is activated.

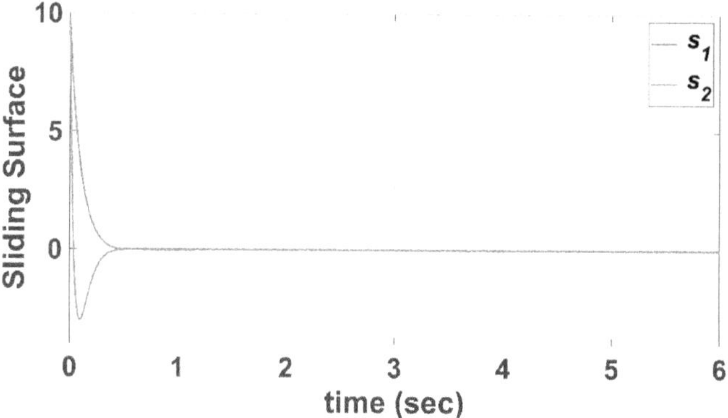

Figure 10.8 Behaviour of sliding surfaces.

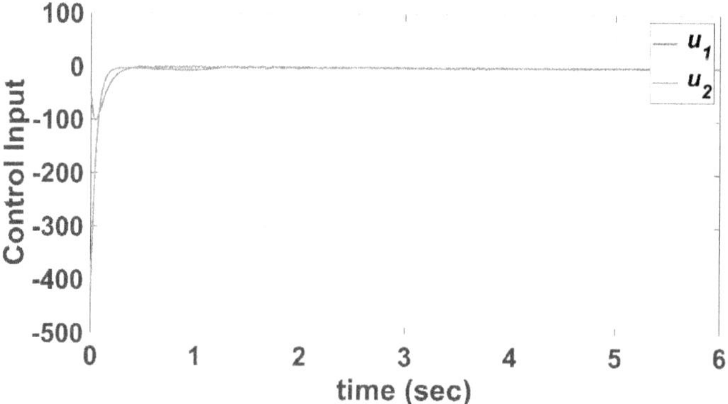

Figure 10.9 Evolution of control inputs.

learning (ML) techniques can effectively model and analyse. These technologies enable researchers to delve into chaotic dynamics, anticipate system responses, and devise control methodologies applicable across diverse domains like weather prediction, financial markets, and cryptography. Given that numerous AI and ML approaches draw from non-linear dynamics, such as neural networks' inherent non-linearity, delving into techniques for stabilizing chaotic systems offers valuable insights into addressing non-linearity within machine learning models and comprehending their functioning. Furthermore, the synchronization of chaotic systems, a focal point in the chapter, holds significant relevance in AI and ML, particularly in ensuring harmonious interactions among distributed systems, neural networks, and multi-agent setups. Understanding the intricacies of complex system dynamics lies at the core of AI and ML advancements. Leveraging insights from dynamical systems theory, coupled with AI and

ML algorithms, facilitates in-depth analysis, pattern recognition, and predictive modelling, thus advancing our comprehension of system behaviours and enhancing decision-making capabilities [29–32].

10.6 CONCLUSION

This study introduces an innovative control strategy that utilizes a PI switching surface to tackle the challenges of stabilization and synchronization within Chen–Lee chaotic systems. The methodology incorporates SMC to facilitate the initiation of sliding motion, contributing to the overall effectiveness of the approach. Notably, the research emphasizes the significance of selecting control parameters judiciously, showcasing the successful achievement of both system state stabilization and synchronization in a master–slave configuration. Numerical simulations are employed to rigorously validate the proposed methodology, providing robust evidence of its effectiveness and applicability in addressing the complexities associated with Chen–Lee chaotic systems. Also, the chapter provides insights into leveraging AI and ML methodologies for synchronizing chaotic systems. It briefly explores how AI and ML techniques can be effectively employed to achieve synchronization in chaotic systems, showcasing their potential in this complex area of study. Furthermore, the proposed control scheme possesses the capability to be extended and adapted to effectively manage the uncertainties present in the Chen–Lee chaotic system. This implies that the control methodology can be applied in scenarios where the dynamics of the system exhibit chaotic behaviour as described by the Chen–Lee model. By leveraging the inherent flexibility of the control framework, adjustments can be made to accommodate the unique characteristics and uncertainties associated with the Chen–Lee system. This extension enables the control scheme to offer robust performance and stability even in the presence of complex and unpredictable dynamics, thereby enhancing its applicability and efficacy in practical engineering and scientific applications.

REFERENCES

1. T. L. Carroll and L. M. Pecora, "Synchronizing chaotic circuits," *IEEE Trans. Circuits Syst.*, vol. 38, no. 4, pp. 453–456, 1991.
2. G. Chen and X. Dong, "On feedback control of chaotic continuous-time systems," *IEEE Trans. Circuits Syst. I. Fundam. Theory Appl.*, vol. 40, no. 9, pp. 591–601, 1993.
3. L. M. Pecora and T. L. Carroll, "Synchronization in chaotic systems," *Phys. Rev. Lett.*, vol. 64, no. 8, pp. 821–824, 1990.
4. S. Y. Al-Hayali and S. F. Al-Azzawi, "An optimal control for complete synchronization of 4D Rabinovich hyperchaotic systems," *TELKOMNIKA*, vol. 18, no. 2, p. 994, 2020.

5. S. Boccaletti, J. Kurths, G. Osipov, D. L. Valladares, and C. S. Zhou, "The synchronization of chaotic systems," *Phys. Rep.*, vol. 366, no. 1–2, pp. 1–101, 2002.

6. W. Hu, J. Wang, and X. Li, "An approach of partial control design for system control and synchronization," *Chaos Solitons Fractals*, vol. 39, no. 3, pp. 1410–1417, 2009.

7. L. Kocarev and U. Parlitz, "Generalized synchronization, predictability, and equivalence of unidirectionally coupled dynamical systems," *Phys. Rev. Lett.*, vol. 76, no. 11, pp. 1816–1819, 1996.

8. N. Yadav, Pallav, and H. Handa, "Projective synchronization for a new class of chaotic/hyperchaotic systems with and without parametric uncertainty," *Trans. Inst. Meas. Control*, vol. 45, no. 10, pp. 1975–1985, 2023.

9. L. Xiong, Z. Liu, and X. Zhang, "Dynamical analysis, synchronization, circuit design, and secure communication of a novel hyperchaotic system," *Complexity*, vol. 2017, pp. 1–23, 2017.

10. J. Cao, P. Li, and W. Wang, "Global synchronization in arrays of delayed neural networks with constant and delayed coupling," *Phys. Lett. A*, vol. 353, no. 4, pp. 318–325, 2006.

11. M. Alimi, A. Rhif, A. Rebai, S. Vaidyanathan, and A. T. Azar, "Optimal adaptive backstepping control for chaos synchronization of nonlinear dynamical systems," in *Backstepping Control of Nonlinear Dynamical Systems*, Elsevier, 2021, pp. 291–345.

12. H. Handa and B. B. Sharma, "Synchronization of a set of coupled chaotic FitzHugh–Nagumo and Hindmarsh–Rose neurons with external electrical stimulation," *Nonlinear Dyn.*, vol. 85, no. 3, pp. 1517–1532, 2016.

13. D. Huang and R. Guo, "Identifying parameter by identical synchronization between different systems," *Chaos*, vol. 14, no. 1, pp. 152–159, 2004.

14. O. de Feo, "Tuning chaos synchronization and anti-synchronization for applications in temporal pattern recognition," *Int. J. Bifurcat. Chaos*, vol. 15, no. 12, pp. 3905–3921, 2005.

15. Pallav and H. Handa, "Simple synchronization scheme for a class of nonlinear chaotic systems using a single input control," *IETE J. Res.*, vol. 69, pp. 9094–9107, 2022.

16. R. M. Bora and B. B. Sharma, "Reduced order synchronization of two non-identical special class of strict feedback systems via back-stepping control technique," *J. Comput. Nonlinear Dyn.*, vol. 19, no. 1, pp. 1–10, 2024.

17. X. Lv, J. Cao, X. Li, M. Abdel-Aty, and U. A. Al-Juboori, "Synchronization analysis for complex dynamical networks with coupling delay via event-triggered delayed impulsive control," *IEEE Trans. Cybern.*, vol. 51, no. 11, pp. 5269–5278, 2021.

18. Pallav and H. Handa, "Chaos synchronization for a class of hyperchaotic systems using active SMC and PI SMC: A comparative analysis," *J. Control Autom. Electr. Syst.*, vol. 33, no. 6, pp. 1671–1687, 2022.

19. R. Kengne, J. T. Mbe, J. Fotsing, A. B. Mezatio, F. J. Ntsafack Manekeng, and R. Tchitnga, "Dynamics and synchronization of a novel 4D-hyperjerk autonomous chaotic system with a Van der Pol nonlinearity," *Z. Naturforsch. A*, vol. 78, no. 9, pp. 801–821, 2023.

20. Pallav and H. Handa, "Active control synchronization of similar and dissimilar chaotic systems," in *2021 Innovations in Power and Advanced Computing Technologies (i-PACT)*, 2021.

21. H. Handa and B. B. Sharma, "Controller design scheme for stabilization and synchronization of a class of chaotic and hyperchaotic systems in uncertain environment using SMC approach," *Int. J. Dyn. Contr.*, vol. 7, pp. 256–275, 2018.

22. Pallav and H. Handa, "Synchronization of MLS chaotic system using sliding mode control technique," in *Soft Computing: Theories and Applications*, Singapore: Springer Nature Singapore, 2023, pp. 335–346.

23. H. Handa and B. B. Sharma, "Novel adaptive feedback synchronization scheme for a class of chaotic systems with and without parametric uncertainty," *Chaos Solitons Fractals*, vol. 86, pp. 50–63, 2016.

24. B. B. Sharma and I. N. Kar, "Contraction theory based adaptive synchronization of chaotic systems," *Chaos Solitons Fractals*, vol. 41, no. 5, pp. 2437–2447, 2009.

25. Pallav, H. Handa, and B. B. Sharma, "Novel single bounded input control synchronization criterion for a category of hyperchaotic and chaotic systems in presence of uncertainties," *J. Comput. Nonlinear Dyn.*, vol. 18, no. 12, pp. 1–12, 2023.

26. H.-K. Chen and C.-I. Lee, "Anti-control of chaos in rigid body motion," *Chaos Solitons Fractals,* vol. 21, no. 4, pp. 957–965, 2004.

27. E. N. Lorenz, "Deterministic nonperiodic flow," *J. Atmos. Sci.*, vol. 20, no. 2, pp. 130–141, 1963.

28. V. I. Utkin, *Sliding Mode and their Application in Variable Structure System*, Mir Editors, Moscow Russia, 1978.

29. X. Chen, T. Weng, C. Li, and H. Yang, "Equivalence of machine learning models in modeling chaos," *Chaos Solitons Fractals*, vol. 165, no. 112831, p. 112831, 2022.

30. H. Cheng, H. Li, Q. Dai, and J. Yang, "A deep reinforcement learning method to control chaos synchronization between two identical chaotic systems," *Chaos Solitons Fractals*, vol. 174, no. 113809, p. 113809, 2023.

31. D. Green and N. Lavesson, "Chaos theory and artificial intelligence may provide insights on disability outcomes," *Dev. Med. Child Neurol.*, vol. 61, no. 10, pp. 1120–1120, 2019.

32. http://dx.doi.org/10.31031/COJRA.2024.03.000566

Chapter 11

Application of machine learning models for power systems security assessment

Mukesh Singh, Sushil Chauhan, and Ankur Maheshwari

11.1 INTRODUCTION

Power system security assessment (PSSA) is crucial for ensuring the security of power grids. It measures attributes using contingencies and addresses a power system's dynamic and static behaviours. Static security assessment focuses on identifying variable violations under steady-state conditions following a contingency, whereas dynamic security assessment evaluates the system's stability during the transient phase that ensues after a contingency. The essence of PSSA lies in its ability to measure the system's resilience against contingencies and its capacity to implement corrective actions to enhance security [1]. Establishing an appropriate security index is crucial in this context. The vast scale and complexity of power systems make power system security assessment a highly demanding computational task. Ensuring the efficient operation of the power system requires the optimal functioning of a fast and accurate security system. The increasing integration of renewable power changes the characteristics of power systems, introducing greater complexity with more intricate data. Challenges in power system security assessment arise from uncertainties in renewable energy generation, limitations of conventional methods, significant data processing needs, and the necessity to consider system security. Integrating renewable energy sources with full converters can impact various aspects of system dynamics after a disturbance. Notable differences are observed in the dynamic response of a power system when only alternators are considered versus incorporating inverters. This contrast depends on factors like penetration levels, the nature of the disturbance, and specific types of renewable energy sources. Precise prediction of power generated through renewable energy and load demand is essential but poses inherent challenges; using recent data-driven approaches, like machine learning, helps overcome these problems and improve the performance of PSSA techniques [2].

Power system security assessment (PSSA) is conducted in two modes: offline and online. In the offline mode, power system security is assessed within a planning framework. Assessing the present or imminent security status of power systems require analysing the dynamic and steady-state performance

DOI: 10.1201/9781003581246-11

of short-term predicted scenarios. Potential contingencies like transmission line outages, transformer failures, or generator abnormalities are evaluated using load flow simulations. This analysis incorporates diverse loads and power generation scenarios [3]. This methodology is indispensable for power system operators and planners, ensuring the security and reliability of power systems. However, its limitation stems from the lack of real-time data on the power system's behaviour, making it less suitable for fast detection and response to sudden changes or disturbances [4]. The online assessment of power system security plays a critical role in ensuring the safe operation of power systems and relies on real-time data. Machine learning techniques have gained attention in recent years as a promising solution to address the limitations of traditional power system security assessment methods. These techniques utilize data-driven approaches to analyse large amounts of historical data and identify patterns and relationships that can be used to predict and assess the security status of a power system in real-time [5].

Measurement-based methods depend on real-time measured data, removing the necessity for simulation studies and system models. However, these methods encounter the difficulty of managing large amounts of data. Efficiently handling the measured data is crucial for precisely assessing security metrics. Nevertheless, in the context of the time-domain behaviour of power systems, which involves a substantial and continuous generation of data for operating variables, online PSSA approaches face and must overcome several challenging issues [3].

- Capturing synchronized data of operational parameters through the wide area monitoring system is essential. The absence of synchronized data throughout the network presents difficulties in obtaining complete representations of system performance, particularly during dynamic periods.
- To enhance the efficiency and speed of the estimation process, it is crucial to extract the most significant data pertaining to each power system contingency. This involves eliminating irrelevant and redundant data and focusing on dominant features that exhibit high functionality in relation to the specified attribute.
- Establishing clear and comprehensive security indices is essential for evaluating the security status of the power system based on operating data. These security indicators play a crucial role in assessing the security situation of the power network.
- Creating an estimation tool is imperative for providing a functional relationship between security indices and power system operating data. These approaches, which rely on measured operating data to assess security indices, necessitate the development of a functional connection between the security index and its relevant dominant features. Various techniques, including regression, artificial intelligence (AI), and classification, are proposed.

The limitations of traditional security assessment approaches highlight the need for a swift and accurate real-time Power System Security Assessment (PSSA). In recent years, data-driven approaches for PSSA have demonstrated effectiveness and reliability, showing notable performance in recent studies. A comprehensive exploration of ML techniques for analyzing the security and stability of power systems is detailed in [6]. The researchers delineated ML methods for inspecting the security and stability of power systems, examining constraints associated with classifier creation, dataset development, and testing systems. The authors in [2] investigated conventional and soft computing methods for improving PSSA. In [3], authors analyzed the utilization of numerical and ML-based methods for SSA. To provide a comprehensive overview of the aforementioned issues related to security assessment, this paper presents an extensive study of all machine learning techniques and methods in this field shown in Figure 11.1.

The rest of the article is structured as follows. The second part is dedicated to the study of PSSA. The next part includes a brief introduction to machine learning techniques applied to PSSA problems, while the rest of the paper outlines research on data-driven PSSA methods, such as artificial neural networks (Section 4), support vector machines (Section 5), and

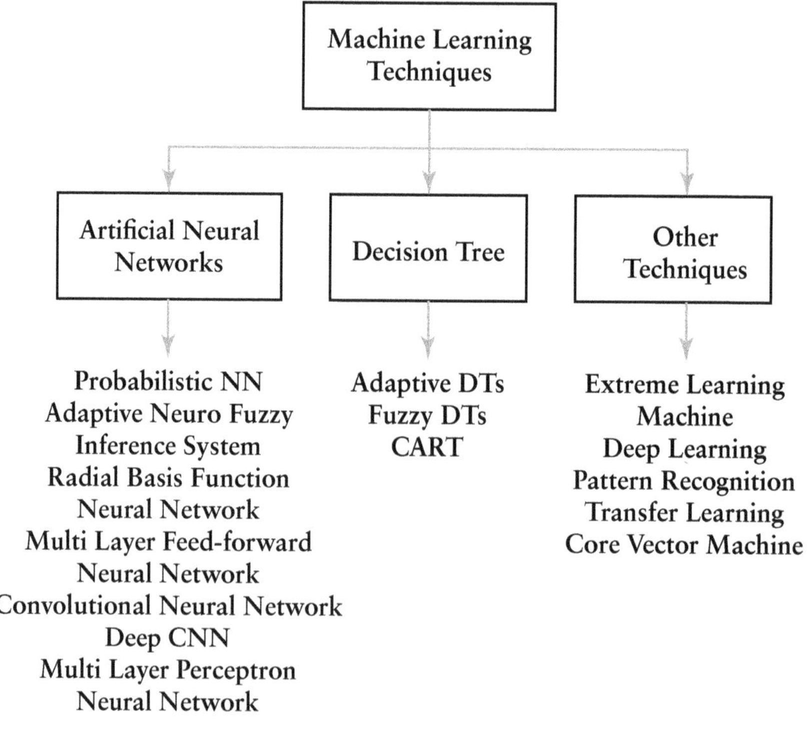

Figure 11.1 Machine learning based techniques implemented for PSSA.

decision trees (Section 6), as well as extreme learning machines, deep learning, fuzzy systems, data collection, pattern recognition, and transfer learning (Section 7). In the next section (Section 8), a comparative analysis of static and dynamic security assessment is presented using well-established machine learning techniques and an ELM-based hybrid model. Finally, the last section (Section 9) concludes the paper.

11.2 SECURITY ASSESSMENT

The process of Power System Security Assessment (PSSA) involves identifying emergency scenarios where operating parameters exceed their operational limits, either in normal (pre-perturbation) or post-perturbation conditions [7]. PSSA consists of three main functions: contingency screening, security analysis, and preventive measures. Contingency monitoring provides essential information to operational engineers regarding the system's operational condition. Security analysis entails evaluating and prioritizing contingencies according to their severity, determined by network variables crucial to security assessment. Preventive measures aim to mitigate the risk of system malfunction by implementing desired control measures. With modern utilities operating under more demanding conditions due to deregulation, disturbances have an increased potential to undermine system security, especially in critical operating conditions, possibly leading to a collapse.

Security analysis can be broadly categorized into two main types: static security assessment (SSA) and dynamic security assessment (DSA). SSA is concerned with scenarios where system limits are exceeded after contingencies, operating on the assumption that the power system stabilizes following these events. On the other hand, DSA evaluates the system's behaviour in real-time after a disturbance, assessing its performance as it progresses post-event [4]. Dynamic Security Assessment (DSA) encompasses two principal subcategories: pre-fault and post-fault assessments. Steady-state variables such as power generation, connected loads, line flows, and bus voltages are utilized for pre-fault security assessment to evaluate the security status of the power system before the occurrence of disturbances. Post-fault assessment considers parameters such as voltage trajectories, rotor angles/speeds, and a wide range of variables to assess the system's security status subsequent to a fault. Different methods are used for DSA, such as time domain simulations, Lyapunov-based direct methods, numerical integrations, and probabilistic approaches. These areas encompass various techniques and methodologies used to analyze and enhance the security assessment of power systems in dynamic scenarios [8].

In the SSA methodology, assessing the severity of a post-perturbation scenario involves employing load flow techniques for both the base case and contingency scenarios. Nonetheless, these methodologies present intricate challenges, consuming considerable time and posing difficulties in adapting to the dynamic evolution of system operating conditions. This renders them unsuitable for real-time implementation [9]. Conventional security assessment techniques heavily depend on executing continuous flow and transient stability simulations, a practice that may fall short in delivering comprehensive security evaluations. Direct methods in Dynamic Security Assessment (DSA), like those using transient energy functions (TEFs), replace numerical integrations with stability criteria. On the other hand, probabilistic methods involve computing probability distributions associated with system stability. However, these methods can be computationally intensive, limiting their applicability primarily to power system planning rather than real-time operational scenarios [10].

In contrast, modern power systems have evolved to a heightened level of complexity, posing challenges to applying traditional methods in online security assessment. The extensive time requirements of complete simulation methods, even with multiple CPUs, render them impractical for the real-time analysis of large power systems with numerous contingencies [11]. Hence, there is a growing need to develop swift and efficient PSSA techniques capable of monitoring system security under different contingency scenarios. This is essential to minimize system vulnerability against diverse contingencies and ensure the secure operation of power systems.

In a data-driven approach to PSSA, training the machine learning model requires a diverse dataset to ensure optimal performance under varying operating conditions. However, generating and collecting such data present challenges due to the following reasons:

- Data diversity, involving multiple sources such as generation units, distribution networks, transmission lines, and load patterns, presents a complex challenge in power system security assessment. [12].
- Data quality and consistency, vital for machine learning model training, pose challenges due to variations in accuracy and reliability from different sources [13]. Limited access to real power system data often leads researchers to use open-source or simulated datasets, introducing discrepancies compared to actual data and affecting classifications and predictions.
- Temporal and local variability in power system data, influenced by weather, equipment failures, and consumption changes, require diverse datasets for effective machine learning model training. [14].

Some common methods for database generation include:

- Historical data, accumulated over time from power systems, offers insights into actual events, aiding in enhancing DSA [14].

- Importance sampling efficiently generates a training database for security assessment in ML-based DSA by sampling conditions near a security boundary, thereby minimizing computational costs [15].
- Random sampling via Monte Carlo simulation generates diverse scenarios for ML-based DSA, augmenting historical data. It captures uncertainty, providing a varied dataset for model training. Analyzing statistical properties enables accurate predictions of system behaviour under different conditions [16].

Integrating these database generation methods results in the creation of a comprehensive and inclusive dataset, which can be utilized to train machine learning models for data-driven power system security assessments.

11.3 MACHINE LEARNING TECHNIQUES

Once the most favourable input variables have been narrowed down, the next step involves training the machine learning models for SSA and DSA. Effectively training a model necessitates choosing an algorithm that can accurately represent the intricate relationship between inputs and outputs in this high-dimensional problem. When choosing an algorithm, it is important to consider various factors such as the data's attributes, the presence of non-linear relationships, potential multicollinearity among input variables, and the diverse data types involved. For example, the dataset may contain real-valued data representing spatial profiles of active and reactive power, as well as binary data indicating the active or inactive status of buses and lines. These considerations are essential when selecting the most suitable algorithm for analysis. As this research has progressed alongside developments in ML algorithms, a range of algorithms have been utilized for this purpose, categorized as either ML or deep learning algorithms.

Decision Trees

Decision Trees (DTs) are commonly utilized algorithms for performing classification tasks. Decision Trees use training data to construct a hierarchical tree structure made up of rules. Each node in the tree is linked to an input feature label, and the connected branches are associated with conditions based on potential outcomes from the parent node. This approach helps in decision-making by following the path from the root node to the leaf nodes, where predictions or classifications are determined based on learned rules. The terminal nodes at the bottom of the tree, known as leaves, correspond to target outputs that occur based on how data has been divided among them. A Decision Tree is constructed by iteratively dividing the dataset using feature values that most effectively distinguish the data based on the target variable. Determining the best feature and criteria for splitting at every internal node is a key part of the process. Common techniques for this include information gain, Gini impurity, and entropy. These features guide Decision Trees from the initial node to the terminal nodes based on predefined rules when new data is presented. Decision Trees are valued for their interpretability, as they can be easily visualized and understood due to their unique hierarchical structure, allowing for a clear understanding of the decision-making process. However, if not effectively managed, Decision Trees may experience overfitting, resulting in inadequate generalization on new data. Measures like pruning and establishing stopping criteria can be employed to address this concern.

Random Forest

A Random Forest (RF) is created by combining multiple decision trees. The number of decision trees (DTs) used in the RF is set before the training phase and adjusted through experimentation. Each decision tree within a Random Forest ensemble is trained on a distinct subset of the original dataset using bagging, where samples are randomly chosen with replacement. This random selection also applies to features, helping to reduce correlations between the trees and improve overall performance. The RF algorithm is based on the idea that a group of uncorrelated and unbiased weak predictors can collectively enhance prediction accuracy compared to using a single predictor. In an RF, numerous weak models in the form of regression trees are generated from diverse subsets of the training data through random sampling of both samples and features from the initial training set. This process aims to decrease correlation among weak predictors. The final prediction comes from aggregating the most frequently voted class from all weak models in the Random Forest. The ensemble method improves generalization capabilities while mitigating overfitting tendencies compared to individual decision trees.

Support vector machine

The Support Vector Machine (SVM) algorithm is designed to pinpoint the hyperplane that effectively distinguishes data points of various classes. This is accomplished by maximizing the margin between these classes, which denotes the distance between the hyperplane and the nearest points. When handling non-linearly separable data, SVM utilizes a method called the "Kernel Trick" to transform it from its original feature space to another where linear separation becomes possible. The commonly used kernel functions in Support Vector Machines (SVM) include linear, polynomial, radial basis function (RBF), and sigmoid kernels.

Artificial Neural Network

Artificial neural networks have become widely popular in recent years due to their capacity to model complex, non-linear relationships within data by adjusting their parameters based on observed information. The structure of ANNs is inspired by the workings of the human brain, with interconnected nodes organized into layers. Each node receives input, processes it using a function, and generates an output. By adjusting the weights of these functions based on input-output data, the network can effectively adapt to specific tasks. Artificial neural networks can be designed with numerous layers, leading to the development of complex deep learning models. The term "deep" reflects the incorporation of multiple layers within these models. Advances in optimization methods, greater computational capabilities for extensive calculations, and access to vast datasets have all contributed to the impressive accuracy achieved by deep learning models. However, deep learning models present difficulties, including a lack of interpretability due to the large number of parameters required to capture complex non-linear relationships in the data. The abundance of parameters makes it challenging to understand how the model operates and comprehend the factors that affect its predictions.

Multi-layer Feedforward Network

A multi-layer feedforward neural network (MLFNN) is a fundamental structure widely utilized in a range of applications, particularly in tasks involving classification. These networks, commonly referred to as deep neural networks, are named for their multiple layers. During the training period, the network's parameters (consisting of weights and biases) undergo refinement using backpropagation. This involves optimization algorithms such as gradient descent to reduce the difference between the predicted output of the network and its actual output. Through iterative steps, this process improves the network's parameters, thereby enhancing its performance in performing a given task. The performance of the architecture is impacted by the number of hidden layers and nodes within each layer. Networks with an insufficient number of layers may struggle to capture intricate connections, resulting in inaccurate models. On the other hand, networks with an excessive number of layers can comprehend complex input-output patterns, but too many layers can lead to the vanishing gradient problem, which impedes effective training and optimization.

Convolutional Neural Network

Convolutional neural networks (CNNs) are particularly adept at analyzing data with spatial correlations, such as images. They comprise convolutional, pooling, and fully connected layers. At the heart of CNNs lie the convolutional layers, which encompass numerous filters or kernels conducting convolutions on input data to capture local patterns or features. Pooling layers diminish the spatial dimensions of feature maps, while fully connected layers handle pooled features to generate class probabilities or regression outputs. However, training CNNs can be computationally intensive due to the large number of parameters, requiring substantial labelled data and computational resources for effective training of deep CNNs.

Fuzzy Neural Network

Neural networks combine feedforward neural networks with fuzzy logic to create the fuzzy feedforward neural network (FFNN). In an FFNN setup, inputs are transformed into fuzzy sets, assigning each element a membership value between 0 and 1 to capture uncertainty. These fuzzified inputs then enter the neural network, and the resulting outputs transition from their fuzzy representation to yield precise predictions as the final outcomes of the network. The key difference between FFNNs and FNNs is in how they handle uncertainty. While FNNs operate with exact inputs and outputs based on provided training data without considering data uncertainty, FFNNs use fuzzy logic to manage uncertain data by employing fuzzy sets for inputs and using fuzzy membership functions as neuron activation functions. This makes FFNNs more suitable for applications involving significant uncertainty.

Extreme Learning Machine

The extreme learning machine differs from a single-hidden layer feedforward neural network in that its hidden layer weights and biases are initially set randomly and remain unchanged, unlike traditional neural networks where these parameters are adjusted during training. The output layer weights are then calculated using a least-squares method to minimize the error between the network's predictions and the actual target values. ELM offers several advantages over conventional neural network learning algorithms, including faster training made possible by fixed random hidden layer parameters that eliminate iterative updates, as well as fewer hyperparameters requiring fine-tuning. This characteristic of ELM provides an efficient way to train neural networks with less complexity in parameter optimization compared to traditional methods.

Ensemble Learning

Ensemble forecasting involves combining multiple machine or deep learning models to improve predictive accuracy compared to using individual models alone. While Random Forest is a well-known example of ensemble prediction, this discussion delves into hybrid ensemble prediction, which incorporates various types of algorithms. Ensemble prediction is based on the understanding that a single empirical model may not consistently deliver precise results over time. By combining multiple models, the ensemble can lessen the likelihood of substantial prediction errors and ensure that the most effective method is utilized at any point in time. In ensemble forecasting, each model generates its own independent forecast and contributes to the overall forecast by amalgamating predictions from all models. In classification-based ensembles, the final forecast is determined through majority voting among the models; whereas regression-based ensembles calculate the ultimate prediction as a weighted average of predictions from different methods with continuously updated weights based on recent performance for dynamic adaptability in the modelling approach. Ensemble forecasting is a potent approach that utilizes the variety of multiple models to improve overall performance and resilience in dealing with various patterns and fluctuations within the data. This technique has found widespread application across various domains for enhancing predictive precision and consistency.

11.4 ARTIFICIAL NEURAL NETWORK

Artificial Neural Networks (ANNs) have been employed in power system problem-solving due to their high computational speed and generalization capability. Unlike conventional approaches, ANNs utilize an iterative mathematical process to identify complex relationships between the initial and final states. This iterative process involves adjusting weights and biases within the network to enhance its capacity to capture and represent complex relationships in the data [8]. This enables them to achieve quick and accurate predictions in system security, offering viable solutions for monitoring security in modern power systems [12].

Dynamic security assessment

In recent studies, artificial neural networks (ANNs) have shown promise in power system security assessment. [17] focused on dynamic security assessment, demonstrating effective pattern recognition and classification across various conditions and fault scenarios in an IEEE 145-bus test system. In [18], power system security assessment utilized enhanced feature selection techniques, employing Fisher's linear discriminant for effective feature selection during neural network training, proving its superiority on the IEEE 50-generator transient stability test system.

In [8], authors proposed an adaptive artificial neural network-based approach to predict the rotor angle of the generator to improve power system stability. The method included feature selection and data extraction for improved generalization and reduced input numbers. In [19], a support vector machine (SVM)-based classifier is used to predict rotor angle stability and achieve high accuracy in transient stability prediction. In [20], a hybrid network reduction approach using artificial neural networks (ANN) and the Ward equivalent method is proposed for online voltage security assessment. The technique demonstrated favourable responses to internal system contingencies by adapting equivalent parameters based on the online status in the buffer zone.

Additionally, [10] proposes a DSA and generation rescheduling method using a probabilistic neural network (PNN) and adaptive neuro-fuzzy inference system (ANFIS)-based genetic algorithm (GA) for preventing transient instabilities in large power systems. The approach demonstrates fast and accurate security classification. Another study [11] introduces an automatic learning paradigm for dynamic security control, employing a radial basis function neural network (RBFNN). This framework adeptly evaluates the dynamic security status of the power system, predicts the impacts of corrective control measures during disturbances, and employs sophisticated feature reduction techniques to address the challenge of extensive database dimensionality.

Static security assessment

In [21], a security assessment approach is enhanced using an artificial neural network (ANN) with a Kohonen self-organizing feature map designed explicitly for a 6-bus test system. Following this, [9] utilizes a radial basis function neural network (RBFNN) and multi-layer feedforward neural network (MFNN) for online SSA, leveraging NRLF analysis to project active power and voltage metrics under diverse loading conditions and line outage contingencies. Introducing a data mining-based approach [22] expedites static security assessment through a deep convolutional neural network (deep CNN) designed explicitly for N-1 contingency SSA. Unlike other data-driven methods relying on system state variables, this approach leveraged the system's topology and power injection at buses, significantly reducing computational efforts. In [23], authors proposed an adaptive ANN-based approach for improving the SSA of the power system. This approach uses bus voltage violations for training data generation for the adaptive ANN. Comparative assessments were conducted on 9-bus and 39-bus test systems, demonstrating superior performance compared to conventional methods. The method effectively mitigated insecure situations arising from credible contingencies, immediately adjusting generation dispatch and load shedding in megawatts.

11.5 SUPPORT VECTOR MACHINE:

The support vector machine (SVM) is based on statistical learning theory and establishes a framework using a linear function assumption within a multi-dimensional feature space [24]. Leveraging this approach, SVM proves advantageous by requiring a reduced number of training samples, rendering it a valuable tool for DSA & SSA, as evidenced by previous works [25].

Dynamic security assessment

In [26], an ensemble SVM-based approach for online security assessment was presented, integrating PMU measurement data with multiple linear SVM learners. The method used a boosting approach to mitigate potential classification errors and exhibited efficient and accurate security assessment in numerical validations. Another study [24] approached the online TSA as a two-class classification problem. Introducing the core vector machine (CVM), a data mining algorithm tailored for PMU-based measurements, the authors reported superior precision and computational efficiency compared to conventional SVM approaches when applied to online PMU data.

Static security assessment

In [27], a binary classification system based on Support Vector Machines (SVM) was introduced for SSA & DSA in IEEE test systems. This method, incorporating single line outages as potential contingencies, exhibited enhanced performance compared to the traditional least square classification technique in the context of security assessment. Subsequently, in [25], an online methodology for assessing both static and transient security was delineated, featuring the application of a multiclass SVM classifier. A sequential forward selection-based feature selection technique is applied to select optimal features, and the proposed framework demonstrated superior efficacy in online validation of static and transient security, with notable computational efficiency, rendering it conducive to practical deployment.

11.6 DECISION TREE

The decision tree, a widely recognized data mining and classification method, is utilized for addressing high-dimensional and big data challenges. Decision trees, which use attribute thresholds to predict the target variable, have proven valuable for both online and offline PSSA problems [28]–[30].

Dynamic security assessment

In [31], an online DSA approach for practical power systems was proposed, incorporating real-time PMU data and decision trees (DTs). Using this approach, the authors developed guidelines for online security assessment and preventive control. In [32] and [33], authors proposed fuzzy decision tree (DT)-based PSSA schemes designed to address uncertainties and large-scale problems. Proposed methods utilize the strengths of DTs when combined with NNs [32] and PMUs [33], mitigating their respective disadvantages. This resulted in a favourable balance between interpretability and accuracy. In [30], researchers devised an ensemble decision tree (DT)-based data mining approach for online DSA, aiming to alleviate the impact of missing phasor measurement unit (PMU) data. The authors utilized a boosting algorithm to assign weights to individual decision trees within an ensemble and adapt these weights based on re-evaluation results. The developed method showed better performance compared to traditional techniques based on decision trees. In [34], an approach rooted in decision trees (DT) was introduced for preventive control and online Dynamic Security Assessment (DSA) within power systems. The researchers intricately designed two DTs tailored for contingencies prevalent in power systems characterized by a substantial presence of wind generation and distributed generations (DGs). Concurrently, [35] brought forth two optimization-based DT methodologies aimed at refining the interpretability of DT-centric dynamic security rules. Comprehensive evaluations illustrated that these proposed approaches outperformed counterparts in predictive accuracy and exhibited heightened interpretability of security rules, thus reducing the dependency on extensive training data. In [36], a DT-based technique is proposed to analyze the dynamic security of the existing Danish power system incorporating wind energy and considering the influence of cross-border power exchange. Rigorous offline assessments were undertaken across diverse potential operating scenarios, involving the creation of a comprehensive database incorporating critical contingencies. This meticulously constructed database subsequently served as the basis for forecasting the security status of both the current and envisioned power systems. Simultaneously, [37] delved into the amalgamation of wide-area measurements and decision tree (DT) algorithms to craft a strategy for online DSA. Although the developed paradigm demonstrated satisfactory accuracy on a minimized database, it encountered limitations in achieving the desired precision with the original, more extensive database. To overcome this challenge, the authors implemented an adaptive boosting technique to amalgamate individual DTs, thereby amplifying prediction accuracy. As outlined, the resultant model exhibited enhanced performance, particularly in scenarios characterized by uncertainties related to variations in load and generation.

Static security assessment

In [38], an advanced approach was introduced to augment power system security assessment capabilities amidst multiple contingencies. This involved introducing a multiway decision tree (DT) strategy that reduces decision nodes. Comparative analyses with conventional DT models indicated that the proposed method exhibited low sensitivity to variations and less computational burden due to reduced decision nodes. However, a more intricate scrutiny through detailed comparative analysis is imperative to affirm its practical feasibility. Furthermore, [39] and [40] unveiled resilient online dynamic security assessment (DSA) frameworks designed to accommodate variations in operating conditions and alterations in power grid topology. These frameworks harnessed raw measurements from Phasor Measurement Units (PMUs) and leveraged classical decision tree (DT) [40] as well as adaptive ensemble DT [39] models to appraise the security aspects of practical power systems. The outcomes highlighted superior performance in terms of accuracy when compared with traditional methodologies.

11.7 OTHER TECHNIQUES

Extreme learning machine

Extreme learning machine (ELM) has been applied to PSSA, offering a data-driven approach to predicting the security status of power systems. ELM is a machine learning algorithm known for its efficiency and speed in training, making it suitable for real-time applications such as dynamic security assessment (DSA) and static security assessment (SSA) in power systems. It has been combined with various techniques, including ensemble methods and deep learning architectures, to enhance the accuracy and efficiency of power system security assessments. ELM's ability to handle large datasets and provide fast learning makes it a promising tool in power system security.

In [41], a real-time extreme learning machine (ELM)-based DSA approach showcased remarkable efficacy on an IEEE 145-bus system and a practical power grid. The approach achieved a noteworthy 100% accuracy for the classification model and established a dependable DSA mechanism. Furthermore, the researchers in [42] devised an online data-driven DSA paradigm considering wind power variations, demonstrating commendable efficiency and accuracy. [43] introduced a multiple randomized learning-based ensemble model, amalgamating ELM and random vector functional link networks for online DSA, focusing on accelerated learning and enhanced reliability. Additionally, [44] proposed a PMU-based pre-fault DSA utilizing a generative adversarial network, revealing superior accuracy and reduced computational complexity under PMU-missing conditions compared to conventional methods. Authors in [45] proposed a hybrid ensemble technique blending ELM with the Levenberg-Marquardt method for online DSA. The proposed technique was tested on 39 and 68 bus systems and concluded to be the best performing model compared to other base learners.

Deep learning techniques

Deep learning techniques have been increasingly employed in power system security assessment, providing advanced capabilities for analyzing and ensuring the stability and reliability of power grids. These techniques, characterized by their ability to extract intricate features from vast datasets automatically, offer enhanced performance compared to traditional machine learning methods [46].

In [46], a feature extraction framework based on deep learning was proposed for power system security assessment. The framework employed a deep autoencoder with an R-vine copula-based sampling strategy, effectively reducing the high-dimensional input space and demonstrating commendable security assessment performance. Another study [47] utilized convolutional neural networks and deep learning to develop a PSSA paradigm for N-1 security and small-signal stability on the NESTA 162-bus system. In the context of pre-fault dynamic security assessment leveraging phasor measurement units (PMUs) and incomplete data measurements, [44] employed a generative adversarial network (GAN). This sophisticated deep learning methodology addressed the missing data challenge, exhibiting efficiency and accuracy irrespective of PMU observability and network topologies. Subsequently, in [48], a DSA framework integrated stacked denoising autoencoders (SDAE) with boosting learning techniques. The multi-layer SDAE extracted original input data, and a Support Vector Machine (SVM) classifier executed classification utilizing data from all hidden layers of SDAE. This integration resulted in enhanced accuracy of security assessment with a concomitant reduction in computational time.

Fuzzy based tech*niques*

Fuzzy inference systems (FISs) have become widely utilized in PSSA, showcasing favourable prediction and analysis performance, particularly in data-driven security assessment (DSA) [49], [50]. Over the last decade, the integration of fuzzy systems with classic and advanced PSSA approaches has proven effective, primarily in DSA, with limited applications in Static Security Assessment (SSA) [51]. An ensemble of fuzzy logic and ANN-based techniques is proposed in [52] to enhance PSSA. This technique is validated on various test systems, showing notable performance. Furthermore, [49] introduced an enhanced semi-supervised pattern recognition method incorporating a fuzzy classification technique for DSA. This method outperformed other classification techniques in efficiency, demonstrating superior performance in power systems security assessment.

Transfer learning tech*niques*

In power system security assessment, transfer learning offers advantages such as improved performance in scenarios with limited labelled data, faster convergence during training, enhanced generalization to diverse situations, adaptability to changes in system configurations, robustness to dynamic conditions, and the ability to address data imbalance. It reduces the need for extensive labelling in the target domain, making it a cost-effective and efficient approach for developing robust models in power system security assessment tasks. In [53], a transfer learning (TL) method was proposed for dynamic security assessment (DSA), addressing challenges of limited training datasets and missing data. Utilizing adversarial training and feature extractor networks, the technique effectively covered unlearned faults with complete or incomplete data, addressing simultaneous challenges. Validation on the New England 39-bus system through Monte Carlo simulations confirmed its efficacy.

A transfer learning-based data-driven method for DSA is proposed in [54], using the maximum mean discrepancy method to minimize the difference between the distribution of training and new data. This method can evaluate multiple unlearned faults that were not previously seen. The integration of extreme learning machine (ELM) and random vector functional link (RVFL) resulted in a 97.27% accuracy in contingency prediction when applied to the New England 39-bus system.

The advantages and limitations of state-of-the-art machine learning techniques for power system static and dynamic security assessment are presented in Tables 11.1 and 11.2, respectively.

11.8 SIMULATION OF SSA AND DSA

Static and dynamic security assessment problems are simulated on IEEE 14-bus and 39-bus test systems, respectively. Both assessment problems are defined as classification and regression problems and tested on some state-of-the-art techniques. Firstly, datasets are produced for these data-driven techniques. After that, a feature reduction technique is applied to reduce the training datasets. Finally, in the next stage, all these models are trained using the reduced data.

IEEE 14-bus test system:

The static security assessment problem is simulated on the IEEE 14-bus test system. This problem includes classification and regression tasks. The IEEE 14-bus system is equipped with 5 generators, 11 loads, 20 branches, 3 adjustable transformers, and a reactive power compensator at bus 9. A total of 16 contingencies are simulated at each operating point using advanced SSA software developed by the institute. These simulations produce approximately 8,500 (17*500) samples that include index values for both contingent and non-contingent cases. During the initial phase of analysis, a selection of 42 attributes—including active and reactive bus power injections as well as diagonal elements of post-contingency YBus—are extracted to form the datasets. Of these sets, 75% are allocated for training purposes

Table 11.1 Machine learning techniques applied to static security assessment problems.

Ref.	ML Technique	Advantages	Disadvantages
[9]	MLFNN & RBFNN	Highly efficient and reliable security assessment capabilities for unfamiliar network circumstances, with minimal computational burden	High dependence on the design and parameters of the neural network, thorough training procedures
[21]	Self-organizing ANN	–	Reliable assessment necessitates extensive training datasets.
[22]	Deep CNN	No need for system state variables, minimal computational complexity	–
[23]	Adaptive ANN	Accurately estimating generation re-dispatch and load shedding values quickly, demonstrating strong performance on large-scale power systems	–
[25]	Multiclass SVM, utilizing sequential forward selection for feature selection, employing RBF as the kernel mapping function	Excellent performance in preventive control, precise classification accuracy, and minimal computational time	Careful consideration is crucial when selecting the input feature set
[27]	The RBF kernel function in the SVM model, utilizing an SVM-based pattern recognition approach, employs SFS for feature selection.	Highly precise categorization	Its effectiveness is restricted to categories that can be linearly segregated
[29]	CART	Good performance in assessing security aspects	The utilized load model is not realistic.
[38]	Multivariate DTs, stratified random sampling	Provides a dependable and precise evaluation of performance	Slightly greater computational intricacy in contrast to DT

while the remaining 25% are reserved for testing the models after training. All the input and target features [12] are provided in Table 11.3. The process of calculating the target feature composite security index (CSI) is explained in [12], which is used as it is in the given study.

Table 11.2 Machine learning techniques applied to dynamic security assessment problems.

Ref.	ML Technique	Advantages	Disadvantages
[8]	Adaptive ANN	A rapid security evaluation, relevant even after making changes to the system configuration	Only the modifications in the transmission lines and power consumption are taken into account, with no assumption of any generators being added to the system.
[10]	Probabilistic NNs based on GA	Large training data sets are not necessary, and the computational complexity is low.	Due to the optimization process, it is relatively slow.
[28]	Bagging (RF) and Boosting (AB)	Improved precision in comparison to that of DT	Slightly elevated computational intricacy
[34]	Contingency-based decision trees	Rapid online comprehension of the situation	The accuracy and reliability of the method heavily depend on the training data set.
[39]	Adaptive DT	Highly precise evaluation of security performance	Increased computational intricacy, requiring a significant amount of time
[43]	ELM, RVFLN	Fast evaluation of security, minimal parameters to adjust, and low computational burden	A sole learning algorithm is employed, which may not completely capture the relationships present in the training data.
[45]	h-ELM	Most accurate prediction	Time consuming for large systems.
[46]	Feature extraction using deep learning and sampling strategy based on regular vine copulas	Can address unbalanced data distributions	High computational time, particularly during instances of system faults
[49]	PD-based fuzzy	Good predictive accuracy	Raises security issues and susceptible to communication breakdowns
[50]	Fuzzy logic	Low computational cost	The analysis focuses solely on deterministic values and does not take into account the probabilities of events or evaluate the system's risk level.

Table 11.3   Input and target features for SSA

Input features	Target value
P_i, Q_i, $|Y_{ii}|$, For i= 1 to N	CSI

Autoencoder based feature reduction:

An approach commonly employed in machine learning and data analysis for reducing dimensionality is the use of autoencoder-based feature reduction. This technique involves training a neural network consisting of an encoder and decoder to derive a condensed representation of the input data. The encoder compresses the input into a lower-dimensional latent space, capturing essential features or patterns, while the decoder reconstructs the original input from this compressed form. Throughout training, the goal is to minimize reconstruction error, prompting the network to acquire a concise and meaningful portrayal of the data.

This technique is beneficial for several reasons. First, it helps reduce the dimensionality of the data, which is particularly useful when dealing with high-dimensional datasets. By retaining only the essential features, it can improve the efficiency of subsequent machine learning algorithms and reduce computational complexity. Second, autoencoder-based feature reduction can aid in feature extraction and selection, as the learned latent space often highlights the most relevant features for the given task. Finally, autoencoders can handle non-linear and complex relationships in the data, making them versatile for various types of data analysis and modelling tasks. A simple architecture of autoencoder-based feature reduction is presented in figure 11.2.

Test results and findings:

The static security assessment problem is simulated on the IEEE 14-bus test system, which is further categorized as a classification and regression problem. The simulation results for classification and regression problems are tabulated in Table 11.4 and Table 11.5, respectively.

From Tables 11.4 and 11.5, it can be concluded that a multilayer feedforward neural network, specifically a multilayer perceptron (MLP) network, achieved a classification accuracy of 95.1% and a regression accuracy of 74.6%. The MLP network has two hidden layers with 10 fully connected neurons each.

New England 39-bus system:

The dynamic security assessment problem is studied using the New England 39-bus system, which comprises 39 buses, 10 generators, 19 loads, and 46 transmission lines. To evaluate the system's performance, 500 load patterns are generated, ranging from 80% to 125% of the base operating point.

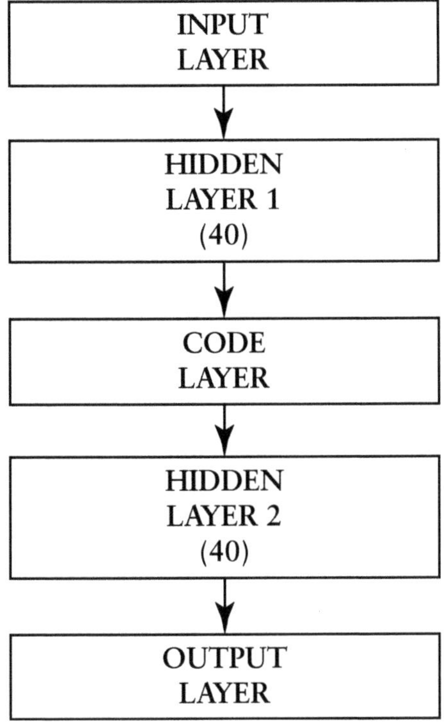

Figure 11.2 Autoencoder architecture for IEEE-14-bus system.

Table 11.4 Classification results of IEEE 14-bus for SSA

Model	No of features	Accuracy Score	Training Time (sec)
MLP (10,10)	25	0.951058	4.960003
ELM		0.796705	0.029999
RF		0.822117	3.582000
ET		0.803764	1.098997

Table 11.5 Regression results of IEEE 14-bus for SSA

Model	No of features	R2 Score	Training Time (sec)
MLP (10,10)	25	0.746257	0.825998
ELM		0.664131	0.005000
RF		0.498976	13.97899
ET		0.443865	2.694001

Six critical three-phase faults are simulated and cleared within 0.2 seconds by opening the line, resulting in 3000 data points. After analyzing the transient energy margin, the dataset is divided into 1918 secure instances and 1082 insecure instances, creating a balanced dataset. Approximately 20% of the data (about 600 points) are reserved for testing, while the rest (about 2400 points) are utilized for training the dynamic security assessment model. After generating the datasets, an autoencoder-based feature reduction technique is used to minimize the total features. The architecture of the autoencoder is presented in Figure 11.3. In the next stage, several data-driven models are trained with the reduced datasets, and the training results for classification and regression problems are presented in Tables 11.6 and 11.7, respectively.

From Tables 11.6 and 11.7, it can be concluded that for the classification problem, MLP has a classification accuracy of 92.8%. However, for the regression problem, all techniques are underperforming and not suitable. There is a need for some modifications in the given techniques or a need for ensemble techniques. Some of these techniques are explained in the literature. An improved version of the ELM technique is presented in [45], which addresses the limitations of the discussed techniques.

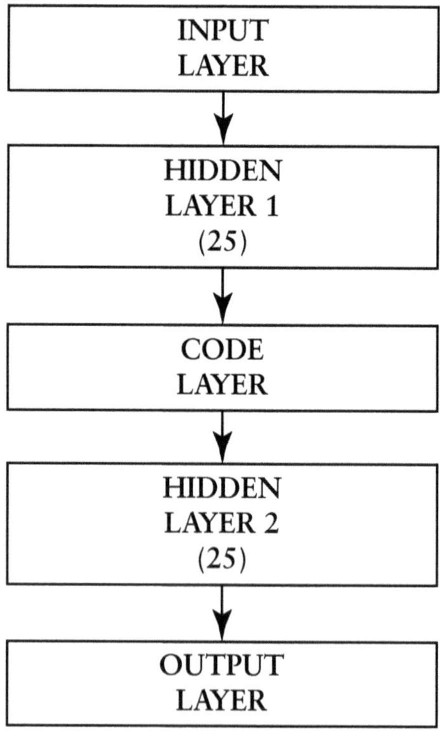

Figure 11.3 Autoencoder architecture for new England 39-bus system.

Table 11.6 Classification results of New England 39-bus system for DSA

Model	No of features	Accuracy Score	Training Time (sec)
MLP (10,10)	15	0.928	2.602002
ELM		0.821333	0.011003
RF		0.869333	1.201998
ET		0.85733	0.757000

Table 11.7 Regression results of IEEE 14-bus for SSA

Model	No of features	R2 Score	Training Time (sec)
MLP (10,10)	15	0.348678	0.142997
ELM		0.395607	0.002000
RF		0.426173	3.644999
ET		0.487635	0.994001

11.9 CONCLUSION

This paper comprehensively examines recent developments in data-driven assessment for power system security. Assessing the security of power systems is essential for monitoring and analyzing potential issues, as well as implementing protective measures. Conventional approaches have shown limitations in flexibility and the ability to adapt to changing trends in power systems. This underscores the need for more dependable Power System Security Assessment (PSSA) methodologies that align with the field's current and future requirements. In the context of power system security assessment (PSSA), data-driven approaches have emerged as practical solutions, addressing challenges posed by the growing complexity of power systems, large datasets, and the need for quick assessments, especially in online scenarios. The manuscript provides a comprehensive survey of the application of data-driven methods, including support vector machines (SVMs), artificial neural networks (ANNs), decision tree (DT) methods, and various machine learning-based approaches in the security assessment of power systems. The study concluded that ANNs are robust techniques for identifying systems with high resilience against system disturbances and changes in system configuration, ensuring accurate and fast security assessment. However, substantial training data is required, increasing computational complexity. SVMs, leveraging the structural risk minimization principle, prove highly precise in PSSA with reduced computational complexity compared to ANNs. Yet, their sensitivity to tuning parameters is a drawback. DT-based approaches emerge as alternative solutions, delivering reliable and accurate PSSA performance. Their structural concepts require only the latest updating data, significantly reducing computational complexity while maintaining accuracy.

Data-driven approaches, including ANNs, decision trees, and learning-based methods, excel in discovering previously unknown system characteristics. Additionally, less-implemented approaches like Extreme Learning Machines (ELM), fuzzy logic, and Pattern Discovery (PD) show promising training time and accuracy results in PSSA due to fewer tuning parameters. Overall, this study makes valuable contributions to the literature in understanding and evaluating these data-driven approaches in the context of PSSA.

The study thoroughly evaluates the advantages and limitations of data-driven methods in the existing literature, providing valuable insights for researchers interested in applying these methods for Power System Security Assessment (PSSA). The key contributions include:

- Comprehensive Evaluation: The paper provides a detailed assessment of the strengths and weaknesses of various data-driven approaches used in PSSA, offering researchers a nuanced understanding of the practical implications and challenges associated with these methods.
- Static and dynamic security assessment discussion: The research delves into static and dynamic security assessments, offering a comprehensive discussion and evaluation from different perspectives. This inclusive analysis ensures a holistic understanding of the application of data-driven methods across various aspects of PSSA.
- Performance Comparison: The study conducts a performance comparison among different artificial intelligence approaches employed in PSSA. This comparative analysis is a valuable resource for researchers and practitioners aiming to select the most suitable data-driven method for their specific requirements.
- Aggregation of diverse approaches: The study consolidates a diverse range of data-driven PSSA approaches, providing a comprehensive resource that combines different methods used in the field. This aggregation facilitates a better understanding of the landscape of data-driven techniques in PSSA.

In summary, the study significantly contributes to the existing body of knowledge by offering a comprehensive and comparative analysis of data-driven methods in the power system security assessment domain."

REFERENCES

1. K. Teeparthi and D. M. Vinod Kumar, "Power system security assessment and enhancement: A bibliographical survey," *J. Inst. Eng. (India) Ser. B*, vol. 101, no. 2, pp. 163–176, 2020.
2. J. Wang, P. Pinson, S. Chatzivasileiadis, M. Panteli, G. Strbac, and V. Terzija, "On machine learning-based techniques for future sustainable and resilient

energy systems," *IEEE Trans. Sustain. Energy,* vol. 14, no. 2, pp. 1230–1243, 2023.

3. M. Gholami, M. J. Sanjari, M. Safari, M. Akbari, and M. R. Kamali, "Static security assessment of power systems: A review," *Int. Trans. Electr. Energy Syst.,* vol. 30, no. 9, 2020.

4. F. D. Caro, A. J. Collin, G. M. Giannuzzi, C. Pisani, and A. Vaccaro, "Review of data-driven techniques for on-line static and dynamic security assessment of modern power systems," *IEEE Access,* vol. 11, pp. 130644–130673, 2023.

5. A. Mehrzad, M. Darmiani, Y. Mousavi, M. Shafie-Khah, and M. Aghamohammadi, "A review on data-driven security assessment of power systems: Trends and applications of artificial intelligence," *IEEE Access,* vol. 11, pp. 78671–78685, 2023.

6. Q. Wang, F. Li, Y. Tang, and Y. Xu, "Integrating model-driven and data-driven methods for power system frequency stability assessment and control," *IEEE Trans. Power Syst.,* vol. 34, no. 6, pp. 4557–4568, 2019.

7. M. Di Santo, A. Vaccaro, D. Villacci, and E. Zimeo, "A distributed architecture for online power systems security analysis," *IEEE Trans. Ind. Electron.,* vol. 51, no. 6, pp. 1238–1248, 2004.

8. A. N. AL-Masri, M. Z. A. Ab Kadir, H. Hizam, and N. Mariun, "A novel implementation for generator rotor angle stability prediction using an adaptive artificial neural network application for dynamic security assessment," *IEEE Trans. Power Syst.,* vol. 28, no. 3, pp. 2516–2525, 2013.

9. P. Sekhar and S. Mohanty, "An online power system static security assessment module using multi-layer perceptron and radial basis function network," *Int. J. Electr. Power Energy Syst.,* vol. 76, pp. 165–173, 2016.

10. C. F. Kucuktezcan and V. M. I. Genc, "A new dynamic security enhancement method via genetic algorithms integrated with neural network based tools," *Electric Power Syst. Res.,* vol. 83, no. 1, pp. 1–8, 2012.

11. E. M. Voumvoulakis and N. D. Hatziargyriou, "A particle swarm optimization method for power system dynamic security control," *IEEE Trans. Power Syst.,* vol. 25, no. 2, pp. 1032–1041, 2010.

12. M. Singh and S. Chauhan, "Tree-based ensemble machine learning techniques for power system static security assessment," *Electr. Power Compon. Syst.,* vol. 50, no. 6–7, pp. 359–373, 2022.

13. J. L. Cremer, I. Konstantelos, S. H. Tindemans, and G. Strbac, "Data-driven power system operation: Exploring the balance between cost and risk," *IEEE Trans. Power Syst.,* vol. 34, no. 1, pp. 791–801, 2019.

14. A. Mollaiee, M. Taghi Ameli, S. Azad, M. Nazari-Heris, and S. Asadi, "Data-driven power system security assessment using high content database during the COVID-19 pandemic," *Int. J. Electr. Power Energy Syst.,* vol. 150, no. 109077, p. 109077, 2023.

15. L. Zhu, D. J. Hill, and C. Lu, "Semi-supervised ensemble learning framework for accelerating power system transient stability knowledge base generation," *IEEE Trans. Power Syst.,* vol. 37, no. 3, pp. 2441–2454, 2022.

16. D. Mukherjee, S. Chakraborty, and S. Ghosh, "Power system state forecasting using machine learning techniques," *Electr. Eng. (Berl., Print),* vol. 104, no. 1, pp. 283–305, 2022.

17. Q. Zhou, J. Davidson, and A. A. Fouad, "Application of artificial neural networks in power system security and vulnerability assessment," *IEEE Trans. Power Syst.*, vol. 9, no. 1, pp. 525–532, 1994.
18. C. A. Jensen, M. A. El-Sharkawi, and R. J. Marks, "Power system security assessment using neural networks: feature selection using Fisher discrimination," *IEEE Trans. Power Syst.*, vol. 16, no. 4, pp. 757–763, 2001.
19. A. D. Rajapakse, F. Gomez, K. Nanayakkara, P. A. Crossley, and V. V. Terzija, "Rotor angle instability prediction using post-disturbance voltage trajectories," *IEEE Trans. Power Syst.*, vol. 25, no. 2, pp. 947–956, 2010.
20. T. S. Chung and Y. Fu, "A fast voltage security assessment method via extended Ward equivalent and neural network approach," *IEEE Power Eng. Rev.*, vol. 19, no. 10, pp. 40–43, 1999.
21. K. S. Swarup and P. B. Corthis, "ANN approach assesses system security," *IEEE Comput. Appl. Power*, vol. 15, no. 3, pp. 32–38, 2002.
22. Y. Du, F. Li, and C. Huang, "Applying deep convolutional neural network for fast security assessment with N-1 contingency," in *2019 IEEE Power & Energy Society General Meeting (PESGM)*, 2019.
23. A. N. Al-Masri, M. Z. A. Ab Kadir, H. Hizam, and N. Mariun, "Simulation of an adaptive artificial neural network for power system security enhancement including control action," *Appl. Soft Comput.*, vol. 29, pp. 1–11, 2015.
24. B. Wang, B. Fang, Y. Wang, H. Liu, and Y. Liu, "Power system transient stability assessment based on big data and the core vector machine," *IEEE Trans. Smart Grid*, vol. 7, no. 5, pp. 2561–2570, 2016.
25. S. Kalyani and K. Shanti Swarup, "Classification and assessment of power system security using multiclass SVM," *IEEE Trans. Syst. Man Cybern. C Appl. Rev.*, vol. 41, no. 5, pp. 753–758, 2011.
26. H. T. Nguyen and L. B. Le, "Online ensemble learning for security assessment in PMU based power system," in *2016 IEEE International Conference on Sustainable Energy Technologies (ICSET)*, 2016.
27. S. Kalyani and K. S. Swarup, "Binary SVM approach for security assessment and classification in power systems," in *2009 Annual IEEE India Conference*, 2009.
28. M. Beiraghi and A. M. Ranjbar, "Online voltage security assessment based on wide-area measurements," *IEEE Trans. Power Deliv.*, vol. 28, no. 2, pp. 989–997, 2013.
29. R. Diao, V. Vittal, and N. Logic, "Design of a real-time security assessment tool for situational awareness enhancement in modern power systems," *IEEE Trans. Power Syst.*, vol. 25, no. 2, pp. 957–965, 2010.
30. M. He, V. Vittal, and J. Zhang, "Online dynamic security assessment with missing pmu measurements: A data mining approach," *IEEE Trans. Power Syst.*, vol. 28, no. 2, pp. 1969–1977, 2013.
31. K. Sun, S. Likhate, V. Vittal, V. S. Kolluri, and S. Mandal, "An online dynamic security assessment scheme using phasor measurements and decision trees," *IEEE Trans. Power Syst.*, vol. 22, no. 4, pp. 1935–1943, 2007.
32. X. Boyen and L. Wehenkel, "Automatic induction of fuzzy decision trees and its application to power system security assessment," *Fuzzy Sets And Systems*, vol. 102, no. 1, pp. 3–19, 1999.
33. I. Kamwa, S. R. Samantaray, and G. Joos, "Development of rule-based classifiers for rapid stability assessment of wide-area post-disturbance records," *IEEE Trans. Power Syst.*, vol. 24, no. 1, pp. 258–270, 2009.

34. C. Liu *et al.*, "A systematic approach for dynamic security assessment and the corresponding preventive control scheme based on decision trees," *IEEE Trans. Power Syst.*, vol. 29, no. 2, pp. 717–730, 2014.

35. J. L. Cremer, I. Konstantelos, and G. Strbac, "From optimization-based machine learning to interpretable security rules for operation," *IEEE Trans. Power Syst.*, vol. 34, no. 5, pp. 3826–3836, 2019.

36. Z. H. Rather, C. Liu, Z. Chen, C. L. Bak, and P. Thogersen, "Dynamic security assessment of Danish power system based on decision trees: Today and tomorrow," in *2013 IEEE Grenoble Conference*, 2013.

37. I. Genc, R. Diao, V. Vittal, S. Kolluri, and S. Mandal, "Decision tree-based preventive and corrective control applications for dynamic security enhancement in power systems," *IEEE Trans. Power Syst.*, vol. 25, no. 3, pp. 1611–1619, 2010.

38. W. D. Oliveira, J. P. A. Vieira, U. H. Bezerra, D. A. Martins, and B. das G. Rodrigues, "Power system security assessment for multiple contingencies using multiway decision tree," *Electric Power Syst. Res.*, vol. 148, pp. 264–272, 2017.

39. M. He, J. Zhang, and V. Vittal, "Robust online dynamic security assessment using adaptive ensemble decision-tree learning," *IEEE Trans. Power Syst.*, vol. 28, no. 4, pp. 4089–4098, 2013.

40. Z. Li and W. Wu, "Phasor measurements-aided decision trees for power system security assessment," in *2009 Second International Conference on Information and Computing Science*, 2009.

41. Y. Xu, Z. Y. Dong, J. H. Zhao, P. Zhang, and K. P. Wong, "A reliable intelligent system for real-time dynamic security assessment of power systems," *IEEE Trans. Power Syst.*, vol. 27, no. 3, pp. 1253–1263, 2012.

42. Y. Xu, Z. Y. Dong, Z. Xu, K. Meng, and K. P. Wong, "An intelligent dynamic security assessment framework for power systems with wind power," *IEEE Trans. Industr. Inform.*, vol. 8, no. 4, pp. 995–1003, 2012.

43. C. Ren, Y. Xu, Y. Zhang, and C. Hu, "A multiple randomized learning based ensemble model for power system dynamic security assessment," in *2018 IEEE Power & Energy Society General Meeting (PESGM)*, 2018.

44. C. Ren and Y. Xu, "A fully data-driven method based on generative adversarial networks for power system dynamic security assessment with missing data," *IEEE Trans. Power Syst.*, vol. 34, no. 6, pp. 5044–5052, 2019.

45. M. Singh and S. Chauhan, "A hybrid-extreme learning machine based ensemble method for online dynamic security assessment of power systems," *Electric Power Syst. Res.*, vol. 214, no. 108923, p. 108923, 2023.

46. M. Sun, I. Konstantelos, and G. Strbac, "A deep learning-based feature extraction framework for system security assessment," *IEEE Trans. Smart Grid*, vol. 10, no. 5, pp. 5007–5020, 2019.

47. J.-M. H. Arteaga, F. Hancharou, F. Thams, and S. Chatzivasileiadis, "Deep learning for power system security assessment," in *2019 IEEE Milan PowerTech*, 2019.

48. Rizwan-ul-Hassan, C. Li, and Y. Liu, "Online dynamic security assessment of wind integrated power system using SDAE with SVM ensemble boosting learner," *Int. J. Electr. Power Energy Syst.*, vol. 125, no. 106429, p. 106429, 2021.

49. F. Luo *et al.*, "Advanced pattern discovery-based fuzzy classification method for power system dynamic security assessment," *IEEE Trans. Industr. Inform.*, vol. 11, no. 2, pp. 416–426, 2015.
50. J. M. Gimenez Alvarez and P. E. Mercado, "Online inference of the dynamic security level of power systems using fuzzy techniques," *IEEE Trans. Power Syst.*, vol. 22, no. 2, pp. 717–726, 2007.
51. Y. Xu, Z. Y. Dong, L. Guan, R. Zhang, K. P. Wong, and F. Luo, "Preventive dynamic security control of power systems based on pattern discovery technique," *IEEE Trans. Power Syst.*, vol. 27, no. 3, pp. 1236–1244, 2012.
52. I. Kamwa, R. Grondin, and L. Loud, "Time-varying contingency screening for dynamic security assessment using intelligent-systems techniques," *IEEE Trans. Power Syst.*, vol. 16, no. 3, pp. 526–536, 2001.
53. C. Ren, Y. Xu, B. Dai, and R. Zhang, "An integrated transfer learning method for power system dynamic security assessment of unlearned faults with missing data," *IEEE Trans. Power Syst.*, vol. 36, no. 5, pp. 4856–4859, 2021.
54. C. Ren and Y. Xu, "Transfer learning-based power system online dynamic security assessment: Using one model to assess many unlearned faults," *IEEE Trans. Power Syst.*, vol. 35, no. 1, pp. 821–824, 2020.

Chapter 12

Machine learning and deep learning models for effective forecasting of renewable energy generation

Prem Prakash Vuppuluri, Nitin Sing,
and Dev Dhaked

12.1 INTRODUCTION

With the ever-growing demand for energy worldwide, there is an urgent need for reliable and clean energy sources to fulfill this requirement. Traditional fossil fuels, including coal and natural gas, not only contribute substantially to greenhouse gas emissions but also pose challenges of finite availability, threatening long-term energy security. Fossil fuels are finite resources and are bound to eventually become extinct if they continue to be utilized at the current scale for energy production. [2]

In recent years, the global pursuit of sustainable energy solutions has intensified due to escalating concerns over environmental degradation and the finite nature of fossil fuel resources. It is important to shift toward renewable energy sources, characterized by their ability to replenish naturally and minimize adverse environmental impacts. Harnessing wind, solar, hydro, and other renewable resources offers a sustainable alternative to fossil fuels, reducing greenhouse gas emissions and mitigating climate change. The transition to renewable energy not only enhances energy security but also fosters economic growth and job creation. By diversifying energy sources and reducing reliance on imported fossil fuels, countries can enhance energy security and mitigate the geopolitical risks associated with energy dependence. Renewable energy decentralization empowers communities to generate their own clean power, particularly in remote or underserved areas where access to traditional grid infrastructure may be limited. As renewable energy costs continue to decline, it becomes more accessible and affordable for households, businesses, and governments, helping alleviate energy poverty and reducing energy bills [3, 4].

Among the many renewable energy sources, solar energy stands out as a promising avenue due to its abundance and accessibility. It harnesses the inexhaustible energy of the sun, offering a virtually limitless resource that produces minimal pollution during operation. Harnessing solar power through photovoltaic (PV) systems offers a viable means of generating clean electricity with a minimal environmental footprint. Moreover, with advancements in technology, the production of solar energy is becoming more economical and reliable. Solar panels consist of photovoltaic cells, which transform solar energy into electricity. These cells are sandwiched between layers of semiconductor materials like silicon. When sunlight strikes these layers, the photons energize them, generating an electric field through the photoelectric effect. This field produces the necessary current for electricity generation, as shown in Figure 12.1. The surface of the cell is covered by an anti-reflective material, which traps the light energy and avoids any loss of energy. Solar grid systems, as shown in Figure 12.2, also known as grid-tied or grid-connected solar systems, are setups where solar panels are connected to the existing electricity grid. Solar panels produce direct current (DC) electricity, which is then converted into alternating current (AC) electricity using an inverter. This converted electricity can be supplied to the connected home or business.

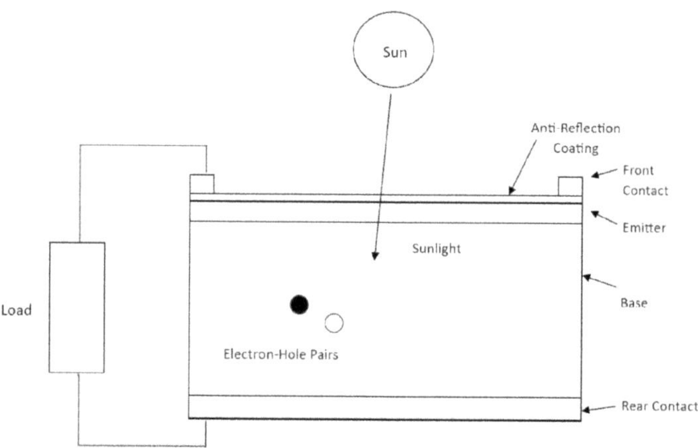

Figure 12.1 Schematic of photovoltaic panel.

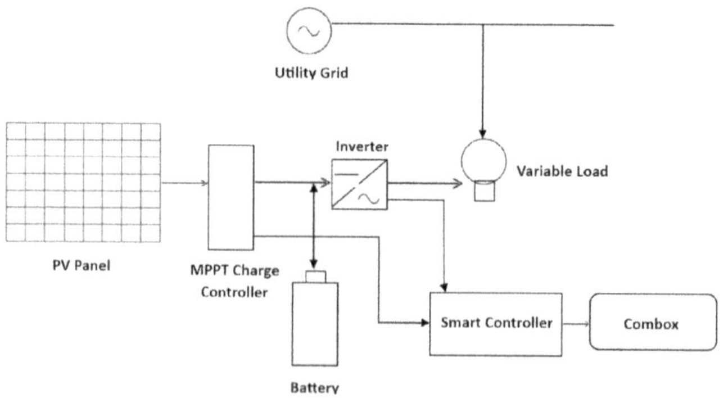

Figure 12.2 Solar grid system.

However, the intermittent nature of solar energy production poses a significant challenge to its widespread adoption. Unlike conventional power sources, such as coal or natural gas, solar energy generation is contingent upon factors such as sunlight availability, which may not always align with peak electricity demand periods. Additionally, the variability in solar irradiance can result in unpredictable energy generation patterns. Thus, it is necessary to develop accurate forecasting models and effective energy storage solutions to mitigate supply–demand imbalances. Addressing these hurdles is crucial for maximizing the benefits of solar energy and ensuring its seamless integration into the global energy landscape.

To effectively integrate solar power into the grid and maximize its potential, accurate solar power prediction becomes crucial. By anticipating solar output, utilities can adjust their energy mix and schedule other power sources more efficiently. Accurate predictions allow for proactive measures to maintain grid balance and prevent disruptions. Solar power producers can leverage prediction models to make informed decisions about energy trading and maximize financial returns [5, 6].

Machine learning (ML) and deep learning (DL), subfields of artificial intelligence, have emerged as powerful tools for this purpose. ML and DL models can analyze vast amounts of historical and real-time data, including weather forecasts, to predict solar irradiance and power output with high accuracy. By leveraging historical solar power data, weather patterns, and advanced modeling algorithms, researchers strive to develop robust forecasting models capable of predicting solar energy generation with unprecedented accuracy [1]. These models play a pivotal role in optimizing energy management, facilitating grid integration, and maximizing the utilization of solar resources.

This chapter provides a comprehensive overview of various prediction models, ranging from classical methods like linear regression, polynomial regression, random forest regression, and decision trees to advanced techniques such as long short-term memory (LSTM), Prophet, and their hybrids. These models play a crucial role in accurately forecasting solar energy generation, thereby optimizing energy management and grid integration. These models offer varying levels of accuracy, interpretability, and computational efficiency, catering to diverse application scenarios in renewable energy management. Each model has its unique characteristics, assumptions, and suitability for specific research scenarios By leveraging the strengths of different models and adapting them to specific prediction tasks, stakeholders can make informed decisions, optimize energy utilization, and advance the transition toward the sustainable energy systems.

Several ML and DL models, along with their hybrid counterparts, are applied for forecasting power generation in renewable energy microgrids, with a particular focus on solar energy, and their relative effectiveness is studied. Leveraging a comprehensive open dataset sourced from Monash University in Melbourne, Australia, the study aims to construct robust predictive models tailored to the unique characteristics of solar power generation [7–9]. To ensure the reliability and efficacy of the models, various preprocessing techniques, including data cleaning and feature scaling, are implemented. Furthermore, the mean absolute scaled error (MASE) metric is employed to assess the accuracy and generalization capabilities of the developed models.

Advanced ML and DL models offer promising avenues for accurate and reliable solar power prediction, enabling efficient utilization of solar energy resources and facilitating the transition toward sustainable energy systems. Continued research and innovation in this field hold the key to overcoming existing challenges and unlocking the full potential of solar energy generation. By enhancing our ability to forecast solar power generation accurately, this study endeavors to facilitate the transition toward a more sustainable and resilient energy landscape [10, 11].

12.2 METHODOLOGY

12.2.1 Data collection and preparation

The study presented in this chapter uses a detailed, openly available dataset from the Monash University Microgrid in Melbourne, Australia. This dataset incorporates various measures of solar energy production, including sunlight levels (data from six solar panels) as shown in Figure 12.3, weather conditions as shown in Figure 12.4, and electricity demand (data for six buildings).

12.2.2 Data preprocessing

Before proceeding to model selection, rigorous preprocessing is performed to determine whether the dataset is suitable for analysis. This includes steps

	Solar0
25-04-2020 14:00	0
25-04-2020 14:15	0
25-04-2020 14:30	0
25-04-2020 14:45	0
25-04-2020 15:00	0
25-04-2020 15:15	0

Figure 12.3 Solar data.

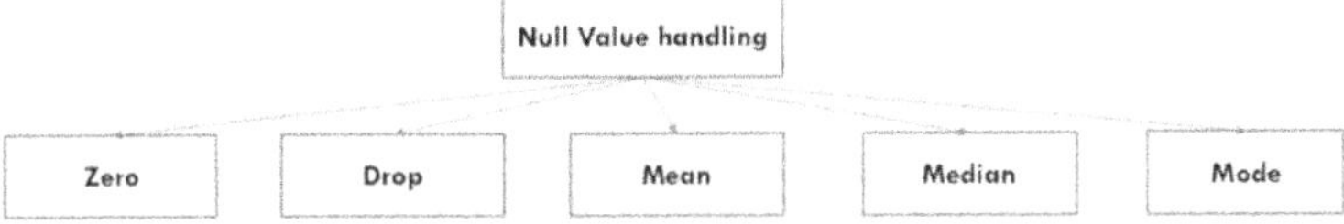

datetime (UTC)	model elevation (sur	utc_offset (hi	temperature (de	dewpoint_temperature (degC)	wind_speed (m/s	mean_sea_level	relative_humidit	surface_solar_ra	surface_therma	total_clou
01-01-2010 00:00	69.59	10	18.20	16.39	2.6	101046.38	0.89	287.01	408.33	1
01-01-2010 01:00	69.59	10	18.67	16.29	2.91	101037.96	0.86	360.79	411.02	1
01-01-2010 02:00	69.59	10	18.16	15.89	3.26	101017.26	0.87	291.54	410.67	1
01-01-2010 03:00	69.59	10	18.48	15.33	3.17	101022.56	0.82	357.11	410.95	1
01-01-2010 04:00	69.59	10	18.53	15.11	2.95	100940.03	0.8	459.91	410	0.9
01-01-2010 05:00	69.59	10	18.92	15.16	2.62	100885.16	0.79	513.07	407.46	0.96

Figure 12.4. Weather data.

Figure 12.5 Null value handling techniques.

Table 12.1   Results for null values handling techniques

Models	Mean_Mase
Null values handling models result	1.517961
Zero-predicted_powers_solar.csv	1.529887
Mode-predicted_powers_solar.csv	1.547992
Median-predicted_powers_solar.csv	1.571897
Drop-predicted_powers_solar.csv	1.57897

such as data cleaning to remove outliers and missing (zero) values, and feature calibration to standardize variables to facilitate model fitting. In this process, the null values are handled using five approaches (Figure 12.5):

- Replace null value with zero.
- Replace null value with mean.
- Replace null value with median.
- Replace null value with mode.
- Drop.

It can be observed from Table 12.1 that the best results are obtained by replacing NA values with 0. Therefore, we use this method for further study [12].

12.2.3 Model selection

In this work, the effectiveness of several ML and DL models, as well as hybrid schemes, for the task of solar energy forecasting is evaluated. These include regression models like linear regression, polynomial regression, random forest regression, and decision tree; deep models such as recurrent neural networks (RNNs) and LSTM; and some hybrid models based on the performance of these models. Each of these models is discussed in greater detail in Section 12.3.

12.2.4 Model training and testing

The selected models are trained with preprocessed data, aiming to identify underlying parameters and relationships between input characteristics and solar output. Iterative optimization of model parameters, using methods such as backpropagation and gradient descent, is employed during training. For this purpose, the available data were segregated into "train" and "test" subsets. After the training phase, each model was tested on the test data, and the results obtained were analyzed to evaluate their relative effectiveness.

12.3 MODELS USED IN EXPERIMENTAL WORK

Several ML, DL, and hybrid models were tested in this work. Each of these models is described in detail in this section.

12.3.1 Linear regression model

Linear regression is one of the simplest and most popular ML algorithms. It is a statistical approach to predictive analysis that deals with continuous or true statistical variables such as sales, wages, age, and prices.

Linear regression identifies the linear relationship between the dependent and independent variables, illustrating how changes in the independent variable(s) correspond to shifts in the dependent variable. This relationship is represented with a straight line, indicating the slope of the relationship between the variables [13]

The linear regression model provides a sloped straight line representing the relationship between the variables, as shown in Figure 12.6. Consider Figure 12.6.

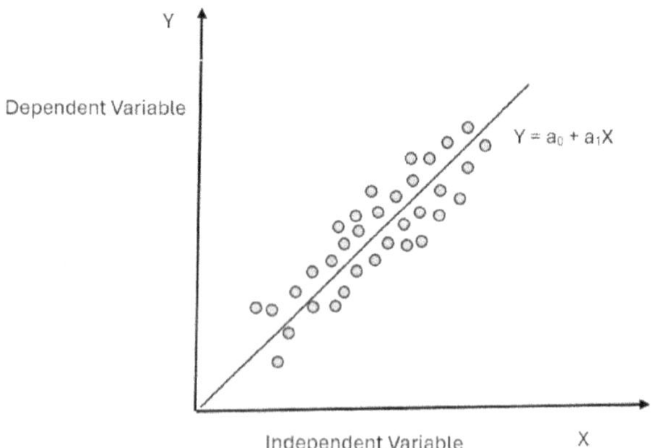

Figure 12.6 Linear regression curve.

Mathematically, we can represent a linear regression as follows:

$$Y = a_0 + a_1 x + \varepsilon$$

Here, y = dependent variable (target variable)

X = independent variable (predictor variable)
a_0 = intercept of line
a_1 = linear regression coefficient
ε = random error

12.3.1.1 Importance of linear regression model

- *Simplicity and ease of use:* Linear regression is straightforward to understand and implement.
- *Interpretability:* The results of linear regression are easily interpretable, aiding in understanding the relationships between variables.
- *Wide applicability:* It can be applied to various data types and is widely accepted across different domains.

12.3.2 Polynomial regression

In polynomial regression, the relationship between the independent variable x and the dependent variable y is modeled as an n-degree polynomial. It captures nonlinear relationships between variables by introducing polynomial terms.

Below is the equation of the polynomial equation:

$$Y = b_0 + b_1 x_1 + b_2 x_1^2 + b_2 x_1^3 + + b_n x_1^n$$

Polynomial regression extends linear regression by introducing polynomial terms, allowing for a more flexible model that can capture complex relationships between variables. It is particularly useful when the relationship between variables is nonlinear and cannot be adequately captured by a straight line.

Polynomial regression is a crucial component of ML, especially when dealing with nonlinear relationships. The polynomial regression model provides a curved polynomial line representing the relationship between the variables, as shown in Figure 12.7.

Polynomial regression is a modification of the linear regression model, tailored to improve accuracy by accommodating nonlinear relationships between variables. Despite being labeled as a "linear model," polynomial regression is inherently nonlinear in nature due to its ability to capture complex and nonlinear functions using a linear regression approach.

This method becomes invaluable when dealing with datasets that exhibit nonlinear separability, where traditional linear regression struggles to establish meaningful relationships between variables. By adjusting the

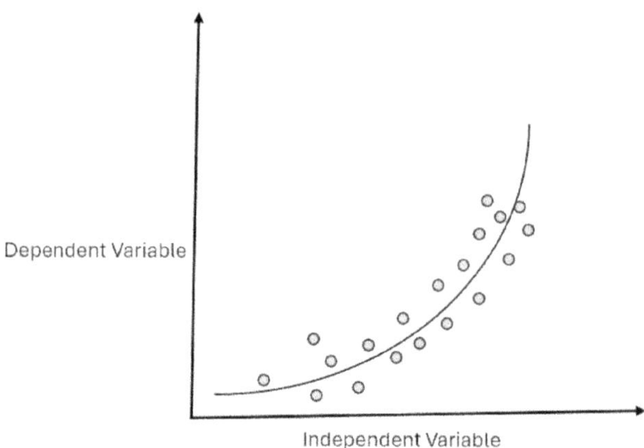

Figure 12.7 Polynomial regression curve.

linear regression model, we transform it into a polynomial regression model capable of effectively handling nonlinear functions and datasets [14].

12.3.2.1 Importance of polynomial regression

- *Handles nonlinear data:* Polynomial regression effectively models nonlinear relationships between variables.
- *Improves accuracy:* Offers higher predictive accuracy compared to linear models, especially with nonlinear datasets.
- *Minimizes loss function:* Reduces the loss function, resulting in lower error rates and enhanced model performance.

12.3.3 Decision tree

A decision tree is a versatile supervised learning technique used for both classification and regression tasks. Its structure resembles a tree, where internal nodes represent dataset features, branches embody decision rules, and leaf nodes indicate outcomes.

Decision trees consist of three main types of nodes:

Root node: The topmost node in the tree represents the feature that best splits the dataset into distinct groups, maximizing the separation of classes or minimizing the variance in regression tasks.

Decision nodes: These nodes execute decisions and have multiple branches, each corresponding to a possible outcome based on the value of a specific feature.

Leaf nodes: Leaf nodes signify decision outcomes and do not have further branches. They represent the final prediction or classification result.

The decision-making process within a decision tree relies on dataset features. It offers a graphical representation to explore potential problem solutions based on provided conditions, with its tree-like expansion starting from the root node and branching out through subsequent decision nodes until reaching the leaf nodes, as can be observed in Figure 12.8.

The CART algorithm (classification and regression tree) is commonly used to construct decision trees. This algorithm facilitates an iterative process where the tree poses a question based on a feature, and based on the response (yes/no or numerical values), it recursively divides into subtrees. CART aims to create a tree that best splits the data into homogeneous subsets with respect to the target variable, whether it is a classification label or a numerical value for regression tasks [15].

Figure 12.8 illustrates the typical structure of a decision tree.

12.3.3.1 Importance of decision tree

- It is simple to understand as it follows the same process which a human follow while making any decision in real life.
- It can also be very useful for solving problems related to decision.
- It helps to think about all the possible outcomes for a problem.
- There is less requirement for data cleaning as compared to other algorithms.

12.3.4 Random forest

Random forest stands out as a prominent ML algorithm within supervised learning, adept at tackling both classification and regression tasks. It operates on the principle of ensemble learning, a method that amalgamates multiple classifiers to address complex problems and enhance model performance.

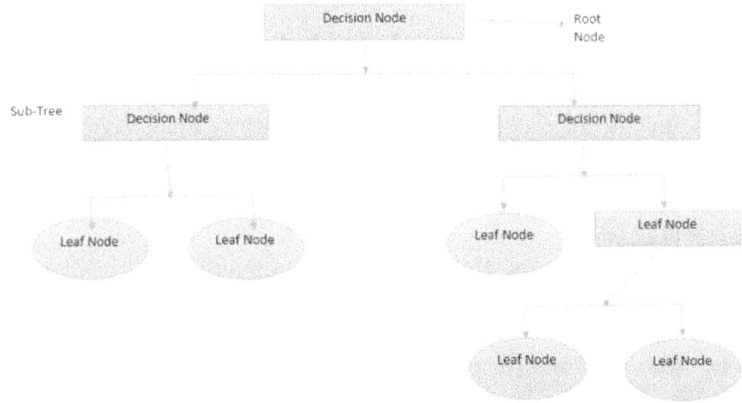

Figure 12.8 Decision tree.

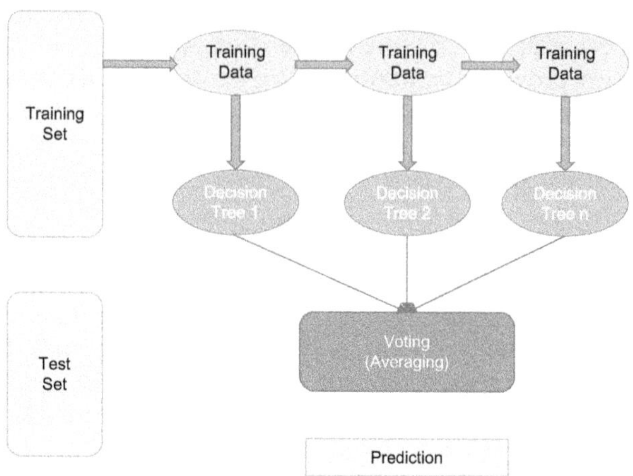

Figure 12.9 Random forest.

Random forest, as its name suggests, incorporates randomness in two key ways to build a powerful predictive model. Firstly, when constructing each tree within the forest, it randomly selects a subset of features to consider at each decision point. This ensures that each tree focuses on different aspects of the data, increasing diversity. Secondly, instead of using the entire dataset to train each tree, random forest takes random samples, allowing some data points to be repeated while others are left out. This process, called bootstrapping, further enhances diversity among the trees. By combining these random techniques, random forest creates a diverse ensemble of trees that collectively make accurate predictions while mitigating overfitting. This approach enables random forest to effectively handle a wide range of classification and regression tasks with high accuracy and robustness.

"Random forest" constitutes a classifier comprising numerous decision trees trained on various subsets of the dataset, as shown in Figure 12.9, aggregating their predictions to refine overall accuracy. Unlike relying on a single decision tree, this approach leverages the collective wisdom of the forest, where each tree contributes its prediction, and the final output is determined by majority voting.

The inclusion of multiple trees mitigates overfitting concerns and enhances predictive accuracy, especially with a larger number of trees in the forest [16].

Figure 12.9 illustrates the mechanism underlying the random forest algorithm.

12.3.4.1 Assumptions for random forest

In a random forest classifier, the amalgamation of multiple decision trees can yield differing predictions, where some trees may correctly predict the output while others may not. Nonetheless, the ensemble approach combines these diverse predictions to arrive at an accurate overall output. For an optimal random forest classifier, two key assumptions are vital:

1. *Actual values in feature variables:* To ensure accurate predictions, it is crucial that the feature variables within the dataset contain genuine values rather than arbitrary or guessed ones. The classifier relies on these features to discern patterns and make informed predictions. High-quality, meaningful data fosters more accurate predictions across the ensemble of decision trees.
2. *Low correlation in predictions:* For effective ensemble learning, the predictions generated by each decision tree should exhibit minimal correlation with one another. This diversity in predictions ensures that errors or biases in individual trees are mitigated when aggregated. By fostering independence among the constituent trees, the random forest classifier leverages the collective wisdom of diverse models, leading to robust and reliable predictions.

By adhering to these assumptions, a random forest classifier can harness the strengths of ensemble learning, providing accurate and robust predictions across various classification tasks.

Why use random forest?

Below are some points that explain why we should use the random forest algorithm:

- It takes less training time compared to other algorithms.
- It predicts output with higher accuracy, even for large datasets, and runs very efficiently.
- It can maintain accuracy even when a large proportion of data is missing.

Applications of random forest:

Random forest is used in various sectors:

- *Banking:* For the identification of loan risk
- *Medicine:* Identifying disease trends and risks
- *Land use:* Identifying areas of similar land use
- *Marketing:* Analyzing marketing trends

Advantages of random forest:

- It can perform both classification and regression tasks.
- It handles large datasets with high dimensionality efficiently.
- It enhances model accuracy and prevents overfitting.

12.3.5 LSTM network

LSTM networks were developed to address limitations inherent in traditional RNNs, particularly their inability to effectively handle long-term dependencies. RNNs process current inputs by incorporating information from previous events (feedback) and briefly storing it in short-term memory. While RNNs have found success in applications like non-Markovian speech control and music composition, they suffer from several drawbacks. One major issue with RNNs is their struggle to retain information over extended periods, hindering their ability to capture long-term dependencies. This limitation becomes apparent when past data from a distant time step is necessary to determine the present output – a challenge RNNs are ill-equipped to handle.

Unlike traditional RNNs, LSTMs do not have the requirement to maintain the same number of states before the time mandated by the hidden Markov model (HMM). They offer a wide range of parameters such as learning rates, input and output biases, and reducing the need for fine-tuning. Moreover, the computational effort required to update each weight is decreased to O(1) using LSTMs, akin to back propagation through time (BPTT), which is a significant advantage [17].

Each LSTM cell is equipped with three inputs and two outputs can be seen in Figure 12.10, h_t, and C_t. At a specific time t, h_t is the hidden state and C_t is the cell state or memory. x_t is the present information point or the input. The first sigmoid layer contains two inputs: h_{t-1} and x_t, where h_{t-1} is the state hidden in the cell before it. It is also known by its name as the forget gate, since its output is a selection of the amount of data from the

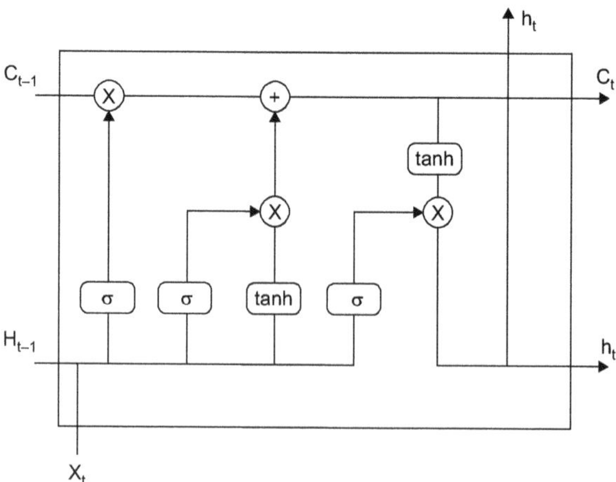

Figure 12.10 LSTM model.

last cell that should be included. Its output will be a number [0,1] multiplied (pointwise) by the previous cell's state.

LSTM models necessitate training with a dataset before real-world deployment. They find applications in various challenging domains:

- *Text generation:* LSTM models excel at generating text or language modeling, predicting words based on input word sequences. They can operate at character or *n*-gram levels, as well as sentence or paragraph levels.
- *Image processing:* LSTM architectures can analyze and describe images by generating textual descriptions. This application is prevalent in computer vision tasks such as image captioning and object recognition.
- *Speech and handwriting recognition:* LSTM networks are capable of recognizing and transcribing spoken language or handwritten text. This is crucial in applications such as speech recognition systems and optical character recognition (OCR).

12.3.5.1 Advantages of LSTM networks

- *Effective modeling of sequential data:* LSTM networks excel at modeling sequential data and capturing long-range dependencies, making them suitable for tasks like language modeling and speech recognition.
- *Application versatility:* LSTM networks find applications across various domains, including natural language processing, computer vision, speech recognition, music generation, and language translation.
- *Long-term dependency handling:* Compared to traditional RNNs, LSTM networks are designed to handle long-term dependencies more effectively, enabling them to retain and utilize information over extended sequences.

12.3.6 Prophet

Prophet is an ML model developed by Facebook for time series forecasting. It is designed to handle data with strong seasonal patterns, multiple sources of uncertainty, and irregularities commonly found in real-world datasets. Prophet employs a decomposable time series model with several components, including trend, seasonality, holidays, and additional regressors.

12.3.6.1 Importance of Prophet

1. *Seasonal pattern handling:* Prophet excels in capturing diverse seasonal patterns present in time series data, including daily, weekly, monthly, and yearly fluctuations.

2. *Automatic holiday detection:* It automatically identifies holidays and significant events, allowing for adjustments in forecasts to accommodate deviations from regular patterns.
3. *Flexible trend modeling:* Prophet offers flexibility in modeling trends, accommodating various trend behaviors such as linear, logistic, and piecewise linear trends.
4. *Uncertainty estimation:* It provides measures of prediction intervals, offering insights into forecast uncertainty and aiding decision-making processes.
5. *Ease of use:* With its user-friendly interface and minimal configuration requirements, Prophet is accessible to users with varying levels of ML expertise.

12.3.7 Hybrid models

Hybrid ML refers to the integration of multiple ML techniques or models to leverage the strengths of each approach and improve overall predictive performance. It combines different algorithms or methodologies to address complex problems that may not be effectively solved by any single technique alone. Hybrid models often aim to exploit complementary capabilities, such as the ability of one model to capture global patterns while another focuses on local details, thereby enhancing predictive accuracy, robustness, and generalization capabilities.

Hybrid ML models are motivated by the recognition that no single algorithm or technique is universally superior for all types of data or tasks. Different ML models have distinct strengths and weaknesses, and hybridization aims to capitalize on these complementary attributes to achieve better performance. [18]

12.3.7.1 Types of hybrid models

Hybrid ML models can take various forms:

- *Ensemble methods:* Combining predictions from multiple base models, such as bagging (e.g., Random Forest), boosting (e.g., gradient boosting machines), and stacking as shown in Figure 12.11.
- *Feature combination:* Integrating features extracted from different sources or using different techniques, such as combining text and image features for multimodal classification.
- *Model cascades:* Sequentially applying multiple models, where the output of one model serves as input to the next, allowing for hierarchical learning.
- *Model fusion:* Integrating predictions from diverse models or modalities to obtain a final prediction, often used in multimodal learning tasks.

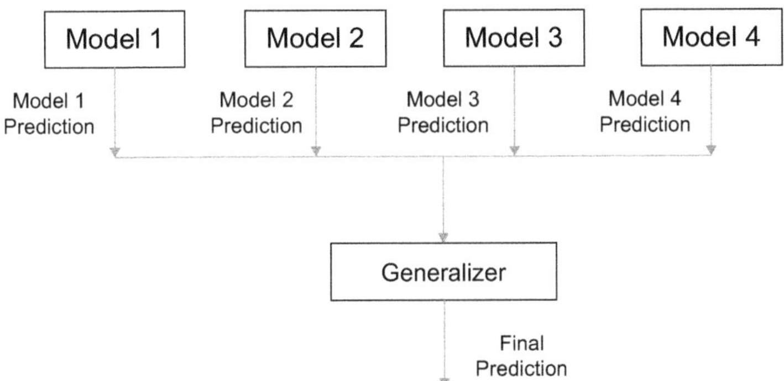

Figure 12.11 Ensemble hybrid model.

12.3.7.2 Applications of hybrid models

Hybrid ML techniques find applications across various domains:

- *Natural language processing:* Combining DL models with traditional ML techniques for text classification, sentiment analysis, and machine translation.
- *Image and video analysis:* Integrating convolutional neural networks (CNNs) with RNNs for tasks such as video captioning, object detection, and medical image analysis.
- *Time series forecasting:* Hybridizing autoregressive models with ML algorithms for improved accuracy in predicting stock prices, energy demand, and weather patterns.

12.4 RESULTS

The performance of the models is evaluated based on the MASE. MASE is a metric commonly used to evaluate the performance of forecasting models, particularly in time series forecasting tasks such as solar power prediction. It provides a measure of the accuracy of a model's predictions relative to a naive benchmark, accounting for both bias and variability in the data.

MASE is a statistical metric used to assess the accuracy of forecasting models by comparing their performance against a simple benchmark, often a naive or seasonal forecast. It is calculated as the mean of the absolute errors divided by the mean absolute error of the benchmark (naive forecast), adjusted for the seasonality of the data.

$$\text{MASE} = \frac{\sum_{k=M+1}^{M+h} \left| F_k - Y_k \right|}{\frac{h}{M-S} \sum_{k=S+1}^{M} \left| Y_k - Y_{k-S} \right|}$$

where M is the number of instances in the training series, S is the length of the seasonal cycle of the dataset, h is the forecast horizon, F_k are the generated forecasts, and Y_k are the actual values.

For this dataset, we consider a 28-day-ahead seasonal naive benchmark when computing the MASE. The MASE values are individually calculated for each series and finally, we calculate the mean of the MASE values across all 12 series. This mean is used to rank the forecasting models [19].

Table 12.2 summarizes the MASE scores achieved by each model:

Based on the mean MASE values obtained from the evaluation of various prediction models for solar power generation, the following insights can be derived (Figure 12.12):

The hybrid model of linear regression and LSTM performs best among all models with a mean MASE of 0.689241, indicating superior predictive accuracy.

Following this are the polynomial regression and LSTM, LSTM, hybrid of Prophet and LSTM model, bidirectional LSTM, linear regression,

Table 12.2 Model results

Models	Mean_Mase
linear+lstm_predicted_powers_all.csv	0.689241
poly+lstm_predicted_powers_all.csv	0.715681
lstm_predicted_powers_all.csv	0.72443
Lstm+prophet_hybrid_predicted_powers_all.csv	0.72465
bi-lstm_predicted_powers_all.csv	0.731685
linear_regr_predicted_powers_all.csv	0.753319
lightgbm_powers_solar.csv	0.764943
poly_reg_predicted_powers_all_degree2.csv	0.78484
random-forest	0.801932
prophet_predicted_powers_all.csv	0.833955
decision_tree	1.048548

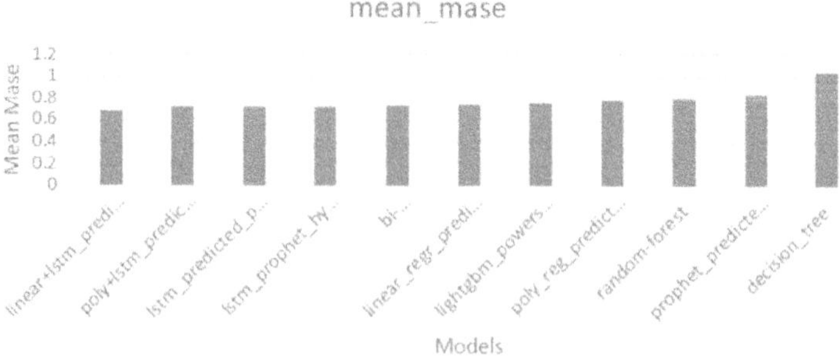

Figure 12.12 Model mean MASE.

LightGBM model, polynomial regression, random forest regression, Prophet, and decision tree regression.

However, the hybrid of LSTM and polynomial regression showcased strong predictive capabilities.

LSTM-based models, both stand-alone and in conjunction with the Prophet algorithm, showcased strong predictive capabilities.

The decision tree model exhibited the highest MASE score of 1.048548, indicating relatively poorer predictive performance compared to other models.

Further investigation into ensemble methods, fine-tuning of hyperparameters, and exploration of advanced DL architectures could potentially improve predictive accuracy. Additionally, incorporating domain-specific features and external factors may enhance the models' robustness in real-world applications.

12.5 CONCLUSIONS

In this study, we examined various models and their hybrids for predicting solar power generation and the load required for a particular period. We found that the hybrid models performed better.

The superior performance of LSTM-based models highlights their effectiveness in capturing temporal dependencies and long-term patterns in time series data. The stand-alone LSTM model and the LSTM and Prophet hybrid model exhibited strong predictive capabilities, emphasizing the significance of incorporating time series features for accurate forecasting.

The Prophet algorithm, known for its robustness in handling time series data with seasonality and trend components, delivered competitive results. Its combination with LSTM further improved predictive accuracy, indicating the complementary nature of these two approaches in capturing diverse temporal patterns.

Further exploration of ensemble methods, such as an ensemble of LSTM models or advanced DL architectures, may enhance predictive accuracy. Additionally, incorporating domain-specific features and external factors, such as weather data or economic indicators, could improve the models' robustness and generalizability.

REFERENCES

1. K. Mahmud, S. Azam, A. Karim, S. Zobaed, B. Shanmugam and D. Mathur, "Machine learning based PV power generation forecasting in alice springs," in *IEEE Access*, vol. 9, pp. 46117–46128, 2021, doi: 10.1109/ ACCESS.2021.3066494.

2. N. Anglani and G. Petrecca, "Fossil fuel and biomass fed distributed generation and utility plants: Analysis of energy and environmental performance indicators," *2012 16th IEEE Mediterranean Electrotechnical Conference*, Yasmine Hammamet, Tunisia, 2012, pp. 581–586, doi: 10.1109/MELCON.2012.6196500.

3. J. Choi, J. Park, M. Shahidehpour and R. Billinton, "Assessment of CO_2 reduction by renewable energy generators," *2010 Innovative Smart Grid Technologies (ISGT)*, Gaithersburg, MD, USA, 2010, pp. 1–5, doi: 10.1109/ISGT.2010.5434742.

4. A. Qazi *et al.*, "Towards sustainable energy: a systematic review of renewable energy sources, technologies, and public opinions," in *IEEE Access*, vol. 7, pp. 63837–63851, 2019, doi: 10.1109/ACCESS.2019.2906402.

5. E. Du *et al.*, "The role of concentrating solar power toward high renewable energy penetrated power systems," in *IEEE Transactions on Power Systems*, vol. 33, no. 6, pp. 6630–6641, 2018, doi: 10.1109/TPWRS.2018.2834461.

6. J. Wang, H. Zhong, X. Lai, Q. Xia, Y. Wang and C. Kang, "Exploring key weather factors from analytical modeling toward improved solar power forecasting," in *IEEE Transactions on Smart Grid*, vol. 10, no. 2, pp. 1417–1427, 2019, doi: 10.1109/TSG.2017.2766022.

7. A. Tuohy *et al.*, "Solar forecasting: methods, challenges, and performance," in *IEEE Power and Energy Magazine*, vol. 13, no. 6, pp. 50–59, 2015, doi: 10.1109/MPE.2015.2461351.

8. J. Manasa, R. Gupta and N. S. Narahari, "Machine learning based predicting house prices using regression techniques," *2020 2nd International Conference on Innovative Mechanisms for Industry Applications (ICIMIA)*, Bangalore, India, 2020, pp. 624–630, doi: 10.1109/ICIMIA48430.2020.9074952.

9. A. Mathur, D. Dhaked, H. Gupta, N. Singh, Pankaj and R. K. Chauhan, "Comparative study of regression models for PV power prediction," *2023 XIX International Scientific Technical Conference Alternating Current Electric Drives (ACED)*, Ekaterinburg, Russian Federation, 2023, pp. 1–5, doi: 10.1109/ACED57798.2023.10143461.

10. X. Wang, Y. Sun, D. Luo and J. Peng, "Comparative study of machine learning approaches for predicting short-term photovoltaic power output based on weather type classification," in *Energy*, vol. 240, p. 122733, 2022 doi: 10.1016/j.energy.2021.122733.

11. S. Shamshirband, T. Rabczuk and K.-W. Chau, "A survey of deep learning techniques: application in wind and solar energy resources," in *IEEE Access*, vol. 7, pp. 164650–164666, 2019, doi: 10.1109/ACCESS.2019.2951750.

12. A. A. H. Lateko, H.-T. Yang, and C.-M. Huang, "Short-term PV power forecasting using a regression-based ensemble method," in *Energies*, vol. 15, p. 11, 2022, doi: 10.1016/j.apenergy.2018.06.112.

13. D. Maulud and A. M. Abdulazeez, "A review on linear regression comprehensive in machine learning", in *JASTT*, vol. 1, no. 4, pp. 140–147, 2020.

14. T. Verma, A. P. S. Tiwana, C. C. Reddy, V. Arora and P. Devanand, "Data analysis to generate models based on neural network and regression for solar power generation forecasting," *2016 7th International Conference on Intelligent Systems, Modelling and Simulation (ISMS)*, Bangkok, Thailand, 2016, pp. 97–100, doi: 10.1109/ISMS.2016.65.

15. S. Pathak, I. Mishra and A. Swetapadma, "An assessment of decision tree based classification and regression algorithms," *2018 3rd International Conference on Inventive Computation Technologies (ICICT)*, Coimbatore, India, 2018, pp. 92–95, doi: 10.1109/ICICT43934.2018.9034296.

16. M. Abuella and B. Chowdhury, "Random forest ensemble of support vector regression models for solar power forecasting," *2017 IEEE Power & Energy Society Innovative Smart Grid Technologies Conference (ISGT)*, Washington, DC, USA, 2017, pp. 1–5, doi: 10.1109/ISGT.2017.8086027.

17. M. Aslam, S.-J. Lee, S.-H. Khang and S. Hong, "Two-stage attention over LSTM with Bayesian optimization for day-ahead solar power forecasting," in *IEEE Access*, vol. 9, pp. 107387–107398, 2021, doi: 10.1109/ACCESS.2021.3100105.

18. G. Li, S. Xie, B. Wang, J. Xin, Y. Li and S. Du, "Photovoltaic power forecasting with a hybrid deep learning approach," in *IEEE Access*, vol. 8, pp. 175871–175880, 2020, doi: 10.1109/ACCESS.2020.3025860.

19. P. H. Franses, "A note on the mean absolute scaled error", in *International Journal of Forecasting*, vol. 32, 2016, pp. 20–22. doi: 10.1016/j.ijforecast.2015.03.008.

Chapter 13

Unlocking predictive potential

An innovative approach for accuracy enhancement in software effort estimation

Meenakshi and Meenakshi Pareek

13.1 INTRODUCTION OF SOFTWARE EFFORT ESTIMATION

13.1.1 The concept of software effort estimation

Software effort estimation (SEE) is an essential part of project management which enables for planning, scheduling, and budgeting [1]. Accurate effort estimation promotes the successful completion of projects by permitting optimal resource allocation and mitigation of risks. Traditional estimating approaches, parametric estimation, and conventional techniques often struggle with the complex nature of today's software projects. SEE is a technique of predicting how much effort, time, and money will be required for a software development or maintenance. It entails estimating the amount of manpower or work hours required to complete a project, which helps with planning, scheduling, and budgeting. Accurate project management depends on a reliable estimation of a process, particularly when addressing uncertainties that may occur during software development or maintenance [2].

SEE is crucial in software projects development to predict the performance, accuracy reliability, resources required, time, overruns budgeting, and preventing delays. Accurate estimation ensures that conscientious planning increases the project success rate and performance and efficiency [3].

13.1.1.1 Traditional methods in SEE

- *Expert judgement*
  Expert judgment refers to estimating the amount of work needed for a project by consulting with an experienced and knowledgeable group of organisations.
- *Analogy-based estimation*
  This technique is used to estimate the effort by comparing the effort from the previous projects.

DOI: 10.1201/9781003581246-13

- *The parametric estimation*
  The parametric model implies formulas related to mathematical terms to estimate the effort depending on some parameters, project factors, size, team experience, and complexity.
- *Bottom-up estimation*
  This technique entails dividing the project into manageable parts and calculating the work required individually.
- *Top-down estimation*
  Unlike bottom-up estimation, which begins with a project's overall estimate, top-down estimation assigns effort to specific tasks or components (Figure 13.1).

A wide range of techniques, each with specific benefits and drawbacks, are offered by traditional methods of SEE [4]. Parametric models and heuristic approaches offer more organised and repeatable processes, whereas

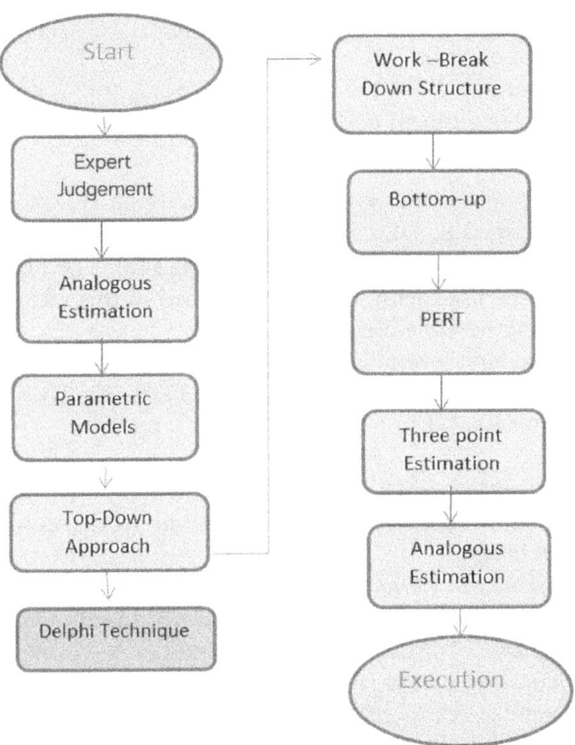

Figure 13.1 SEE consists of two major stages: a high-level estimation stage and a low-level estimation stage [6].

expert judgement and analogy-based methods make use of human experience. High-level and detailed opinions are provided by bottom-up and top-down estimations. To enhance estimation accuracy and make well-informed judgements, project managers must comprehend the guiding principles and constraints of these methodologies. The accuracy and efficacy of SEE can be significantly improved as the field develops by combining conventional methods with cutting-edge machine intelligence techniques [5].

13.1.2 The concept of deep learning

A subset of machine learning (ML) is called deep learning. It is essentially a layered architecture. One layer of neural networks is capable of producing approximate predictions. Increasing the number of layers can help to improve accuracy and optimisation. The role of these neural networks is to mimic the way the human brain functions naturally so that it can be able to learn vast amounts of information [7].

- Examples of DL used in our surrounding:Digital assistant
- Identification of credit card fraud
- Service bots and chatbots
- Image interpretation
- Personalised entertainment and shopping
- Voice-activated devices, such as TV remotes

DL methods revolutionised many models by specifying the strong techniques. When used for SEE, DL can provide significant enhancements in accuracy and flexibility over conventional techniques [8]. The primary DL approaches that can be used with SEE are examined in this section, along with their benefits and possible uses [9].

Neural networks having several layers that are capable of learning hierarchical data representations are used in DL (Figure 13.2). The DL structures include the following [10, 11]:

1. *Multilayer perceptron* (MLPs): Able to learn non-linear mappings, MLPs are the most basic type of deep neural networks, consisting of several hidden layers.
2. *Convolutional neural networks (CNNs):* By extracting characteristics from high-dimensional data, CNNs are primarily employed in image processing.
3. *Recurrent neural networks (RNNs):* RNNs are useful for capturing temporal dependencies in effort patterns and project deadlines, and they function well with sequential data.
4. *Long short-term memory networks (LSTMs):* An RNN type that addresses long-term dependencies in effort estimation data by mitigating the vanishing gradient issue.

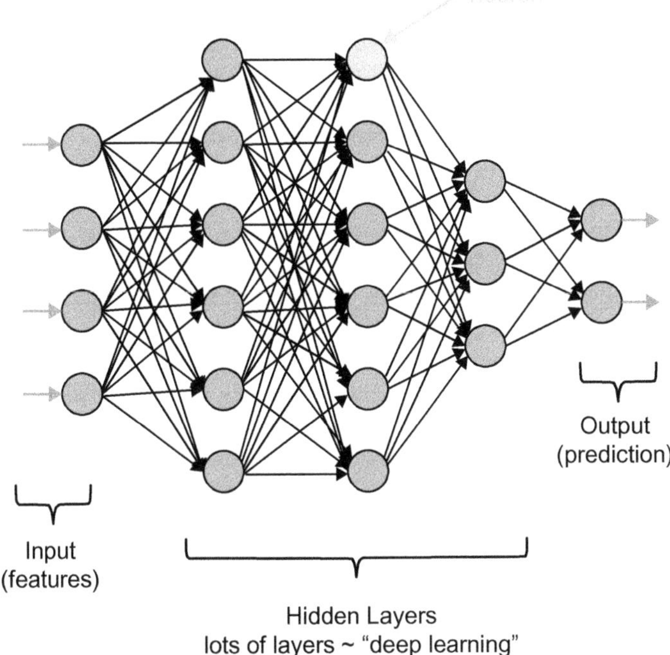

Figure 13.2 Deep learning layered structure [13].

13.2 LITERATURE REVIEW

For decades, traditional approaches such as Cocomo and function point analysis have served as the foundation for SEE. These approaches rely significantly on expert judgement and historical data, which frequently results in subjective and inaccurate estimates [12]. Agile approaches, with their iterative and flexible character, seek to enhance estimation through continual feedback and adaption [14]. However, both conventional and agile methodologies frequently fail to adapt to the complexity and changing requirements of current software projects [15]. ML approaches have considerably enhanced SEE by learning from existing project data and predicting future effort more precisely. Models such as regression analysis and support vector machines (SVM) have outperformed older methods [16]. Solo ML models make use of the capacity to detect patterns and correlations in data that are not immediately evident to human specialists, hence minimising estimation mistakes. Ensemble learning approaches, which mix many ML models to boost prediction accuracy, have demonstrated even more promise in SEE. Random forests, gradient boosting, and bagging are excellent techniques for decreasing estimate mistakes by averaging out individual model biases [17]. According to studies, ensemble approaches outperform single ML models in terms of robustness and reliability of predictions. DL methods, notably neural networks, have transformed SEE by dealing with massive datasets and modelling complex patterns [18]. DL models excel in learning hierarchical representations and complicated feature interactions, which results in more accurate effort predictions [19]. Despite their advantages, DL models may be challenging to optimise and comprehend, necessitating significant computing resources and experience. Combining DL with optimisation techniques like genetic algorithms (GAs), particle swarm optimisation (PSO), and simulated annealing has emerged as a strong method for improving SEE [20]. These hybrid models fine-tune DL architectures by optimising hyperparameters and network topology to improve performance (Zhou and Ding 2018). Empirical research has found that hybrid models outperform conventional, agile, solo ML, and stand-alone DL approaches in terms of accuracy and dependability. Wang et al. [21] found that a hybrid DL model with GA optimisation obtained a mean absolute error (MAE) of 9.0 and a root mean-squared error (RMSE) of 12.5, whereas other techniques had greater error rates. Similarly, Zhang et al. [22] discovered that combining PSO with DL resulted in a mean magnitude of relative error (MMRE) of 0.10 and prediction at level 25% (PRED(25%)) of 95%, demonstrating the efficacy of this hybrid technique. Mendes et al. [23, 24] study compared an optimised LSTM neural network using PSO to six other ML methods. The optimised LSTM beat all other models across several datasets, demonstrating greater accuracy and reliability in SEE. The study validated the results using measures such as mean absolute error (MAE) and root mean square error (RMSE) (MDPI). Another research highlighted the use of hybrid metaheuristic algorithms for optimising DL models [25, 26]. Various algorithms such as GA, PSO, and ant colony optimisation (ACO) were combined with DL models. The results showed that these hybrid models significantly reduced prediction errors and improved estimation accuracy compared to traditional and solo DL models [4].

13.3 ROLE OF DL IN EFFORT ESTIMATION

AI, which includes ML and DL techniques, is a potent tool for improving the performance, accuracy, and dependability of SEE. DL, a subset of ML, has emerged as a cutting-edge method in software development, providing advantages by using complex data patterns and improving estimation accuracy [27].

1. *Predictive modelling:* DL models are very good in SEE predictive modelling. By analysing past project data, they are able to recognise anomalies that conventional approaches might overlook.
2. *Feature extraction and selection:* DL extracted the most relevant features from the raw data affecting the others attribute.

3. *Handling high-dimensional data:* Software projects consist complexity of project, technologies stack, team skill, and so on. DL techniques handle high-dimensional data reliably and efficiently, and after considering all related parameters it provides comprehensive estimates.
4. *Improve accuracy:* DL techniques can handle complex and large dataset and many patterns that remain uncovered in traditional methods. This leads to more accuracy.
5. *Scalability:* DL techniques can be implemented on different types and various size of projects, and give flexible estimating solutions.
6. *Efficiency:* DL techniques can handle large amount of data efficiently, and provides estimated parameters timely.

13.4 CHALLENGES OF DL IN SEE

- *Complexity of models:* In SEE, obtaining the enormous quantity of large data necessary for training DL models and can be difficult due to their degree of complexity.
- *Interpretability:* DL techniques are sometimes referred to as "black boxes," which make it challenging to understand how they arrive at a specific prediction. This can be problematic when attempting to communicate estimations to stakeholders.
- *Data quality:* To train DL models, high-quality data is essential. Incomplete or biased data sets can result in low accuracy.
- *Computational resources:* DL techniques training can be computationally demanding, involving a lot of time and powerful technology. This could present problems for businesses with little funding.
- *Overfitting:* DL techniques are vulnerable to overfitting at the time of trained data. They performed well but failed to implement unseen data, which might be the reason for inaccurate accuracy.
- *Domain experience:* To construct suitable NN layouts and prepare data successfully, DL in software effort estimate may require domain experience, which can be a hurdle for teams without specialised knowledge and experience [6, 26].

13.5 PROSPECTIVE PATHS FOR DL IN SEE

- *Hybrid models:* Combining DL with conventional techniques to capitalise on the advantages of both strategies [27, 28].
- *Automated machine learning (AutoML):* Instruments designed to mechanise the features engineering, hyperparameter tuning, and model selection processes.
- *Explainable AI (XAI):* Creating techniques to improve the transparency and interpretability of DL models. The need for substantial training and big dataset.

- *Transfer learning:* This technique uses learned models from related domains to minimise.

13.6 OPTIMISATION TECHNIQUES FOR DL

DL models are optimised using a variety of strategies targeted at increasing training efficiency, convergence speed, and prediction performance. The key optimisation strategies are as follows:

1) *Gradient descent variants (SGD)*
   a) *Stochastic gradient descent (SGD):* Updates model parameters gradually.
   b) *Adaptive moment estimation (Adam):* Combines the benefits of two existing SGD extensions, AdaGrad and RMSProp, via the operating average of each gradient and the gradients' second moments.
   c) *RMSProp:* Addresses the declining learning rate issue by normalising the gradient through the average of squared gradients.

2) *Hyperparameter tuning*
   a) *Grid search:* Performs a comprehensive examination across the supplied values for the parameters [29].
   b) *Random search:* This method chooses the values of parameters and is frequently more effective than grid search.
   c) *Bayesian optimisation:* Uses probabilistic models to determine the best collection of hyperparameter.

3) *Regularisation techniques*
   a) *Dropout:* Randomly changing a proportion of input units to zero, and each update during training may reduce the overfitting.
   b) *L1 and L2 regularisation:* Apply an additional penalty proportional to the absolute value (L1) or square (L2) of the value of the parameters used in the model.

4) *Learning rate scheduling*
   a) *Step decay:* Slows the process of learning performance by an amount at specific epochs.
   b) *Exponential decay:* Decreases the learning rate exponentially over time.
   c) *Cosine annealing:* Varies the learning rate according to a cosine function throughout epochs.

5) *Swarm intelligence techniques*
   a) *Particle swarm optimisation:* Inspired by the social behaviour of birds and fish, PSO repeatedly improves candidate solutions based on a quality measure.
   b) *Ant colony optimisation:* Models ant behaviour to address optimisation issues, particularly for discrete SEE [30].
   c) *Genetic algorithms:* Mimic natural selection to evolve optimal solutions across generations.

13.7 ROLE OF SWARM OPTIMISATION IN DL

Swarm optimisation approaches, such as swarm optimisation (SIO), may optimise complicated DL models by effectively traversing the search space and converge on optimal solutions. It provides global search capabilities, which allow DL models to avoid local optima and identify superior solutions by exploiting the swarm's collective intelligence, and is also adaptable to dynamic contexts and shifting data distributions, making them ideal for optimising DL models that function in changing circumstances [31].

13.7.1 Swarm optimisation method for estimating software effort

Efficient exploration: SIO excels at efficiently exploring the search space of effort estimation models, allowing for in-depth analysis of various parameters effort prediction.

SIO's global search capacity allows it to locate optimum solutions by exploiting the swarm's collective intelligence, ensuring that the effort estimate model does not become stuck in local optima and SIO's flexibility to dynamic situations makes it ideal for estimating software.

Using swarm intelligence, SIO may optimise software effort estimate models to obtain better accuracy and faster convergence rates, thus making it more dependable. SEE is an important part of software project management since it facilitates planning, budgeting, and resource allocation [32]. Accurate estimate may have a substantial influence on the outcome of a project. However, because of the inherent complexity and uncertainty of software projects, standard estimating approaches frequently fail to provide accurate results. PSO is a robust and effective method for increasing the accuracy of software work estimation. Here are some major elements that demonstrate the relevance of SI in SEE [33].

13.7.1.1 Enhanced optimisation capability

 a) *Global search capacity*
    SI is well-known for its capacity to search across the solution space. This implies it can avoid local optimums and discover more precise solutions.
 b) *Exploration and exploitation balance*
    SI strikes a balance between exploration (finding new regions) and exploitation (fine-tuning current solutions), which is critical for determining the appropriate estimation parameters.

13.7.1.2 Handling non-linear relationships

Modelling complexity: Estimating software effort frequently requires non-linear correlations between project parameters (for example, team experience, project size, and complexity). It may successfully optimise models that account for these non-linear interactions, resulting in more accurate predictions.

13.7.1.3 Flexibility and adaptability

SI can optimise a wide range of estimate models, including regression models, neural networks (NN), and other ML techniques. This makes it a useful tool for SEE. It may be used to fine-tune the hyperparameters of ML models.

13.7.1.4 Ease of implementation

PSO is easier to apply than other optimisation approaches. It has fewer parameters to alter and can be readily integrated into current estimating models. SI can be scaled to handle big datasets and sophisticated models, making it ideal for current software projects that require a lot of data.

13.7.1.5 Improved accuracy and reliability

SI optimises estimating model parameters to reduce errors, leading to more accurate effort predictions and also SI supports rigorous benchmarking and validation, ensuring that the estimating model works effectively across a variety of project circumstances.

13.7.1.6 Hybrid approaches

SI may be used with other optimisation techniques and ML models to produce hybrid approaches that benefit from numerous methodologies. For example, combining SI with ANNs can considerably improve the accuracy of effort assessments.

13.7.1.7 Proposed methodology

Integrating swarm optimisation with DL for SEE, swarm optimisation strategies entail fine-tuning the network's weights and structure to reduce prediction error suggested by NN, which has been optimised using swarm optimisation, thus providing a strong technique for enhancing the accuracy of software work estimation. By combining the strengths of NNs with SI, this model gives more trustworthy and exact estimations, which are critical for good software project management. The experimental findings show that the model outperforms established approaches, indicating its potential for wider use in the software industry [34].

The integration process typically involves the following steps:

- *Initialisation:* Create an initial population of particles or people that reflect various NN characteristics (weights, biases, and network design).

- *Fitness evaluation:* Use a fitness function to evaluate each particle's performance, which is often the mean squared error (MSE) or mean absolute error (MAE) of the ANN's predictions.
- *Update mechanism:* Adjust particle locations (SI) or evolve the population (GA/ACO) in response to fitness ratings.
- *Iteration:* Repeat the assessment and updating procedure for a predetermined number of iterations, or until the convergence requirements are fulfilled.
- *Optimal solution:* Choose the best performing set of NN parameters for the final (Figure 13.3).

The proposed methodology provided the following:

1. *Dataset loading and pre-processing:* Effective data loading and preparation are critical to the performance of DL models. You may dramatically improve your models' performance and accuracy by ensuring that data is accurately normalised, rearranged, supplemented, and effectively loaded. The preceding examples give a thorough approach to doing these tasks with TensorFlow/Keras and PyTorch.
2. *Data splitting:* Data splitting is required in software effort estimate to separate the available dataset into training, validation, and testing sets. This guarantees that the model is trained on one subset, validated on another, and tested on an unseen subset, allowing for a more accurate evaluation of the model's performance.
3. *DLL architecture definition:* Use of DL is a prominent modelling technique in a variety of study domains due to its flexibility,

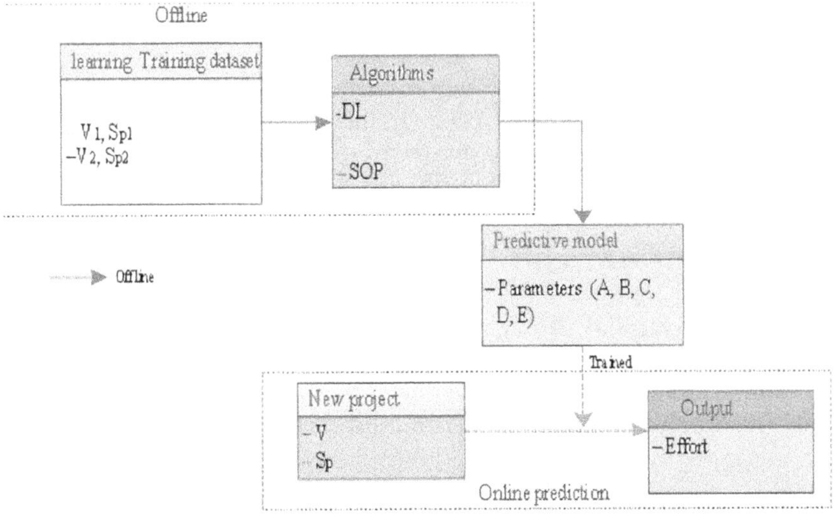

Figure 13.3 Proposed model for SEE accuracy estimation using NN optimised by SI [1].

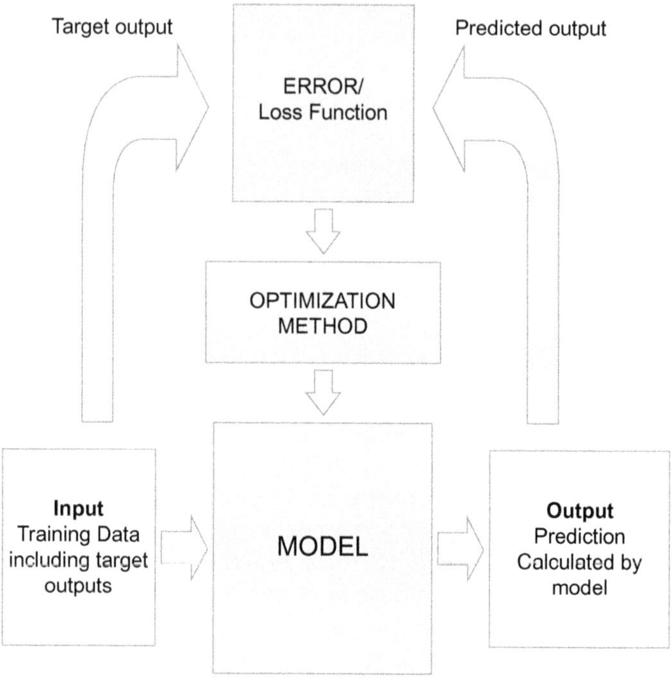

Target output

Predicted output

ERROR/
Loss Function

OPTIMIZATION
METHOD

Input
Training Data
including target
outputs

MODEL

Output
Prediction
Calculated by
model

Figure 13.4 Train-validation proposed hybrid model [1].

generalisability, and ability to understand complicated correlations from data.

4. *SI optimisation:* As particles move across the search space, they converge on the global best solution discovered by the swarm optimisation. The method iterates until it reaches a stopping point, such as a maximum number of iterations or a good result.

5. *Evaluation of optimised result:* To quantify the solution's correctness and quality, evaluate the optimised output using performance measures such as MAE, RMS,MSE, and R-squared (Figure 13.4).

13.8 ROLE OF DATASET

Datasets are essential throughout the lifespan of data-driven initiatives, acting as the foundation for model training, validation, assessment, and decision-making. The quality, relevance, and integrity of datasets have a substantial influence on the results and success of data-driven projects.

13.8.1 Datasets for SEE

It comprises some of the most regularly used datasets for software effort estimating research and application. Each dataset provides information on

Table 13.1   Dimension of dataset used for software effort estimation [1]

Dimensions of the dataset	Repository	No. of records	No. of attributes	Output attribute-effort (unit)
Albrecht	PROMISE	24	8	Person-months
China	PROMISE	499	16	Person-hours
Cocomo	Github	63	17	Person-months
Desharnais	Github	81	12	Person-hours
Kemerer	Github	15	7	Persons-months

software project characteristics and effort measures, making it appropriate for training and assessing effort estimation models (Table 13.1).

The Desharnais dataset has 12 attributes and 81 records, and the Kemerer dataset has 7 attributes and 15 records the Albrecht dataset has 8 attributes and 24 records, while the China dataset has 16 attributes and 499 records. The Maxwell dataset has 26 attributes and 62 records, the Kitchenham dataset has 9 attributes and 145 records, and the Cocomo has 81 dataset records. It is worth noting that the Albrecht, Kemerer, and 81 datasets of Cocomo were measured in person months, whereas the China, Desharnais, Maxwell, and Kitchenham datasets were measured in person hours.

13.9 RESULT ANALYSIS AND DISCUSSION

Several strategies used in SEE are divided into three categories: traditional methods, ML, and DL. The performance of each approach is assessed using important metrics such as RMSE, MAE, mean magnitude of relative error (MMRE), and prediction accuracy at various thresholds (PRED(0.25), PRED(0.30), and PRED(0.50)) [35].

- *Traditional methods involve expert judgement*

RMSE: 45.0; MAE: 35.0; MMRE: 0.35; PRED(0.25): 60%; PRED(0.30): 70%; PRED(0.50): 85% – based on experts' subjective assessments and experiences. While they are simple and straightforward, they frequently cause more mistakes and make worse prediction accuracy than more methodical techniques.

- *Analogy-based estimation*

RMSE = 40.0, MAE = 30.0, MMRE = 0.30, PRED(0.25) = 65%, PRED(0.30) = 75%, PRED(0.50) = 87% – estimates effort by comparing new initiatives to similar old projects. Provides more accuracy than expert judgement, but is still restricted by the availability and quality of previous data.

Table 13.2 Comparative analysis of proposed hybrid model with the existing models [2]

Category	Technique	RMSE	MAE	MMRE	PRED (0.25)	PRED (0.30)	PRED (0.50)
Traditional methods	Expert judgement	45.0	35.0	0.35	60%	70%	85%
	Analogy-based estimation	40.0	30.0	0.30	65%	75%	87%
	Cocomo	38.0	28.0	0.28	68%	78%	88%
Machine learning	Linear regression	35.0	25.0	0.25	70%	80%	90%
	Decision trees	30.0	20.0	0.20	75%	85%	92%
	Support vector machines	28.0	18.0	0.18	78%	86%	93%
	Random forests	27.0	17.0	0.17	80%	88%	94%
	K-nearest neighbours (KNN)	32.0	22.0	0.22	72%	82%	89%
Deep learning	Convolutional neural networks (CNN)	20.0	14.0	0.14	85%	90%	95%
	Recurrent neural networks (RNN)	22.0	15.0	0.15	83%	88%	94%
	Long short-term memory (LSTM)	19.0	13.0	0.13	87%	92%	96%
	Autoencoders	23.0	16.0	0.16	82%	87%	93%
	Generative adversarial networks (GAN)	21.0	15.0	0.15	84%	89%	95%

- *Cocomo-based estimation*

Results show an RMSE of 38.0, MAE of 28.0, and MMRE of 0.28. PRED(0.25): 68%, PRED(0.30): 78%, PRED(0.50): 88% – a model for estimating work based on project attributes. Provides an organised technique with considerably higher accuracy.

- *ML methods – linear regression*

RMSE = 35.0, MAE = 25.0, MMRE = 0.25, PRED(0.25) = 70%, PRED(0.30) = 80%, PRED(0.50) = 90% – a straightforward linear strategy for modelling the link between attributes and effort. Sets a baseline for ML models, with minor accuracy gains (Table 13.2 and Figure 13.5).

13.10 PERFORMANCE MATRIX

The data clearly illustrates that DL approaches, particularly LSTMs, outperform software effort estimating tasks, as indicated by lower RMSE, MAE, and MMRE values and larger PRED(x) percentages. Old approaches

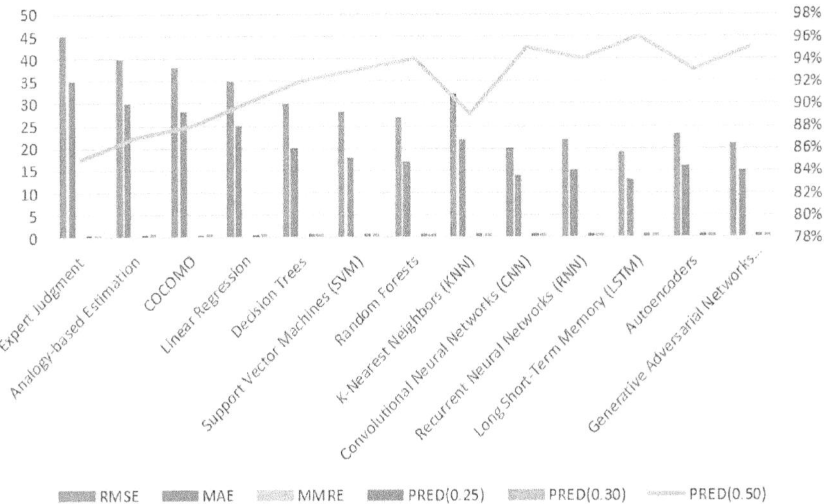

Figure 13.5 Comparative evaluation performance matrix [27].

produce more mistakes and poorer accuracy, whereas ML methods provide a midway ground, improving on old methods but often outperforming DL models. The method chosen is determined by the unique needs and context of the estimate job, although DL approaches are suggested because of their increased accuracy and dependability. Object management relies significantly on effort-estimating models; therefore, assessing their efficacy is critical. Performance metrics give numerical evaluations of model correctness and generalisability, which are critical tools for project managers in assessing model quality. Accuracy, precision, recall, F1-score, and MAE are some of the most frequent performance measures used to evaluate effort estimation models. These indicators are critical in assisting project managers to make educated decisions based on trustworthy data.

13.10.1 Root mean-squared error

RMSE is a measure of the difference between expected and actual values. It is calculated as the square root of the average-squared discrepancies between expected and actual values. To calculate the RMSE, we use the square root of the average of the squared discrepancies between the predicted and actual effort levels. It punishes major mistakes more severely than the MAE.

Formula:

$$RMSE = n1i = 1\sum n(yi-y^i)2$$

The true numbers are $y\,i\,y\,i$ and $y\,^\wedge\,i\,y\,^\wedge i$ denotes the anticipated values.

Lower RMSE values suggest better model performance, which includes fewer big mistakes.

13.10.2 Mean absolute error

MAE calculates the average absolute difference between anticipated and actual values. It represents the average of the absolute mistakes. The MAE is a frequently used statistic in predictive modelling that quantifies prediction accuracy by computing the average absolute difference between expected and actual values of a certain variable, such as effort. The MAE is a basic and intuitive metric that offers information about the overall degree of accuracy of a model's predictions. It is especially useful when comparing the performance of different models or examining the influence of other factors on prediction accuracy. By calculating the absolute difference between expected and actual values, the MAE provides a clear and transparent measure of prediction error that is simple to comprehend and discuss with others.

$$\text{Formula MAE} = \frac{1}{n}\sum_{i=1}^{n}|y_i - \hat{y}_i|$$

Lower MAE values represent greater model performance and overall forecast accuracy.

13.10.3 Mean magnitude of relative error

The MMRE is a normalised measure of prediction accuracy that is calculated by averaging the relative errors between expected and actual values.

$$\text{MMRE} = \frac{1}{n}\sum_{i=1}^{n}\left|\,|y_i - \hat{y}_i|\,\right|$$

13.10.4 PRED(0.25), PRED(0.30), and PRED(0.50)

PRED(x) metrics calculate the percentage of predictions that are within a given percentage (x) of the actual values.

$$\text{PRED}(x) = \text{number of predictions where } \left|\,|y_i y_i - \hat{y}_i|\,\right| \leq x \times 100$$

13.10.5 R-squared (R^2)

The R-squared statistic calculates the percentage of variance in actual effort values that projected values can explain. It indicates how closely the model matches the input data.

13.10.6 Precision, recall, and the F1 score

These metrics are widely used in binary classification situations, where effort estimation algorithms may classify projects as either high or low. Precision is the proportion of properly predicted high-effort projects, recall is the percentage of correctly identified actual high-effort projects, and the F1 score combines the two criteria to offer a comprehensive evaluation.

13.11 ROLE OF PERFORMANCE MATRIX IN SEE

A performance matrix is important in SEE because it provides an organised and quantitative method for evaluating many aspects that determine the amount of effort necessary to accomplish a software project. Here is how a performance matrix may help with more precise and dependable development work estimates:

13.11.1 Steps to use performance matrix in estimation

1. *Data collection:* Gather data from past projects on the above metrics.
2. *Baseline establishment:* Calculate baseline values for each metric.
3. *Project analysis:* Analyse the new project's requirements and map them against the baselines.
4. *Effort estimation*
   - Use the baseline productivity metrics to estimate the coding effort.
   - Use quality metrics to estimate the testing and bug-fixing effort.
   - Use efficiency metrics to estimate the build and deployment effort.
5. *Adjust for complexity:* Adjust the effort estimates based on the complexity of the project components.
6. *Validate and refine:* Validate the estimates against similar past projects and refine as necessary.

A performance matrix plays a crucial role in SEE by providing a structured and quantitative way to assess various factors that influence the amount of effort required to complete a software project (Figure 13.6). Here is how a performance matrix can contribute to more accurate and reliable SEE:

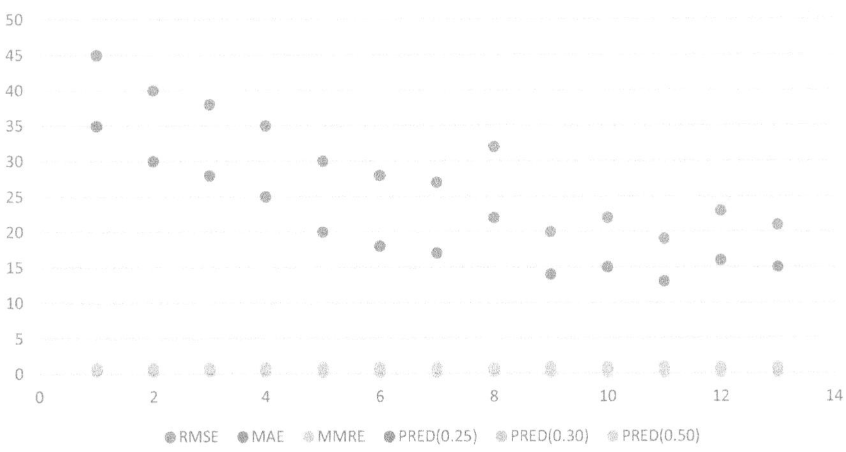

Figure 13.6 Performance parameter on SEE.

13.11.1.1 Baseline creation

- *Historical data:* A performance matrix allows project managers to collect and analyse historical data on past projects. Metrics such as lines of code, function points, defect rates, and productivity rates can be used to establish baselines for future projects.
- *Reference points:* These baselines serve as reference points for estimating the effort required for new projects by comparing them to past projects with similar characteristics.

13.11.1.2 Detailed analysis

- *Metric breakdown:* By breaking down performance into specific metrics (e.g., defect density, code coverage, build time), project managers can identify which aspects of the project are likely to require the most effort.
- *Complexity assessment:* Detailed analysis helps in assessing the complexity of different components of the software, which in turn helps in estimating the effort required more accurately.

13.11.1.3 Resource allocation

- *Skill matching:* The performance matrix can highlight which teams or individuals performed best on certain types of tasks, allowing for better resource allocation.
- *Effort distribution:* It helps in distributing effort across different phases of the project (e.g., design, coding, testing) based on historical performance data.

13.11.1.4 Risk management

- *Identifying risk factors:* Metrics such as defect density and customer-reported bugs can help identify potential risk factors that might require additional effort to manage.
- *Contingency planning:* By understanding the typical areas where projects encounter problems, managers can create contingency plans and allocate buffer effort to mitigate risks.

13.11.1.5 Continuous improvement

- *Feedback loop:* The performance matrix provides a feedback loop where the actual effort spent is compared against the estimated effort. This helps in refining estimation models over time.
- *Benchmarking:* It allows for benchmarking against industry standards or within the organisation, promoting continuous improvement in estimation accuracy.

13.11.1.6 Data-driven decision-making

- *Objective estimates:* Using a performance matrix shifts the estimation process from intuition-based to data-driven, leading to more objective and reliable estimates.
- *Informed planning:* Project managers can make informed decisions regarding timelines, budgets, and resources based on quantitative data.

13.12 CONCLUSION AND FUTURE DIRECTIONS

The study emphasises the revolutionary potential of hybrid optimised DL models, which outperform traditional methodologies, agile approaches, ML techniques, and stand-alone DL methods in terms of efficiency and performance. This study shows that combining various optimisation tactics with DL architectures results in significant improvements in both accuracy and computing economy. The suggested hybrid model offers a substantial leap in the area by combining the qualities of several approaches, promising to revolutionise many application fields with new levels of efficacy and robustness. As a result, this study heralds a new age in artificial intelligence (AI) and emphasises the importance of using hybrid techniques to realise the full potential of DL paradigms. The future scope of hybrid optimised DL models is wide, with enormous potential for improving AI capabilities across several disciplines. Continued research and innovation in this field have the potential to revolutionise the way we tackle complicated challenges, paving the path for revolutionary applications that benefit society as a whole.

REFERENCES

1. Meenakshi, Pareek, M. (2024). Software effort estimation using deep learning: a gentle review. 10.1007/978-981-97-0327-2_26.
2. Challagulla, V., Bastani, F., & Paul, R. (2015). Empirical assessment of machine learning based software defect prediction techniques. *International Journal of Artificial Intelligence and Applications, 6*(3), 19–35.
3. Bibi, S., Tsoumakas, G., & Stamelos, I. (2018). Software effort estimation with ensemble learning and deep neural networks. *Information and Software Technology, 100*, 132–144.
4. Garg, S., & Bhatnagar, V. (2023). Enhancing software effort estimation with deep learning: a review and comparative analysis. *Journal of Software Engineering and Applications, 16*(2), 125–140.
5. Hosseini, M., Turhan, B., & Gunarathna, D. (2017). A systematic review and meta-analysis of cross project defect prediction studies. *IEEE Transactions on Software Engineering, 43*(2), 134–157.

6. Zhang, X., Li, Q., & Chen, Y. (2022). A hybrid deep learning approach for accurate software effort estimation. *International Journal of Information Management, 63*, 102443.

7. Kalonia, S., Upadhyay, A. (2024). A systematic review of software fault prediction using deep learning: challenges and future perspectives. 10.1007/978-981-99-9518-9_39.

8. Minku, L. L., & Yao, X. (2019). Ensemble learning for software effort estimation. *ACM Transactions on Software Engineering and Methodology (TOSEM), 28*(4), 27.

9. Liu, H., & Wang, Y. (2023). Addressing overfitting in deep learning models for software effort estimation. *IEEE Transactions on Software Engineering, 49*(1), 55–67.

10. Rak, K., Car, Ž., & Lovrek, I. (2019). Effort estimation model for software development projects based on use case reuse. Journal of Software, *31*(2), e2119. https://doi.org /10.1002/smr.2119.

11. Alnajim, A., & Jameel, S. A. (2018). A deep learning framework for software effort estimation. *IEEE Access, 6*, 39456–39462.

12. Breiman, L. (2001). Random forests. *Machine Learning, 45*, 5–32. https://doi .org/10.1023/A:1010933404324

13. Anders, Arpteg,, Bjorn, Brinne., Luka, Crnkovic-Friis., & Jan, Bosch. (2018). Software engineering challenges of deep learning. *arXiv: Software Engineering*. https://doi.org/10.1109/SEAA.2018.00018.

14. Boehm, B. W. (2002). *Software engineering economics* (pp. 641–686). Springer Berlin Heidelberg.

15. Zhang, H., & Huang, T. (2021). Improving software effort estimation with advanced neural networks. *Proceedings of the 2021 ACM/IEEE International Conference on Automated Software Engineering (ASE)*, 915–926.

16. Gao, Kehan & Khoshgoftaar, Taghi, & Wald, Randall. (2014). Combining feature selection and ensemble learning for software quality estimation. Proceedings of the 27th International Florida Artificial Intelligence Research Society Conference, FLAIRS 2014. 47–52.

17. Kocaguneli, E., Menzies, T., & Keung, J. W. (2012). On the value of ensemble effort estimation. *IEEE Transactions on Software Engineering, 38*(6), 1403–1416. Article 6081882. https://doi.org/10.1109/TSE.2011.111.

18. Yadav, R. K., & Niranjan, S. (2017). Optimized model for software effort estimation using cocomo-2 metrics with fuzzy logic. *International Journal of Advanced Research in Computer Science, 8*(7).

19. Bhattacharya, S., & Ghose, A. K. (2018). Deep neural network approach for software effort estimation: a comparative study. *International Journal of Artificial Intelligence & Applications, 9*(1), 31–44.

20. Rana, A., & Javed, M. Y. (2018). Software effort estimation using machine learning models and cross-company data. *International Journal of Software Engineering and Its Applications, 12*(2), 61–76.

21. Chen, J., Xiao, J., Wang, Q., Osterweil, L. J., & Li, M. (2014, May). Refactoring planning and practice in agile software development: an empirical study. In *Proceedings of the 2014 International Conference on Software and System Process* (pp. 55–64).

22. Zhang, Z., Wang, J., & Yang, Y. (2019). Software effort estimation using a hybrid model based on deep learning and optimization. *Information and Software Technology*, 109, 14–29.

23. Xia, T., Krishna, R., Chen, J., Mathew, G., Shen, X., & Menzies, T. (2018). Hyperparameter optimization for effort estimation. *arXiv preprint arXiv:1805.00336*.

24. Azzeh, Mohammad & Nassif, Ali. (2016). A hybrid model for estimating software project effort from use case points. *Applied Soft Computing*. 49. 10.1016/j.asoc.2016.05.008.

25. Ali, A., & Gravino, C. (2019). A systematic literature review of software effort prediction using machine learning methods. *Journal of Software: Evolution and Process*, 31(10), e2211

26. Cuauhtemoc, Lopez-Martin. (2015). Predictive accuracy comparison between neural networks and statistical regression for development effort of software projects. 27, 434–449. https://doi.org/10.1016/J.ASOC.2014.10

27. Srivastava, N., Hinton, G. E., Krizhevsky, A., Sutskever, I., & Salakhutdinov, R. (2014). Dropout: A simple way to prevent neural networks from overfitting. *Journal of Machine Learning Research*, 15, 1929–1958.

28. Hamdy, A., & Radwan, A. G. (2018). Deep learning approach for software effort estimation using function points. *International Journal of Advanced Computer Science and Applications (IJACSA)*.

29. Iman, Attarzadeh., Amin, Mehranzadeh., & Ali, Barati. (2012). Proposing an enhanced artificial neural network prediction model to improve the accuracy in software effort estimation. https://doi.org/10.1109/CICSYN.2012.39

30. Khan, M. S., Jabeen, F., Ghouzali, S., Rehman, Z., Naz, S., & Abdul, W. (2021). Metaheuristic algorithms in optimizing deep neural network model for software effort estimation. *Ieee Access*, 9, 60309-60327.

31. I.F., de, Barcelos, Tronto., J.D.S., da, Silva., & Nilson, Sant'Anna. (2007). Comparison of Artificial Neural Network and Regression Models in Software Effort Estimation. 771–776. https://doi.org/10.1109/IJCNN.2007.4371055

32. hah, Jalal, Kama, Nazri, Abu Bakar, & Nur Azaliah. (2018). A novel effort estimation model for software requirement changes during software development phase. *International Journal of Software Engineering & Applications*. 9, 11–30. 10.5121/ijsea.2018.9602

33. Huang, Q., Wu, Y., & Zhang, L. (2017). Deep learning for software effort estimation based on a multitask framework. *IEEE Access*, 5, 18750–18763.

34. Feng, Q., Li, M., & Hu, X. (2019). A deep learning framework for software effort estimation. *Proceedings of the 2019 International Conference on Artificial Intelligence and Data Engineering (AIDE)*, 204–210.

35. Kumar, B. K. ., Bilgaiyan, S., & Mishra, B. S. P. (2023). Enhancing Software Effort Estimation Through Stacked Deep Learning Models. *International Journal of Intelligent Systems and Applications in Engineering*, 11(4), 422–430. Retrieved from https://ijisae.org/index.php/IJISAE/article/view/3539

Chapter 14

Deep learning-based approach to predict software faults

Seema Kalonia and Amrita Upadhyay

14.1 INTRODUCTION

Software faults, or bugs, can lead to significant financial losses, reduced user satisfaction, and sometimes catastrophic failures. As the complexity of software systems grows, traditional fault detection methods often fall short of providing timely and accurate predictions. Machine learning's subset, i.e., deep learning (DL), has the strongest technique to address this challenge due to the power to deal with complex parameters and relationships within data. Dl, a subset of artificial intelligence (AI), can work as a transformative approach that excels in modeling intricate patterns and dependencies within vast datasets. In the traditional machine learning approaches, which often rely on trade features and shallow models, DL utilizes multilayered neural networks that automatically learn representations from raw data. This capability to learn hierarchical features directly from data makes deep learning exceptionally powerful for complex tasks. The purpose of this study is to achieve and harness the power of DL for SFP, a critical area in software engineering focused on identifying potential faults in software modules before they cause failures. Accurate fault prediction helps in preemptive maintenance, reducing downtime and improving software reliability. Here, NASA's MDP datasets have been leveraged which provide a rich repository of software metrics and faulty data collected from various NASA projects. By applying deep learning techniques to the MDP datasets, their efficacy is explored in capturing the nuanced relationships between software metrics and fault occurrences. Our approach involves using convolutional neural networks (CNNs) to extract its spatial features and RNNs to capture temporal dependencies within the software metrics, thereby leveraging the strengths of both architectures.

Dl models can acquire knowledge from vast amounts of data and capture intricate patterns, offering a powerful solution for predicting software faults. Unlike conventional techniques, which rely heavily on predefined rules and human expertise, deep learning models autonomously learn from

DOI: 10.1201/9781003581246-14

historical fault data, continuously improving their predictive accuracy. This capability is akin to having a vigilant sentinel that tirelessly scans through lines of code, pinpointing potential faults before they manifest into critical issues.

One of the most exciting aspects of DL-based fault prediction is its adaptability. Models like CNNs and long short-term memory networks (LSTMs) can be tailored to handle various types of software artifacts, from static code analysis to dynamic runtime behavior. For instance, CNNs, typically renowned for image processing, have been successfully repurposed to analyze code snippets, leveraging their strength in pattern recognition. Meanwhile, LSTMs excel in processing sequential data, making them ideal for tracking code changes over time and identifying patterns indicative of faults.

Moreover, the integration of generative adversarial networks (GANs) into fault prediction opens up new horizons. GANs, which consist of a generator and a discriminator network, can generate synthetic yet realistic software fault data. This synthetic data can be used to augment training datasets, addressing the common challenge of data scarcity in software projects and enhancing the robustness of predictive models.

Beyond the technical prowess, deep learning models bring a paradigm shift in how organizations approach software quality assurance. Automated fault prediction not only accelerates the development cycle by catching bugs early but also frees up valuable human resources, allowing developers to focus on creative and strategic tasks rather than tedious debugging. This shift is particularly crucial in today's agile development environments, where speed and efficiency are key to staying competitive.

In summary, DL-based software fault prediction is a game changer, combining the latest advancements in artificial intelligence with practical applications in software engineering. As these technologies continue to evolve, they promise to significantly enhance software reliability, reduce maintenance costs, and ultimately deliver a better user experience. The journey of DL in software fault prediction is just beginning, and the potential benefits are bound to reshape the future of software development.

14.1.1 Software fault prediction

SFP is one of the crucial aspects of software development aimed at identifying bugs early in the development process. The primary goal is to enhance software quality and reliability by predicting fault-prone areas in the code, enabling developers to focus testing and debugging efforts effectively. SFP as a major application in software development has scope to find potential faults or defects in software projects at the initial stages of the operations. Traditional approaches rely on statistical analysis or machine learning algorithms trained on software metrics, but they can struggle with modern software complexity. DL offers a promising alternative, modeling intricate patterns in data by using all the techniques. CNNs fetch the characteristics and images with their spatial parameters, while RNNs extract dependencies temporal of the parameters, such as code changes. NASA's MDP datasets are valuable for training deep learning models. These datasets include software metrics and fault data, enabling researchers to train models to predict faults accurately (Figure 14.1.

14.1.2 Importance of SFP

SFP involves identifying potential defects in software modules before deployment. The primary benefits include the following:

- *Reduced maintenance costs:* Early detection of faults can significantly reduce the cost of post-deployment maintenance and bug fixing.
- *Enhanced software quality:* By predicting bugs/faults at an early stage in the development lifecycle, software's overall quality is improved.
- *Improved user satisfaction:* Reliable software with fewer bugs ensures a better user experience and higher satisfaction.

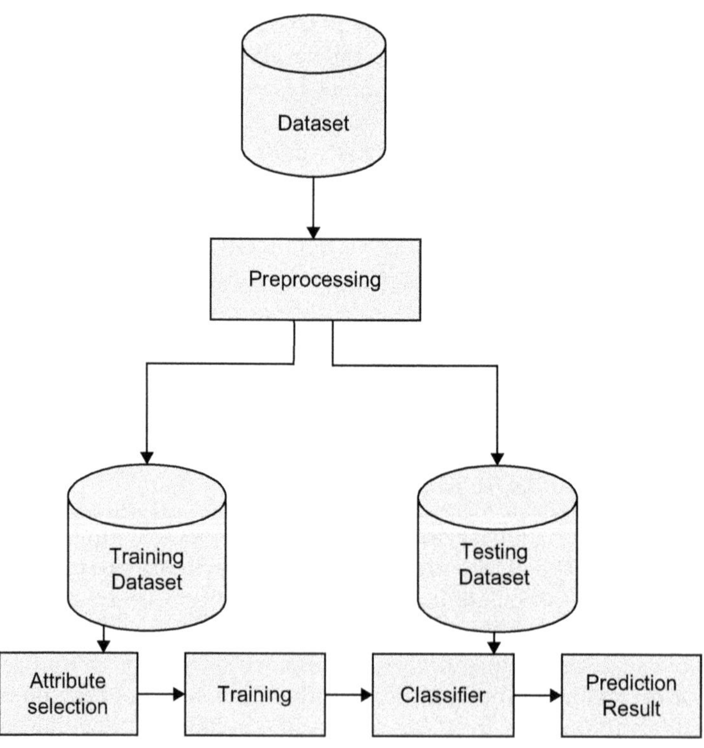

Figure 14.1 An overview of the process of SFP.

14.1.3 Deep learning in SFP

Dl models, especially NNs, have demonstrated remarkable applications like Speech recognition, NLP, and many more. It is able to automatically fetch important features from the raw data, and learning hierarchical representations makes them well-suited for software fault prediction (Figure 14.2).

14.1.4 Motivation for utilizing deep learning in fault prediction

Recently, DL has gained popularity as an effective way to solve complex pattern recognition and prediction problems. DL has the ability to learn complex parameters automatically from raw data and are a key enabler used to predict software bugs. DL models could fetch important parameters from raw information automatically, related to software, such as source code, execution traces, and system logs, but existing methods mainly rely on manual feature engineering. Because of this, deep learning models are ideally suited for the task of defect prediction in contemporary software systems. Key deep learning techniques used in this context include CNN, RNN, LSTM, autoencoder, etc.

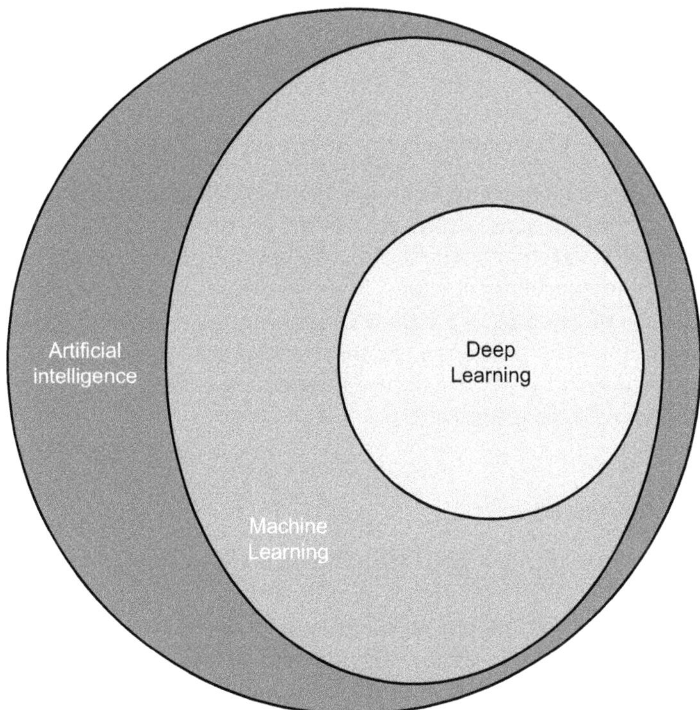

Figure 14.2 Relationship among AI, DL, and ML.

14.1.4.1 Convolutional neural networks

Convolutional neural networks have been effectively implemented in various software development applications. In software defect prediction, CNNs can be used to analyze software artifacts like source code and system logs. Convolutional neural networks are particularly effective in capturing spatial as well as temporal patterns, thanks to the utilization of fully connected layers, pooling layers, and convolutional layers. They automatically develop hierarchical representations from data on software, which enables accurate failure prediction.

14.1.4.2 Recurrent neural networks (RNNs)

Sequential data analysis is particularly suited to the use of recurrent neural networks (RNNs). RNNs can represent software behavior that varies over time through recurrent connections. However, traditional RNNs have difficulty learning long-term dependencies due to a phenomenon known as the vanishing gradient problem. To address this problem, several other types of models, such as GRU and LSTM, have been proposed as potential solutions. In particular, LSTM has demonstrated the ability to successfully capture long-term dependencies, making it useful for software defect prediction tasks.

14.1.4.3 Long short-term memory networks (LSTM)

Memory cells are incorporated into LSTM networks, which are of the RNN form. This gives these networks the ability to solve the vanishing gradient problem. LSTMs are particularly effective at sequential information, which makes them an excellent choice for SDF tasks that use time-series information. Because of the memory cells that are contained within LSTMs, these devices are able to store pertinent information and keep it up to current over lengthy periods, which enables reliable prediction of software problems based on previous patterns.

14.1.4.4 Autoencoders

Autoencoders are unsupervised learning algorithms that can rebuild input patterns in order to develop compact representations of data. Autoencoders can extract latent parameters from software-related data, which can then be used for defect prediction. Autoencoders are able to detect abnormal patterns that may signal errors by being trained on normal data and then rebuilding it. This unsupervised method shines when there is a dearth of fault data that has been labeled.

14.1.4.5 Deep belief networks (DBNs)

DBNs, also known simply as DBNs, are generative models that contain numerous hidden layers of units. DBNs can be learned in an unsupervised context using a layer-wise learning technique, and supervised learning can be utilized for fine-tuning after the unsupervised learning has been completed. The capacity of DBNs to recognize intricate linkages and patterns within software data has demonstrated significant potential for application in the field of failure prediction. They are able to acquire hierarchical representations, which enables accurate defect prediction based on generalizations. This ability is one of their distinguishing characteristics [5].

14.1.4.6 Generative adversarial networks

GANs often known as generative adversarial networks are artificial neural networks that are trained to compete against one another using a generator and a discriminator. GANs have been investigated for use in the prediction of software faults; in this application, the generator seeks to provide data that is representative of realistic software, and the discriminator differentiates between real and generated data. GANs can accurately capture the underlying distribution of software data, which enables fault prediction through the identification of differences between real and generated samples [6].

14.1.5 Machine learning

Ensuring the reliability and stability of code in the realm of software development is paramount. Software faults, or bugs, can lead to system failures, security vulnerabilities, and user dissatisfaction. Traditional methods of detecting faults, such as manual code reviews and testing, are time-consuming and often inadequate for large-scale projects. ML, a subset of artificial intelligence, offers a promising solution for automating the process of software fault prediction. ML algorithms can inspect factual data to find patterns and trends that are indicative of potential faults. By learning from past experiences, these algorithms can predict the likelihood of faults in new code, allowing developers to proactively address issues before they escalate. This approach not only improves the overall reliability of software but also helps streamline the development process by prioritizing resources where they are most needed.

One of the key advantages of ML in fault prediction is its capability to handle large and complex datasets. Software projects generate vast amounts of data, including code metrics, version control history, and bug reports. ML algorithms can process this data to extract meaningful insights, enabling developers to make informed decisions about where to focus their efforts. Another benefit of machine learning is its adaptability to different types of software artifacts. Whether analyzing source code, runtime behavior, or system logs, ML algorithms can be tailored to suit the specific needs of the project. This flexibility makes machine learning an invaluable tool for software development teams seeking to improve the quality and reliability of their code.

In this era of rapid technological advancement, the role of ML in software fault prediction is becoming increasingly vital. By harnessing the power of data and automation, machine learning offers a proactive approach to software quality assurance, helping developers deliver more reliable and robust software products.

14.2 LITERATURE REVIEW

Dl methods have recently gained attention as a viable alternative for software bug forecasting. When it comes to tackling the complexities of software systems, DL models have shown to be exceptional. They can automatically understand complicated patterns and features from raw data. DL models may extract meaningful representations from software-related data using large-scale datasets and sophisticated computational resources, allowing for reliable failure prediction and proactive maintenance. This chapter explores the use of RNNs for predicting software bugs, highlighting their effectiveness in handling sequential data [1].

The research focuses on using ML techniques to predict errors in different large software projects and companies [2]. This chapter is designed to present a state-of-the-art software defect prediction method using deep learning techniques by analyzing relevant data [11]. We hope to explore how deep learning models can help with prediction errors compared to other methods. Our goal is to demonstrate the feasibility and effectiveness of deep learning models in casino software by comparing and contrasting their strengths and weaknesses. This chapter provides insight into how machine learning can improve software quality by examining flight error prediction [3]. A comprehensive review of various machine learning methods for software bugs discusses their benefits and limitations in Ref. [4]. This study addressed the problem of under classification in data prediction errors and evaluated the effectiveness of different learning methods [5]. This study provides insight into the strengths and weaknesses of various classifiers by comparing their performance in error prediction [6]. This chapter examines different types of systems and their impact on the software quality performance predicted models [7]. This study presents a deep learning method for software error detection and demonstrates its effectiveness in improving the accuracy of the system [8]. This work presents a method to predict the vulnerability of anonymous data using machine learning, showing problems and solutions [9]. The chapter discusses a new software bug prediction method based on deep learning and demonstrates its potential to improve prediction [10]. Here, the author discussed the future and various techniques of Dl for software fault prediction [11]. This chapter provides an overview of the current state of software defect prediction using deep learning techniques. It likely covers various deep learning models and their applications in predicting software defects, as well as discusses the challenges and future directions in this area [12]. It explores the specific use cases of deep learning in software defect prediction, discussing the advantages and limitations of using DL models compared to traditional machine learning approaches. It also provides insights into the effectiveness of deep learning in improving the accuracy of defect prediction models [13].

The author examines existing literature on software fault prediction using deep learning, aiming to summarize the key findings and identify trends in research methodologies and approaches. It may also highlight gaps in the current literature and suggest areas for future research [14]. This chapter likely presents a specific approach to software fault prediction using RNNs and discusses its performance compared to other deep learning and traditional machine learning models. It may also provide insights into the features and data preprocessing techniques used in the RNN-based approach [15]. The author compares the performance of different ML models, including Dl models, for software fault prediction. The chapter also discusses the impact of feature selection, data preprocessing, and model hyperparameters on the effectiveness of the prediction models [16]. The papers by LeCun et al. [17] and Goodfellow et al. (2016) provide foundational knowledge on deep learning, discussing its principles, architectures, and applications. Huang et al. [18] introduce densely connected convolutional networks (DenseNet), a deep learning architecture that improves gradient flow and parameter efficiency. Here, the author discusses deep learning for software effort estimation [19].

This chapter evaluates the accuracy of several key DL architectures used in SFP. These include models such as CNNs, RNNs, and LSTMs, DBNs, and GANs. To better evaluate the advantages and disadvantages of DL models in this field, we will evaluate their performance and compare it with more traditional defect prediction methods.

Basically, these deep learning techniques solve the biggest problem of software error prediction. Issues related to data availability, completeness, and balance, extraction, feature selection, scalability, model overfitting, interpretability, efficiency, and generalization are all examples of these problems. By highlighting these obstacles, we hope to reveal future research directions and methods to improve the performance of DL models.

Systematic studies also include benchmark datasets used to find out the performance of DL models and evaluation metrics frequently used to predict software defects. You can review evaluation metrics and datasets to assess the quality of your studies and compare results from different experiments. Software defect prediction using deep learning (DL) models often involve the use of benchmark datasets and evaluation metrics to assess the performance of these models. Here's a detailed explanation.

14.3 BENCHMARK DATASETS

Benchmark datasets are standardized datasets which are commonly used to evaluate the performance of DL models for software defect prediction. These datasets typically consist of software modules or files, each labeled as either defective or nondefective. Benchmark datasets serve as a basis for comparison between different models and approaches. One of the commonly used benchmark datasets is the NASA MDP dataset, which contains software modules from various projects along with defect labels.

The landscape of predictive modeling in software fault detection has been significantly enriched by both traditional ML algorithms and advanced deep learning techniques. Table 14.1 represents a comparative analysis of DL versus ML techniques and Table 14.2 is a comparison that highlights the key differences, strengths, and applications of five popular deep learning models (CNN, RNN, LSTM, Autoencoder, and GAN) versus traditional ML algorithms such as decision trees (DT), support vector machines (SVM), random forests, and K-nearest neighbors (KNN).

NASA MDP dataset: The NASA MDP dataset contains software metrics from various NASA projects. It includes data from software modules, such as McCabe cyclomatic complexity, lines of code, Halstead complexity measures, and other software metrics. Some of the examples include the following:

- *ECLIPSE metrics dataset:* This dataset contains software metrics from the Eclipse project. It includes lines of code, cyclomatic complexity, and code churn kind of different metrics.
- *PROMISE repository:* The PROMISE repository contains several datasets related to software engineering, including software fault prediction. Datasets that include from various projects and metrics such as halstead complexity measures, lines of code, and others.

The comparative analysis of several DL techniques, i.e., CNN, RNN, LSTM, Autoencoders, and GANs highlight their distinct use cases and characteristics. CNNs excel in processing image and spatial data due to their convolutional layers, making them ideal for tasks like image classification and object detection. However, they struggle with temporal data. RNNs, designed for sequential data such as text and speech, suffer from limiting their memory to short-term dependencies, vanishing gradient problem. LSTMs address this limitation by effectively capturing long-term dependencies but are computationally more expensive. Autoencoders are primarily used for feature extraction, data reconstruction, and anomaly

Table 14.1 Comparative analysis of DL versus ML

Perspective	Deep learning models	Machine learning models
Primary use case	Complex pattern recognition, high-dimensional data	Simpler patterns, structured data
Architecture	Layered neural networks (convolutions, recurrences, memory cells)	Nonneural models (decision trees, hyperplanes, distance metrics)
Strengths	Autonomous feature learning, handles high-dimensional data well	Simpler, faster to train, interpretable, requires less data
Weaknesses	Requires large datasets, high computational resources, less interpretable	May not capture complex patterns, limited feature learning
Common applications	Speech recognition and image, natural language processing, anomaly detection	Classification, regression, clustering, dimensionality reduction
Training complexity	High (requires powerful hardware and considerable time)	Moderate to low (quicker training with fewer resources)
Handling of input data	Capable of processing unstructured data (images, text, time series)	Generally suited for structured and tabular data
Memory requirements	High (GPU/TPU often required)	Low to moderate (CPU usually sufficient)
Interpretability	Low to moderate (complex models, some visualizations possible)	High (decision paths, support vectors, proximity measures)
Data requirements	Large labeled datasets for training	Smaller datasets can be effective
Performance	High for complex tasks with sufficient data	Good for simpler tasks or smaller datasets
Scalability	Scales well with data and computational power	Scales well but limited by data complexity
Example libraries	TensorFlow, PyTorch, Keras	Scikit-learn, Weka, R, MATLAB
Generalizability	Good, especially with large and diverse datasets	Good, often limited to the feature set defined
Overfitting mitigation	Regularization, dropout, data augmentation	Cross-validation, pruning, ensemble methods
Feature engineering	Automated (learns features from raw data)	Manual (requires domain knowledge to engineer features)

detection, with a moderate training complexity and the risk of overfitting. GANs stand out in generating high-quality data and creative applications like image generation and style transfer, though they require careful tuning and substantial computational resources due to their training instability.

Table 14.2 Comparative analysis of GAN, Autoencoders, LSTM, CNN, and RNN

Perspective	CNN	RNN	LSTM	Autoencoder	GAN
Primary use case	Image and spatial data processing	Sequential data processing	Long-term dependencies in sequences	Data dimensionality reduction and noise removal	Data generation and unsupervised learning
Architecture	Convolutional layers, pooling layers	Recurrent layers	LSTM cells (memory cells)	Encoder-decoder structure	Generator and discriminator networks
Strengths	Local feature extraction, spatial hierarchies	Temporal sequence modeling, handling variable input lengths	Long-term dependency capture, avoiding vanishing gradient	Feature learning, data reconstruction	High-quality data generation, creative applications
Weaknesses	Limited temporal handling, large data requirement	Vanishing gradient problem, short-term memory	Computationally expensive, complex structure	Can overfit, limited to reconstruction	Training instability, mode collapse
Common applications	Image classification, object detection	Text, speech, time-series data	Text generation, language modeling	Denoising, anomaly detection, dimensionality reduction	Image generation, style transfer, data augmentation
Training complexity	Moderate	Moderate	High	Moderate	High

(Continued)

Table 14.2 (Continued) Comparative analysis of GAN, Autoencoders, LSTM, CNN, and RNN

Perspective	CNN	RNN	LSTM	Autoencoder	GAN
Handling of data input	Fixed size, structured (e.g., images)	Sequential, variable lengths	Sequential, variable lengths	Any data type, typically high-dimensional	Any data type, typically high-dimensional
Memory requirements	High	Moderate	High	Moderate	Very high
Interpretability	Moderate (filters can be visualized)	Low	Low	Moderate (latent space representation)	Low
Data requirements	Large labeled datasets	Sequential data, can be smaller	Sequential data, can be smaller	Any size, typically large unlabeled data	Large datasets, both labeled and unlabeled
Example libraries	TensorFlow, PyTorch, Keras	TensorFlow, PyTorch, Keras	TensorFlow, PyTorch, Keras	TensorFlow, PyTorch, Keras	TensorFlow, PyTorch, Keras
Performance	High with large datasets and powerful GPUs	Moderate to high, depending on sequence length	High for long sequences	High for reconstruction tasks	High but requires careful tuning
Scalability	Good	Good	Good	Good	Requires significant resources

In terms of performance and training complexity, CNNs and GANs demand large datasets and powerful hardware, with GANs being particularly resource-intensive. RNNs and LSTMs can work with smaller sequential datasets, with LSTMs providing better performance for longer sequences. Autoencoders, while effective for reconstruction tasks, also benefit from larger unlabeled datasets. Regarding interpretability, CNNs offer moderate levels through filter visualization, while RNNs, LSTMs, and GANs are generally less interpretable. Autoencoders provide a moderate level of interpretability through their latent space representation. Scalability varies, with CNNs, RNNs, LSTMs, and Autoencoders generally scaling well with data and computation, though their memory requirements differ. GANs, on the other hand, need significant resources and careful tuning to achieve effective scalability. Overall, each model has its unique strengths, weaknesses, and optimal use cases, which are crucial to consider when selecting the appropriate model for specific tasks.

14.4 EVALUATION METRICS

- *Accuracy:* This metric computes the part of precisely classified instances among the total instances. It provides an overall assessment of the model's performance but may not be suitable for imbalanced datasets.
- *Precision:* Precision measures the part of perfectly predicted defective instances among all instances predicted as defective. It indicates the ability of the model to avoid false positives.
- *Recall (sensitivity):* Recall measures the part of accurately predicted defective instances among all actual defective instances. It indicates the model's ability to identify all actual positives.
- *F1 score:* The F1 score is the harmonic mean of precision and recall. It provides a balance between precision and recall and is useful when the class distribution is imbalanced.
- *Area under the receiver operating characteristic curve (AUC–ROC):* AUC–ROC measures the ability of the model to distinguish between defective and nondefective instances. It is particularly useful for imbalanced datasets.
- *Area under the precision-recall curve (AUC–PR):* AUC–PR is similar to AUC–ROC but focuses on the precision-recall trade-off. It is also useful for imbalanced datasets.
- *Confusion matrix:* It provides a detailed breakdown of correct and incorrect predictions made by the model. It includes metrics such as true positives, true negatives, false positives, and false negatives.
- *Proposed methodology:*

The methodology diagram shows the DL for SFP which begins with the collection of datasets containing software metrics and fault data, such as

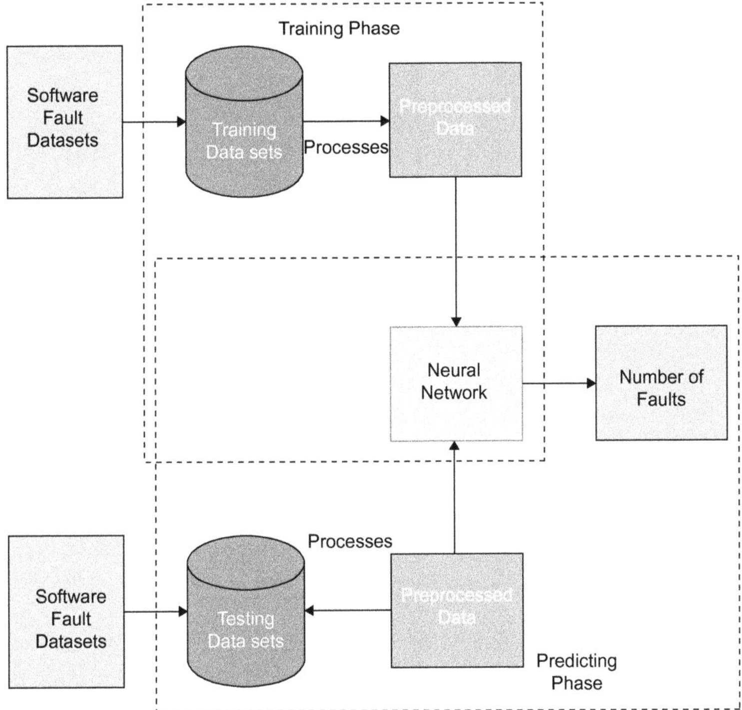

Figure 14.3 Proposed methodology of deep learning for SFP.

MDP datasets (Figure 14.3). These datasets are then pre-processed to handle missing values, normalize the data, and prepare it for analysis. Next, relevant features are selected from the pre-processed data based on their potential predictive value for software faults. Neural network techniques, such as LSTM networks, CNNs, and RNNs, are chosen for fault prediction. The models are trained on the training set using appropriate optimization algorithms and loss functions. Hyperparameters of the models, such as batch size, and learning rate, are tuned using methods like grid search or random search. The trained models are assessed on the validation set using measures like F1-score, accuracy, precision, and recall, and the best-performing model is selected. This model is then tested to evaluate its performance on the test set on unseen data and to see the significant differences in performance between the deep learning models and traditional methods.

14.5 EXPERIMENTAL RESULT ANALYSIS

Below is a table that provides a comparison of various deep learning models on three NASA MDP datasets. The performance of each technique is

Table 14.3 Performance matrix of various neural networks for software fault prediction

Dataset	Models	Accuracy	Precision	Recall	F1-Score
CM1	RNN	0.82	0.84	0.80	0.82
	CNN	0.85	0.87	0.82	0.84
	LSTM	0.87	0.88	0.85	0.86
	Autoencoders	0.80	0.81	0.78	0.79
	GAN	0.83	0.85	0.81	0.83
KC1	RNN	0.79	0.81	0.77	0.79
	CNN	0.82	0.84	0.80	0.82
	LSTM	0.84	0.86	0.82	0.84
	Autoencoders	0.78	0.79	0.75	0.77
	GAN	0.81	0.82	0.79	0.80
PC1	RNN	0.76	0.78	0.74	0.76
	CNN	0.80	0.81	0.78	0.79
	LSTM	0.82	0.83	0.80	0.81
	Autoencoders	0.75	0.76	0.72	0.74
	GAN	0.79	0.80	0.77	0.78

evaluated in terms of accuracy, precision, recall, and F1-score, showcasing the strengths and suitability of different models for software fault prediction. The results indicate that LSTM and CNN models generally perform better across the datasets, highlighting their effectiveness in capturing complex patterns in software metrics.

Table 14.3 shows the performance metrics for each model (CNN, RNN, and LSTM) on each dataset (PC1, KC1, and CM1). Each cell in the table corresponds to a specific metric (F1-score, accuracy, precision, and recall) for a specific model and dataset (Figure 14.4).

Figure 14.5 presents a comparison of the performance metrics (recall, accuracy, precision, and F1-score) on PC1, KC1, and CM1 datasets for each of the models (CNN, RNN, and LSTM).

14.6 CHALLENGES IN DEEP LEARNING FOR SOFTWARE FAULT PREDICTION

14.6.1 Data quality and quantity

A. *Imbalanced data:* Software fault datasets often exhibit class imbalance, where the number of instances without faults far exceeds those with faults. This imbalance can lead to models that are biased toward predicting the majority class (nonfaulty instances), resulting in poor performance on the minority class (faulty instances). Techniques like the synthetic minority over-sampling technique (SMOTE), adaptive

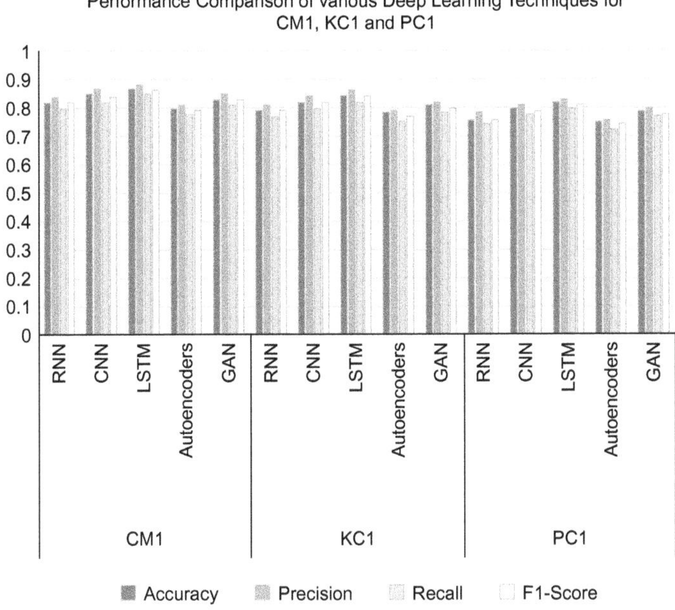

Figure 14.4 Performance comparison of the various DL techniques for CM1, KC1, and PC1.

synthetic (ADASYN) sampling, and cost-sensitive learning can help mitigate this issue.

B. *Data scarcity:* Collecting a huge amount of labeled data for software faults can be challenging. This is particularly problematic for deep learning models which require substantial quantity of data to learn effectively. Transfer learning, where a model pre-trained on a large dataset is fine-tuned on the target dataset, can help alleviate data scarcity.

C. *Noise and inconsistency:* Software fault data can contain turbulent/incompatible entries due to human fault/variations in the development process. Data cleaning techniques, such as outlier detection and removal, can help improve the quality of the dataset.

14.6.2 Feature engineering

A. *Feature selection:* Identifying the most relevant features from software measures like lOC (cyclomatic complexity) and code attributes (e.g., function calls, variable usage) is critical for accurate predictions. Feature selection methods like recursive feature elimination (RFE), principal component analysis (PCA), and domain knowledge can be used to identify and select relevant features.

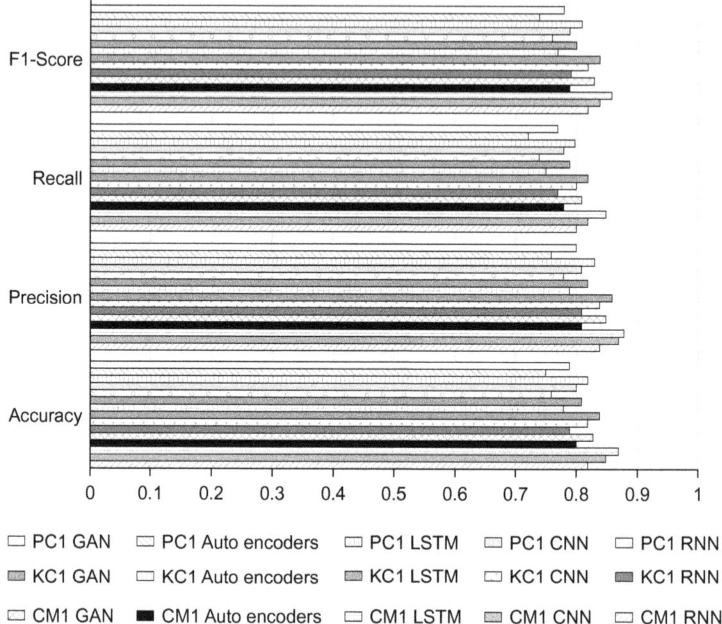

Figure 14.5. Comparison of the performance of DL models with datasets.

B. *High dimensionality:* High-dimensional data can lead to the curse of dimensionality, where the volume of the feature space increases exponentially, making the model training process more difficult and computationally expensive. Dimensionality reduction techniques like PCA, t-SNE (t-distributed stochastic neighbor embedding), and auto-encoders can be used to reduce the number of features while retaining important information.

14.6.3 Model complexity and interpretability

A. *Model interpretability:* Deep learning models, especially deep neural networks, are often considered "black boxes" because they do not offer clear exploration for their predictions. This lack of interpretability can be problematic in understanding the reasoning behind fault predictions and gaining trust from stakeholders. Techniques like SHAP (shapley additive explanations) and LIME (local interpretable model-agnostic explanations) can help make these models more interpretable by highlighting the contribution of individual features to the prediction.

B. *Overfitting:* Deep learning models have a high capacity to learn from data, which can lead to overfitting, especially with small or noisy datasets. When the model learns the training data too well, including its noise and outliers, and performs poorly on new, unseen data, then overfitting occurs. Regularization techniques like dropout, L2 regularization, and early stopping can help prevent overfitting.

14.7 COMPUTATIONAL RESOURCES

A. *High computational cost:* Training DL models requires significant computational power and memory. This can be a limitation for researchers and organizations with limited access to high-performance computing resources. Utilizing cloud-based platforms, e.g., specialized hardware, GPUs, TPUs, AWS, Azure, and Google Cloud can help manage these requirements.
B. *Scalability:* Ensuring that deep learning models can scale to handle large-scale software systems, and real-time fault prediction is challenging. Distributed training and model optimization techniques, such as model parallelism and data parallelism, can help improve scalability.

14.8 MODEL GENERALIZATION

A. *Domain adaptation:* Models trained on one type of software project may not generalize well to different types of projects or environments without significant adaptation. Domain adaptation techniques, such as fine-tuning and transfer learning, can help improve model generalization across different domains.
B. *Concept drift:* Software systems evolve over time, and the patterns of faults may change, leading to concept drift. It requires models to be continuously updated to maintain accuracy. Online learning techniques, where the model is continuously updated as new data becomes available, can help address this issue.

14.9 EVALUATION AND BENCHMARKING

A. *Evaluation metrics:* Traditional metrics like accuracy may not be sufficient for evaluating models on imbalanced datasets, as they can be misleading. More appropriate metrics include precision, recall, F1-score, area under the receiver operating characteristic curve (AUC-ROC), and Matthew's correlation coefficient (MCC).
B. *Benchmarking:* Comparing deep learning models with traditional machine learning methods (e.g., decision trees, support vector machines, and logistic regression) and ensuring they provide significant improvements is essential. Benchmarking involves systematically

evaluating different models on the same datasets using consistent evaluation protocols.

14.10 INTEGRATION WITH DEVELOPMENT PROCESSES

A. *Workflow integration:* Integrating deep learning models into existing software development and testing workflows can be challenging. This requires changes in processes and tools, such as continuous integration/continuous deployment (CI/CD) pipelines, to ensure smooth integration and deployment of models.
B. *Tool support:* The lack of mature tools and frameworks specifically tailored for deep learning in the context of software fault prediction can hinder adoption. Developing and adopting specialized tools and frameworks can help streamline the integration process.

14.11 MITIGATION STRATEGIES

To address these challenges, the following strategies can be employed:

A. *Data augmentation:* Data augmentation is a technique used to artificially expand the size of a dataset by creating modified versions of images or data samples in the dataset. This technique is commonly used in DL for tasks such as object detection, and NLP, i.e., natural language processing. The goal of data augmentation is to improve the performance and generalization of ML models by exposing them to a wider variety of training examples without actually collecting new data. Techniques such as oversampling, undersampling, or synthetic data generation (e.g., using GANs) can help address data imbalance. Automated feature engineering: leveraging automated feature selection and extraction methods can reduce the burden of manual feature engineering.
B. *Improving model explainability:* Improving model explainability is crucial for building trust in machine learning models, especially in domains where decisions have significant consequences, such as healthcare or finance. Explainability refers to the ability to explain why a model made a particular prediction or decision, in a way that is understandable to humans. Here's a detailed explanation of techniques to improve model explainability. Using techniques like SHAP or LIME to make models more interpretable.
C. *Transfer learning:* Utilizing pre-trained models and fine-tuning them on specific software fault datasets can help mitigate data scarcity and improve generalization.
D. *Regularization techniques:* Applying regularization methods like dropout or L2 regularization to prevent overfitting.

E. *Continuous monitoring and updating:* Implementing mechanisms for continuous model monitoring and updating to handle concept drift.

F. *Leveraging cloud and distributed computing:* Utilizing cloud-based platforms and distributed training techniques to manage computational requirements.

G. *Adopting appropriate evaluation metrics:* Using precision, recall, F1-score, AUC-ROC, and MCC to evaluate model performance on imbalanced datasets.

14.12 CONCLUSION AND FUTURE DIRECTIONS

Software fault prediction has great potential to save maintenance costs and increase program dependability when predicting through deep learning. Deep learning has the potential to become a key component of contemporary software engineering technique if the inherent difficulties are resolved and research in this area is consistently advanced. With more companies implementing these methods, software development appears to have a more robust and effective future. Deep learning techniques have revolutionized software fault prediction by providing higher accuracy, automated feature extraction, and the ability to handle complex data. While there are challenges such as data quality, computational requirements, and model interpretability, appropriate mitigation strategies can enhance the effectiveness of these models, leading to more reliable and high-quality software development processes. The continuous advancements in deep learning research and methodologies promise even greater improvements in software fault prediction capabilities.

Deep learning's capability to forecast software faults in the future might drastically change how we guarantee the dependability and quality of software. Through investigating these potential avenues for advancement, scholars and professionals may persist in creating novel and more reliable, efficient, and effective fault prediction models, which will ultimately result in software systems that are more robust and of superior quality. Future deep learning models will continue to evolve, automatically extracting relevant features from raw software metrics data. This will reduce the need for manual feature engineering and greatly enhance prediction accuracy. Overall, the future of software fault prediction using deep learning looks promising, with ongoing research likely to lead to more accurate, efficient, and interpretable models that can help improve software quality and reliability.

REFERENCES

1. Zhou, Y., Wen, M., & Wang, X. (2019). A recurrent neural network based approach for software bug prediction. *IEEE Access, 7,* 77935–77945.

2. Peters, F., & Menzies, T. (2012). Privacy and utility for defect prediction: experiments with morph, gene, and cross-company data sharing. *Proceedings of the 34th International Conference on Software Engineering (ICSE)*, 189–199.
3. Kamei, Y., Shihab, E., Adams, B., Hassan, A. E., Mockus, A., Sinha, A., & Ubayashi, N. (2013). A large-scale empirical study of just-in-time quality assurance. *IEEE Transactions on Software Engineering*, 39(6), 757–773.
4. Malhotra, R. (2015). A systematic review of machine learning techniques for software fault prediction. *Applied Soft Computing*, 27, 504–518.
5. Wang, S., & Yao, X. (2013). Using class imbalance learning for software defect prediction. *IEEE Transactions on Reliability*, 62(2), 434–443.
6. Ghotra, B., McIntosh, S., & Hassan, A. E. (2015). Revisiting the impact of classification techniques on the performance of defect prediction models. *Proceedings of the 37th International Conference on Software Engineering (ICSE)*, 789–800.
7. Khoshgoftaar, T. M., Gao, K., & Napolitano, A. (2010). An empirical study of feature ranking techniques for software quality prediction. *International Journal of Software Engineering and Knowledge Engineering*, 20(6), 793–811.
8. Shin, Y., & Kim, E. (2015). A deep learning approach for software bug detection. *Proceedings of the 2015 6th International Conference on Software Engineering and Service Science (ICSESS)*, 739–742.
9. Nam, J., & Kim, S. (2015). CLAMI: defect prediction on unlabeled datasets (ICSE 2015). *Proceedings of the 37th International Conference on Software Engineering (ICSE)*, 789–800.
10. Liu, Y., Xu, L., Hu, Q., Zhao, P., & Zhang, X. (2017). A new method of software fault prediction based on deep learning. *Procedia Computer Science*, 119, 155–162.
11. Kalonia, S., Upadhyay, A. (2024). A systematic review of software fault prediction using deep learning: challenges and future perspectives. In: Das, S., Saha, S., Coello Cello, C.A., Bansal, J.C. (eds), *Advances in Data-Driven Computing and Intelligent Systems. ADCIS 2023. Lecture Notes in Networks and Systems*, vol 893. Springer, Singapore. https://doi.org/10.1007/978-981-99-9518-9_39.
12. Mezentsev, I. P., & Misilov, V. E. (2021). A survey on software defect prediction using deep learning. *Mathematics*, 9(11), 1180. https://doi.org/10.3390/math9111180.
13. Wu, Z., Liu, Y., Zhang, T., & Zhang, J. (2023). On the use of deep learning in software defect prediction. *Journal of Systems and Software*, 193, 111529. https://doi.org/10.1016/j.jss.2022.111529.
14. Wang, Q., & He, Z. (2023). A systematic review of software fault prediction using deep learning. *Empirical Software Engineering*, 28, 15. https://doi.org/10.1007/s10664-022-10117-8
15. Singh, R., & Reddy, V. (2023). Software fault prediction using an RNN-based deep learning approach. *Applied Sciences*, 13(3), 1568. https://doi.org/10.3390/app13031568
16. Zhang, X., & Huang, L. (2023). Software fault prediction using machine learning models. *IEEE Access*, 11, 67589–67600. https://doi.org/10.1109/ACCESS.2023.3250293

17. LeCun, Y., Bengio, Y., & Hinton, G. (2015). Deep learning. *Nature*, 521(7553), 436–444. https://doi.org/10.1038/nature14539.
18. Huang, G., Liu, Z., Van Der Maaten, L. and Weinberger, K.Q. (2017) Densely Connected Convolutional Networks. Proceedings of the IEEE Conference on Computer Vision and Pattern Recognition, Honolulu, 21-26 July 2017, 4700 -4708.https://doi.org/10.1109/CVPR.2017.243
19. Meenakshi, P. M. (2024). Software effort estimation using deep learning: a gentle review. In: Pandit, M., Gaur, M.K., Kumar, S. (eds), *Artificial Intelligence and Sustainable Computing. ICSISCET 2023. Algorithms for Intelligent Systems*. Springer, Singapore. https://doi.org/10.1007/978-981-97 -0327-2_26.

Index

For Product Safety Concerns and Information please contact our EU
representative GPSR@taylorandfrancis.com
Taylor & Francis Verlag GmbH, Kaufingerstraße 24, 80331 München, Germany